Biotechnology of Ectomycorrhizae

Molecular Approaches

Biotechnology of Ectomycorrhizae

Molecular Approaches

Edited by

Vilberto Stocchi

University of Urbino
Urbino, Italy

Paola Bonfante

University of Turin and Center of Mycology, CNR
Turin, Italy

and

Marco Nuti

University of Padua
Padua, Italy

Springer Science+Business Media, LLC

Library of Congress Cataloging-in-Publication Data

International Symposium on Biotechnology of Ectomycorrihizae:
   Molecular Approaches (1994: Urbino, Italy)
     Biotechnology of ectomycorrhizae : molecular approaches / edited
   by Vilberto Stocchi, Paola Bonfante, Marco Nuti.
        p. cm.
     Includes bibliographical references and index.
     ISBN 978-1-4613-5770-4   ISBN 978-1-4615-1889-1 (eBook)
     DOI 10.1007/978-1-4615-1889-1

     1. Ectomycorrhizas--Biotechnology--Congresses.   I. Stocchi,
   Vilberto.   II. Bonfante, Paola.   III. Nuti, Marco P.   IV. Title.
   TP248.27.F86I57 1994
   589.2--dc20   95-35586
                                                         CIP

Proceedings of an International Symposium on Biotechnology of Ectomycorrhizae: Molecular
Approaches, held November 9–11, 1994, in Urbino, Italy

ISBN 978-1-4613-5770-4

© 1995 Springer Science+Business Media New York
Originally published by Plenum Press, New York in 1995
Softcover reprint of the hardcover 1st edition 1995

10 9 8 7 6 5 4 3 2 1

PREFACE

Forty years after the discovery of the helix nature of DNA and more than twenty after the first applications of recombinant DNA technology to the pharmaceutical industry, the Pandora's vase of biotechnology seems far from being empty. New products for agriculture and the food industry are constantly being placed on the market, and powerful monitoring techniques have been developed to track non-modified and genetically modified vaccines, viruses, microbes and plants released into the environment. Molecular approaches for taxonomic purposes, which might also be useful for quality control and assurance, have been successfully developed and used for taxonomic purposes in the last decade for both prokaryotic and eukaryotic cells, including yeasts and filamentous fungi.

Mycorrhizae are one example of a traditional biotechnology that can greatly benefit from the latest molecular approaches. These universal symbioses between soil fungi and plant roots play a central role in most of the natural and agricultural ecosystems in such key processes as nutrient cycling, soil structural conservation and plant health. For these reasons, mycorrhizae have been successfully used to improve the quality of forest and agricultural seedlings, to produce high-quality micropropagated plants and to increase the production of edible mushrooms of high economic value, such as truffles. However, although controlled inoculation of oak and hazel seedlings with ectomycorrhizal truffles has been carried out for decades in France and Italy, and is still expanding commercially, several technological gaps remain to be filled. These include:

- the difficulty in evaluating mycorrhizal status after inoculation with truffles or other ectomycorrhizal symbionts, which are morphologically very similar, such as *Tuber uncinatum* and *Tuber aestivum,* or *Tuber borchii* and *Tuber magnatum*;
- the difficulty in rapidly verifying that an inoculated species is effectively and efficiently colonizing the soil where indigenous populations of minor economic importance can compete with the inoculated ones, and
- the difficulty in assessing genetic drifts that might occur in the population of inoculated strains as a consequence of management practices or environmental pressure.

Ultimately, scientifically sound legislative frameworks will be adopted only if the scientific community is able to offer reliable, reproducible methods, enabling the problem of frauds to be overcome in this economically relevant international market.

Yet biotechnology offers a range of opportunities in the field of mycorrhiza research that are of course not limited to truffles and to ectomycorrhizal fungi. Under the auspices of the National Research Council of Italy, the Symposium was deemed timely, in that it offered a wider forum for the exchange of scientific information and discussion of new and emerging concepts in this area. By comparing conventional and innovative approaches, and using reliable experimental systems, such as *Neurospora,* along with the more complex and less understood symbiotic associations, we believe that the gaps in current knowledge can be filled, and old experimental limitations will be overcome.

The Editors

CONTENTS

ECTOMYCORRHIZAS IN THE ECOSYSTEM

Structural, Functional and Community Aspects

D. J. Read

Department of Animal and Plant Sciences
The University of Sheffield
Sheffield
United Kingdom

INTRODUCTION

One hundred years have elapsed since Frank (1894) used a combination of careful field observations and laboratory growth studies to develop the first detailed hypothesis concerning the significance of the ectomycorrhizal symbiosis. He envisaged that mycorrhizal fungi absorbed organic nitrogen (N) from the superficial layers of forrest soil, passing this element to the trees, at the same time obtaining carbon to sustain themselves. It is in many ways regrettable that in the ensuing century, research on the symbiosis has not been based on the same integrated approach, but followed two distinct and often divergent paths, one leading to analysis of function under simplified laboratory conditions, the other to evaluation, in the field, of the relationships between the fungi involved and plant roots. In order to achieve any understanding of the ecological significance of the ectomycorrhizal symbiosis, a synthesis of progress made in the two separate lines of advance is essential. This paper, while attempting to provide such a synthesis, also contains a plea that in the next century of research the two hitherto largely independent paths will converge so that the diverse biological attributes of the symbiosis can be viewed in a more realistic context.

Faced with the complexities of the soil environment, some have concluded that progress towards understanding of mycorrhizal biology can only be achieved under simplified and strictly controlled conditions in the laboratory. This approach was exemplified by the pioneering studies of Melin (1925), who first examined in culture, the nutritional requirements of mycorrhizal fungi then synthesised mycorrhiza on pine in sterilised sand and demonstrated that the mycelia of these fungi could capture phosphate (Melin & Nilsson 1950), as well as nitrogen in both mineral (Melin & Nilsson 1952), and amino-acid (Melin & Nilsson 1953) forms and facilitate their transfer to the host plant.

Simplification was carried further by Harley and his co-workers (Harley & Smith 1983) who, using excised mycorrhiza of *Fagus sylvatica* showed that the sheath, consisting of up to 40% by weight of the colonised root, increased the effectiveness of capture of phosphate and ammonium N from a bathing medium, both ions being stored in the fungus

Biotechnology of Ectomycorrhizae, Edited by Vilberto Stocchi et al.
Plenum Press, New York, 1995

and subsequently released to the plant. The significance of these types of study lay in the fact that they demonstrated the *potential* of mycorrhiza to play a major role in soil. However, they suffered the inevitable weakness of all laboratory experiments that they do not simulate the realities of the natural environment. Of particular concern is that the substrates employed bear little relationship, either qualitatively or quantitatively, to those in which mycorrhizal roots proliferate in the field.

Those approaching the question of mycorrhizal function by the second path, rather than ignoring natural complexity, have grappled with it and attempted, perhaps inevitably at a more descriptive level, to seek patterns. Thus, the extent of the distribution of this type of symbiosis between (Meyer 1973, Trappe 1987, Alexander & Hogberg 1986, Newman & Reddell 1987, Hogberg 1992) and within (Hesselman 1900, Fontana 1977, Haselwandter & Read 1980) plant families, has been described, the enormous diversity of morphological types (Peyronel 1922, Dominik 1969, Agerer 1991) and of fungal species involved in the association, emphasised (Trappe 1977, Molina *et al* 1992) and the geographical distribution of predominantly ectomycorrhizal ecosystems charted (Moser, 1967, Read 1991) (Fig 1). These types of study have enabled us to define, in broad terms, the physico-chemical nature of the environments in which ectomycorrhizal systems become dominant and hence to bring into sharper focus the questions which need to be addressed in the next hundred years of research on mycorrhiza. They confirm and extend Frank's original observation that ectomycorrhizal systems are associated with particular soil conditions in which, for reasons of climate or chemistry of the substrate, frequently a combination of both, a significant proportion of the nutrient fund of the ecosystem is sequestered in organic form near the soil

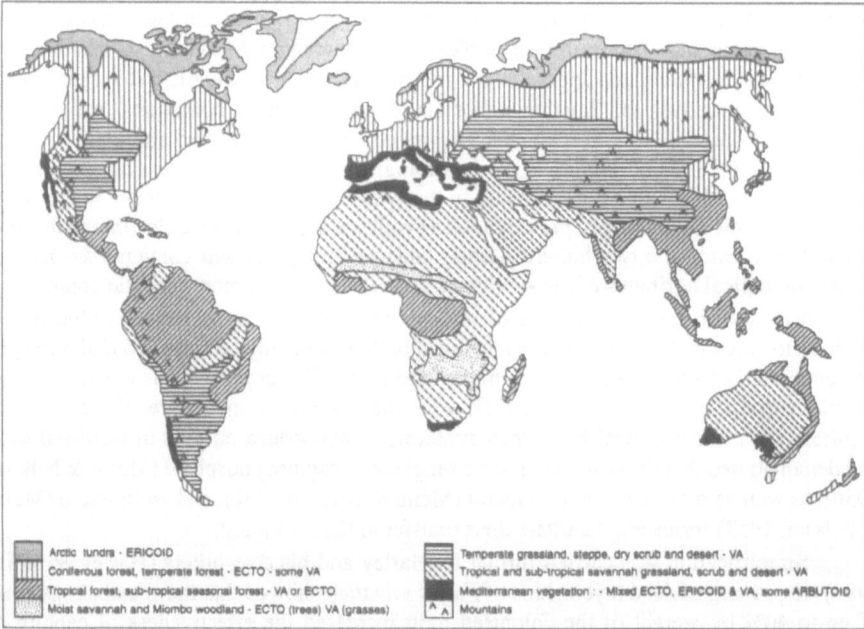

Figure 1. The distribution of the major climatically determined terrestrial biomes of the world showing the importance of systems dominated by ectomycorrhizal species across the land masses of Eurasia and North America. Activities of man have reduced the extent of these ectomycorrhizal forests in many parts, where they have been replaced by herbaceous vegetation with VA mycorrhiza. Systems dominated by ectomycorrhizal trees may have understorey vegetation of ericoid or VA mycorrhizal species. (Modified from Read 1991).

surface. Here, often but not necessarily in accordance with the poor base status of the lower horizons of the soil, the ectomycorrhizal roots proliferate over 80% of the mycorrhizal types being typically localised in the top 10 cm of the soil profile (Chilvers & Pryor 1965, Meyer, 1973, Harvey *et al* 1976). This is by nature an acidic medium where carboxyl groups associated with organic detritus predominate and which is characterised by its extremely low *available* nutrient status. Even in the relatively unusual circumstance in which ectomycorrhizal trees or shrubs are found on calcareous substrates (Clement *et al* 1977, Lapeyrie & Chilvers 1985) it appears that the colonised roots are associated with organically enriched surface horizons or shallow rankers. One of the most obvious patterns to emerge from exposure of these patterns of distribution concerns the nature of the relationship both in structural and functional terms between mycorrhizal roots and the organic substrates in which, as observed by Frank (1894) they preferentially proliferate.

STRUCTURAL FEATURES OF ECTOMYCORRHIZAL ROOTS

There is now little dispute over the nature of the complex fungal structure, the Hartig-net, which provides the intimate contact between the mycorrhizal partners. Indeed, it appears to be constructed in a remarkably uniform manner across a whole range of plant-fungus partnerships. The hyphae of the fungal partner, in penetrating between the radial walls of the outer cells of the root cortex, produce a largely unseptate but much branched and compact fan-like development(Fig 2) in which the individual walls of the hyphal branches provide a very large surface area and hence a structure which was recognised by Dudderidge & Read (1984a) as being analogous to that of 'transfer cells' of plants. These hyphae are multinucleate, coenocytic, and contain numerous mitochondria as well as extensive rough endoplasmic reticulum (Duddridge & Read 1984a, b, Massicotte *et al* 1986, Blasius *et al* 1986, Kottke & Oberwinkler 1986a, b 1987) (Fig 3). These are all features suggestive of intensive physiological activity.

In contrast to the apparent uniformity of structures seen in the Hartig net, the fungal mantles which envelope the root show a wide range of morphological and anatomical features which are sufficiently distinctive to allow not only conventional classification of 'type' based upon shape or color such as that provided by Dominik (1969) but, of much greater value, the identification of fungal genera and even species involved in the symbiosis (Agerer 1987, 1991, 1992). Mantle scrapings reveal characteristic patterns of cell arrangement in parenchymatous and plectenchymatous matrices which are diagnostic at the generic, and sometimes species, levels. Thus amongst the important fungal genera the mycorrhiza of which can now with some experience be unequivocally recognised with the aid of a microscope, are *Lactarius, Russula* and *Amanita*. Some commonly occurring types, the fungal partners of which are as yet unidentified, are categorised according to the host genera on which they occur. Hence *Piceirhiza,* and *Fagirhiza* refer to unidentified types on spruce and beech respectively. So long as it is recognised that some of the fungi involved in these associations are not necessarily restricted to these genera, this appears to be a satisfactory arrangement for the present. Advances in molecular technology promise progress towards identification of many of these so far unresolved taxa in the near future.

The most intractable yet perhaps most vital component of the ectomycorrhizal system is the vegetative mycelium, which not only provides contact between the root and its surrounding environment but also fulfills the essential role of exploration and exploitation of the nutrient resources of the soil and their transport to the mantle. This function is dependent upon the integrity of the mycelial network, the inherent fragility of which, combined with the microscopic nature of its individual elements and their inaccessibility in soil make non-destructive analysis problematical. In addition to the requirement for a

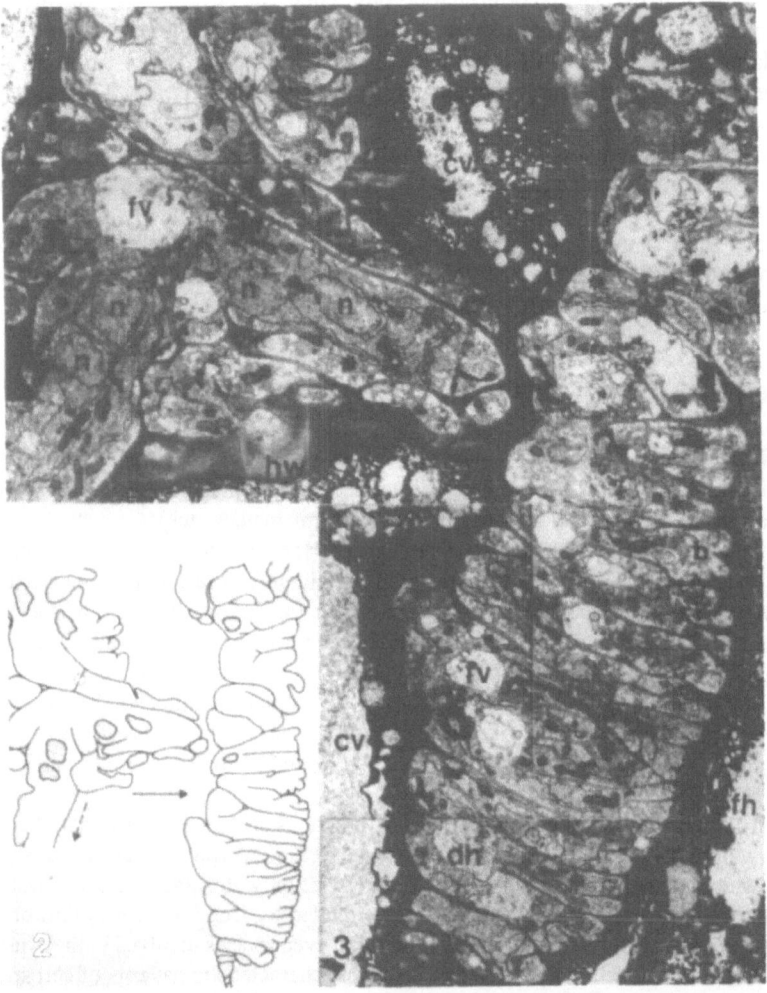

Figure 2. Pattern of branching of hyphal elements of the Hartig net interpreted from a tangential section (see Fig. 3) through mycorrhiza formed by *Amanita muscaria* on *Picea abies*. The main growth direction of the hyphae is transverse to the root axis (solid arrow) except in one case where it is longitudinal to the axis (dashed arrow). Their repeated branching over the surface of the host cell produces an extensive proliferation of hyphal walls and hence a structure with large internal surface area that is analogous to a transfer cell. (Reproduced from Kottke and Oberwinkler 1987 with permission).

Figure 3. Tangential section through the Hartig net in a mature region of an ectomycorrhiza formed between *Picea abies* and *Amanita muscaria*. The fungal structure is interpreted (see Fig. 2) as being a much branched and closely packed series of finger-like hyphal projections which penetrate between the radial walls of the vacuolate cortical cells (cv) of the host. Old hyphae are dilated (dh) and vacuolate (fv) while younger hyphae at the advancing edge of the net are narrow and finely vacuolate (fh). Two pairs of nuclei (n) indicate the dicaryotic and coenocytic nature of the mycelium. Numerous mitochondria (m) and extensive rough endoplasmic reticulum* suggest intense physiological activity. (Reproduced from Kottke and Oberwinkler, 1987, with permission).

non-destructive method of analysis, other prerequisites for a realistic evaluation of the structure and function of the mycorrhizal mycelium include the use of natural substrates free of additional exogenous carbon sources and the maintenance of a microbial community which is as representative as possible of that occurring in nature.

With these constraints in mind Brownlee *et al.* (1983), Read *et al.* (1985), Read (1991, 1992) and Coutts and Nicoll (1990) have investigated the structure of ectomycorrhizal mycelia as they develope from ectomycorrhizal plants over non-sterile forest soil or peat in transparent observation chambers. These studies have enabled the processes of extension and differentiation of the mycelia to be visualised and quantified.

Two types of growth pattern are observed. In one, hyphae emanating from individual colonised roots show finite growth largely as undifferentiated single hyphae producing fans which occupy the soil immediately around that root. If colonised laterals occur in clusters these fans overlap. This pattern is exemplified by such fungi as *Piloderma croceum* and *Cenococcum geophilum*, and appears to be the norm in ascomycetous ectomycorrhiza, as well as in some basidiomycetes.

In the second type, growth of hyphae into uncolonised soil takes place continuously. As a result, the hyphae, which may commence their growth as individuals emanating from single roots as in the previous type, produce much larger fans which advance at a rate of 2-3 mm/day through the substrate. A feature of these fans is that while at the advancing front the hyphae retain their individuality, behind it they aggregate to form compact linear organs which have been variously called mycelial 'cords', 'strands' and 'rhizomorphs'. Because they do not have an organised apex consisting of a meristem-like cluster of hyphal tips, these are not structurally the same as the true rhizomorphs of fungi such as *Armillaria* and the term strand or cord is preferred. In many ectomycorrhizal fungi cords are internally differentiated, wide 'vessel' hyphae having few or no septa and diameters up to 30μm, being ensheathed in finer densely cytoplasmic elements. These strands connect the individual hyphae of the advancing mycelial front with the mantle.

The vessel hyphae do not however enter the mantle. Instead, they branch repeatedly outside the mantle and enter it, again in the form of a fan-like structure of narrower elements, seen by Agerer (1992) as being analogous to a river delta. The zone of contact of the delta-like mycelial system with the mantle varies with mycorrhizal type. In detached roots these hyphal aggregates are normally seen only as disorganised tassel-like structures, but in intact systems it is most common to see them attached basipetally to the mycorrhizal mantle of the individual root, or the cluster of rootlets. (Fig 4a, b). Here they appear to provide the main if not the only point of communication between the mantle and the soil. This is confirmed in observation chambers and in pictures taken of mycorrhizas formed in the field behind root 'windows' (Egli & Kälin 1990) which demonstrate that, whether occurring as individuals or as clusters infected rootlets most commonly are found in air pockets and hence have little or no direct contact with surrounding soil.

Mycelial chords entering the mantle are thus the pipe-lines that connect the colonised rootlets to soil resources which may be discontinuously distributed over considerable distances from the root. The view that the mantle has a storage rather than nutrient capture function is encouraged by the observation that in representatives of many of the most important mycorrhizal genera, such as *Amanita, Rhizopogon, Russula* and *Suillus,* the structure is strongly hydrophobic (Unestam, 1991). This attribute, while being advantageous in terms of providing for retention of acquired resources, would be a considerable disadvantage for an absorptive structure. Some types of mantle, for example, those formed by *Cenococcum* and *Thelephora,* appear not to be hydrophobic indicating that there may be a range of hydrophobicities according to the fungi involved.

At the end distal ends of the mycelial system the hyphae advance as a broad front (Finlay & Read 1986a, b, Coutts & Nicoll 1990a) which provides optimal efficiency of

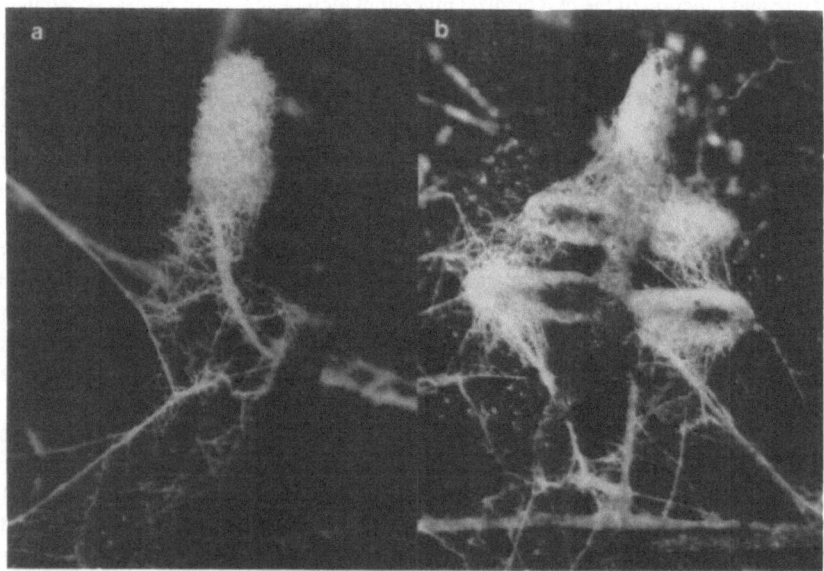

Figure 4. (a) Individual mycorrhizal fine root of *Larix* colonised by *Boletinus cavipes*. (b) Branched mycorrhizal lateral root of *Picea* colonised by *Paxillus involutus*. In both cases colonisation has occurred while the root developed in an air-filled soil pore. Contact between the colonised root and the soil is maintained by mycelial chords through which most of the transport to and from the mycorrhizal sheath must occur. (From Read 1992).

exploration in a system containing patchily distributed resources. The explored territory behind the advancing fan of individual hyphae is occupied by only two types of structure, the compact linear organs which connect the 'front' to the colonised root, and areas in which dense proliferation of individual hyphae is retained in discrete areas. These were termed 'patches' by (Finlay & Read 1986a, b), who hypothesised on the basis of their observation that carbon was preferentially allocated to the mycelium in those areas, that they were localised sites of nutrient enrichment in which intensive foraging by the fungus was occuring. Subsequent studies (Read 1991, Carleton & Read 1991, Bending & Read 1995) have confirmed, by placing organic residues collected from the fermentation (FH) horizon of forest soil in the path of the advancing front, that formation of such patches can be induced and that they persist as discrete units after the front has passed. The patches are to some extent comparable, but on a much smaller scale, with the extensive mycelial mats formed by fungi of genera such as *Hysterangium* and *Gautieria* in soils under *Pseudotsuga* and *Eucalyptus* (Griffiths et al 1990, 1991, Griffiths & Caldwell 1992).

If the intensive colonisation of FH material seen in laboratory microcosms occurs, as would be predicted, when these fungi encounter the same types of resource in the field, it seems that a considerable proportion of the fungal biomass observed to occupy the FH region of forest soils, and which has hitherto been regarded as being largely made up of decomposers (Miller 1974, Visser & Parkinson 1975) is, in fact, attributable to ectomycorrhizal species. Such observations challenge us to re-investigate the functional basis of the ectomycorrhizal symbiosis, in terms of the relationships of the fungi involved both with the host plant and the rest of the soil community.

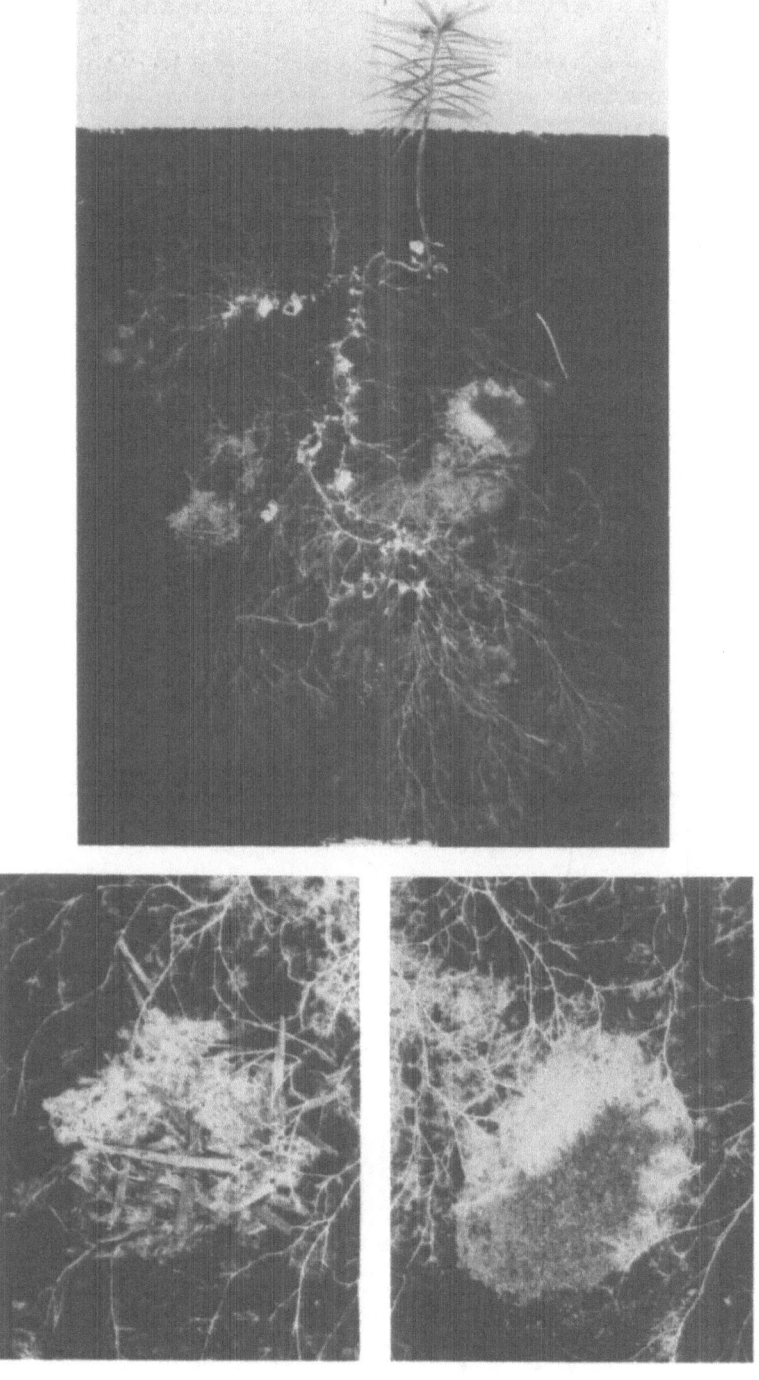

Figure 5. *Larix leptolepis* colonised by *Boletinus cavipes* the mycelium of which are extending across non sterile humified forest soil as an exploratory fan. Intensive development of mycelium to form a patch is induced by introduction of litter, in the uncomminuted (left inset) or comminuted (right inset) form, which had been freshly collected from the fermentation horizon (FH) of forest soil (from Read 1992).

FUNCTIONAL FEATURES OF ECTOMYCORRHIZAL ROOTS

Despite Frank's observation that ectomycorrhizal roots proliferated preferentially in organic soil horizons, and his suggestion that they were here involved in the mobilisation of nitrogen and carbon from these resources, the view has prevailed that these fungi are in fact scavenging largely for mineral ions. It has done so partly because influential work, notably that of Lindeberg (1944), Norkrans (1950) and Lundeberg (1970), suggested that ectomycorrhizal fungi had little or no ability to use complex organic sources of the major nutrients nitrogen and phosphorus. The elegant laboratory studies of Melin and Harley referred to earlier may inadvertently have added credence to this view. Their results clearly demonstrate

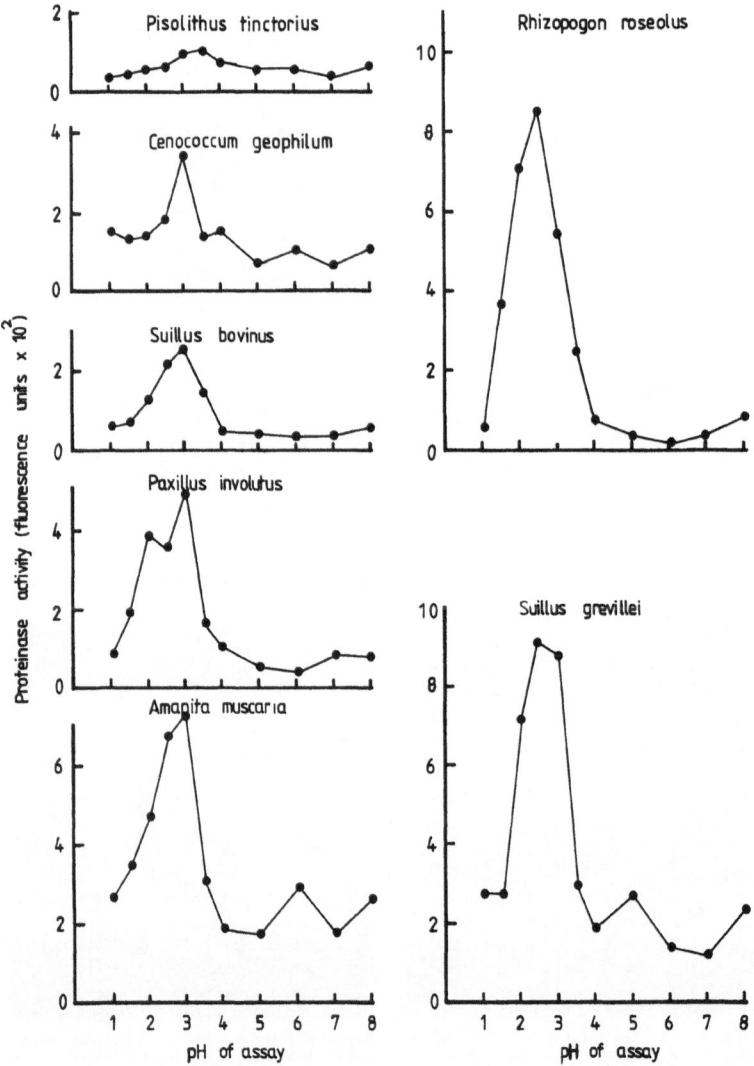

Figure 6. Relative amounts of proteinase activity of a series of ecologically distinctive ectomycorrhizal fungi all grown and assayed under the same conditions (described in Leak & Read 1990). Note the characteristically low pH for activity of the enzyme and its pronounced inhibition above pH 4.0.

that ectomycorrhizal colonisation can enhance the ability of the plant to capture phosphate and ammonium ions from a bathing medium. They do not, however, answer or even address questions concerning the qualitative nature of the source materials being exploited in the real world.

Forests in which trees are predominantly ectomycorrhizal characteristically occur when low rates of mineralisation lead to accumulation of nitrogen in surface organic residues (Post *et al* 1985, Read 1991a). It is a feature of these ecosystems that, in part because the nitrogen requirements of plant tissues are approximately 10 times those of phosphorus, N becomes the most important growth limiting nutrient. This being the case, selection favouring mechanisms which enhance access to the element were to be expected, and Frank's hypothesis that mycorrhizal roots accumulate in the organic residues in order to mobilise N appears to be not unreasonable particularly because we now know that the FH layers of the soil profile in which their roots selectively proliferate are those in which N mobilisation is greatest (Staaf & Berg 1977). The difficulty arises when we come to test the hypothesis.

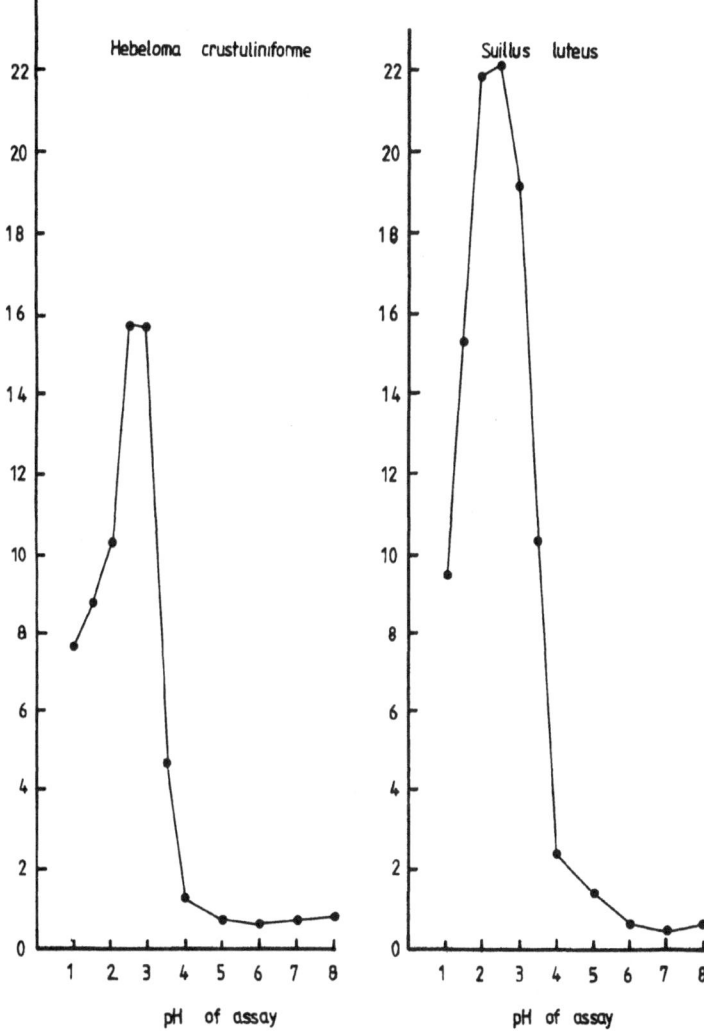

Complex organic polymers, when presented as potential substrates to mycorrhizal fungi in sterile form, are unlikely to be qualitatively representative of those which will be encountered by the fungi in nature. The sterilisation process itself produces structural charges. Alternatively, natural substrates presented in the non-sterile condition contain a mixed microbial population of potential decomposers which makes quantification of activity specifically of the ectomycorrhizal fungi impossible. It is in this situation that a combination of the simplified approach in which the biochemical potential of the fungi can be examined, with that using complex natural substrates, can yield advances.

Laboratory studies using soluble protein as high molecular weight nitrogenous substrates under aseptic conditions have revealed (Melin 1925, Fontana, Abuzinadah & Read 1986a) that some ectomycorrhizal fungi referred to as being 'protein-fungi' can readily use such polymers as sole source of nitrogen. Importantly, when such fungi are grown in mycorrhizal association and supplied with protein as the sole N source, a significant proportion of the nitrogen, which they assimilate largely in the form of amino compounds (Finlay et al 1988; Read, et al 1989), is transferred to the host plant (Abuzinadah and Read, 1986b, 1989). In pure culture, when grown with protein as sole source of N, and supplied with carbon as they would be by their host plants in nature, the protein fungi release ammonium only when the exogenous carbon sources are exhausted. (Read et al 1989). It seems inappropriate in view of these observations to regard mineralised forms as being the only, or even the most important sources of 'plant available' nitrogen in ectomycorrhizal forests.

It has been confirmed that 'protein fungi' produce an acid carboxyproteinase. Moreover, under a given set of standardised conditions, the amounts of activity expressed by a range of such fungi (Fig. 6), while being generally lower than those produced by ericoid fungi, can be broadly related to some of their known ecological attributes. Relationships are apparent on the one hand between enzyme activity and the nature of the substrates characteristically exploited by these fungi in nature, and, on the other, between proteolysis and the experimentally observed patterns of nitrogen transfer from protein to the host plants which they infect. In terms of substrate, *Suillus luteus* and *S. grevillei*, which are amongst the most active producers of protease and associates of pine and larch, respectively, in some of the most recalcitrant litters. At the other extreme *Pisolithus tinctorius*, which produces little enzyme, is a fungus of mineralising environments, as, often, is *Cenococcum geophilum*, which can accompany pioneering hosts on virgin soils at the tree line in alpine environments (Trappe, 1988). Experimental determinations of nitrogen transfer from fungus to host (Abuzinadah and Read, 1989) indicate effectiveness in the order *Hebeloma crustuliniforme*, *Amanita muscaria*, *Paxillus involutus*, which again accords with the order observed in the enzyme assays.

There is a need to treat the results of such screening experiments with caution because the ectomycorrhizal fungi differ in their sensitivities to the culture conditions employed to induce enzyme activities. However, there is little reason to question the basic observation that some ectomycorrhizal fungi have the *potential* to obtain their nitrogen requirements from the primary sources of the element which are the proteins of plant and microbial residues in the horizons occupied by their host roots. There remains the question of whether when presented with *natural* substrates these capabilities are expressed.

The structural affinities of these fungi for natural substrates are not, as shown earlier, in doubt, since their intensive proliferation over FH materials indicates an ideal spatial relationship. An earlier observation that formation of mycelial patches over introduced organic matter from the FH zone was associated with enhanced N content of the foliage of the host (Read 1991a) led to experiments in which the functional aspects of these exploitative modes of growth were examined (Bending & Read 1995). Radioactively labelled carbon was fed to the shoots of mycorrhizal plants, from which ectomycorrhizal mycelia of *Suillus*

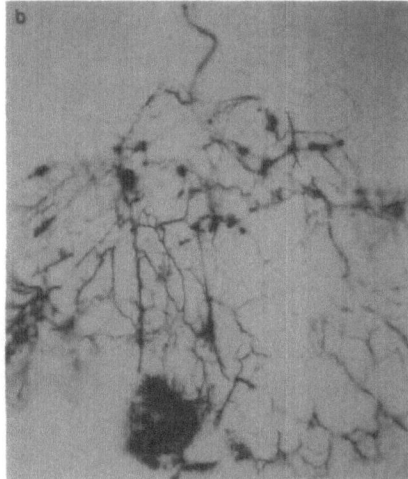

Figure 7. (a) Observation chamber with mycelium of *S. bovinus* with trays of fermentation horizon organic matter (FH) at different stages of colonisation. (b) Autrodiograph of chamber shown in a) after feeding of $^{14}CO_2$ to shoot. Assimilate is allocated only to the tray colonised for less then 40 days (asterisk).

bovinus were extending to form 'patches' of different known ages and stages of development in blocks of introduced organic matter from a pine soil FH horizon. Allocation of carbon to these patches, as revealed by autoradiography, took place over a relatively short period from the time of initial colonisation (time zero) to approximately 40 days later (Fig 7a, b). Sequential visual and microscopic examination of the patches (Fig 8a, b, c, d) indicated initially vigorous colonisation, patch formation, followed by decline and change of color of the ectomycorrhizal mycelium, as well as extensive sporulation of saprotrophic fungi. This pattern of events suggests that the exploitative phase, which appears to be associated with selective allocation of carbon to patches, is confined to a period during which specific resources are available for mobilisation.

By determining the concentrations of the major nutrients N, P & K before and after colonisation of the introduced FH material it was confirmed (Bending & Read 1995) that their exploitation by *S. bovinus* did indeed lead to export of these elements, the concentrations of which declined by 23, 22 and 30% respectively over the c40 day period during which the patches were selectively importing carbon. Equivalent values for substrates colonised by *T. terrestris* in a parallel series of chambers were 13 and 21% for N & K respectively, while no change in P concentration was observed with this fungus (Figs 9a, b, c). As pointed out by Leake (1995) the values of N & P release by *S. bovinus* are comparable with the estimates, 32 & 33% respectively, reported by Entry *et al* (1991) to occur over a growing season in *Pseudotsuga* litter colonised by naturally occurring ectomycorrhizal mat forming fungi. The values reported by Bending and Read convert to fluxes of 78, 4.1 & 6.1 μg gm FH/day for N P & K respectively over the 40 days of active occupation of the introduced

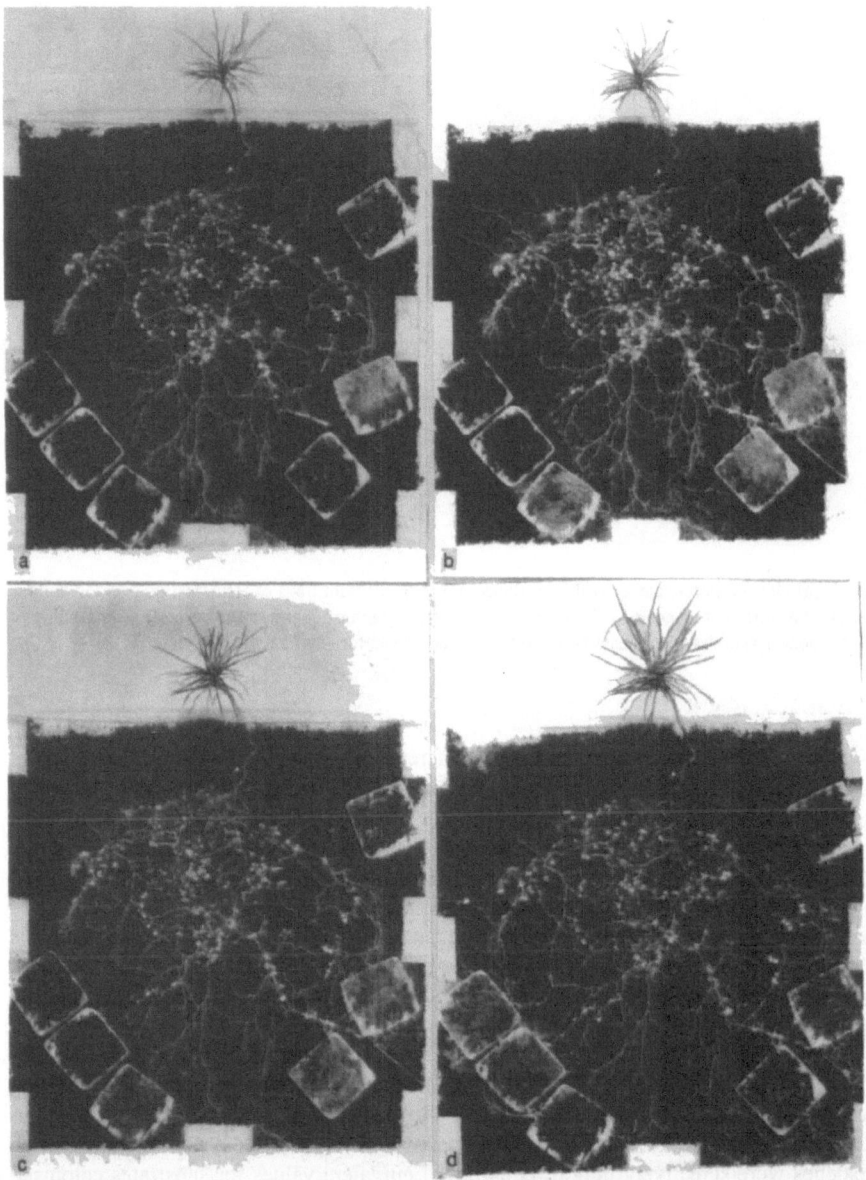

Figure 8. Chronosequence of stages of development of the mycelium of *S bovinus* growing from mycorrhizal roots of *P sylvestris* in a transparent observation chamber to colonise surrounding homogeneous peat and trays of fermentation horizon organic matter (FH) a) Day 0 b) Day 14 c) Day 28 d) Day 42

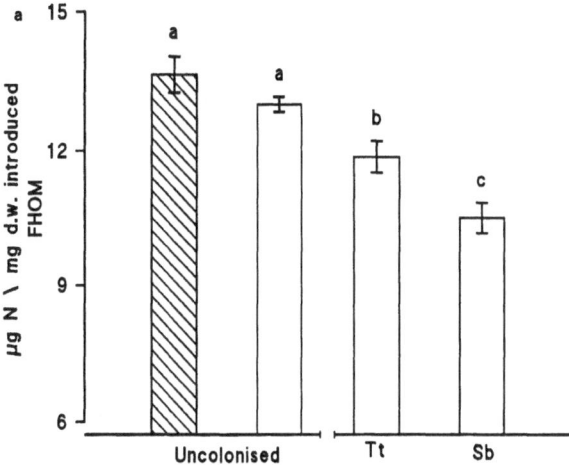

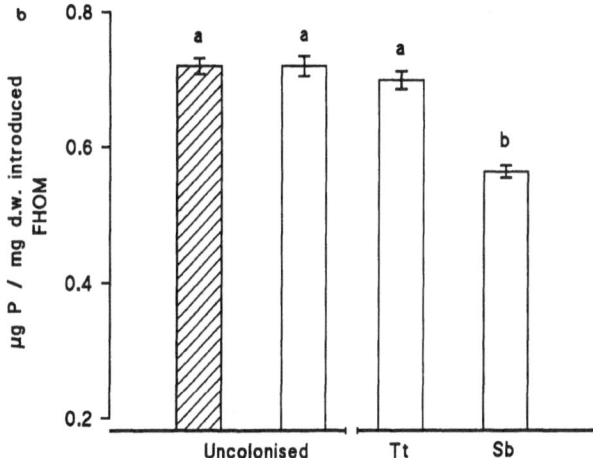

Figure 9. (a) Initial (hatched) and final (open) nitrogen concentrations in uncolonised FH relative to that after colonisation by *S. bovinus* and *T. terrestris*. (b) Initial (hatched) and final (open) phosphorus concentrations in uncolonised FH relative to that after colonisation by *S. bovinus* and *T. terrestris*. (c) Initial (hatched) and final (open) potassium concentrations in uncolonised FH relative to that after colonisation by *S. bovinus* and *T. terrestris*. Bars represent +/- standard error of the mean. Significance of differences in nitrogen (a), phosphorus (b), and potassium (c) concentration, as determined by 1-way analysis of variance. (Treatments possessing different letters are significantly different p<0.01).

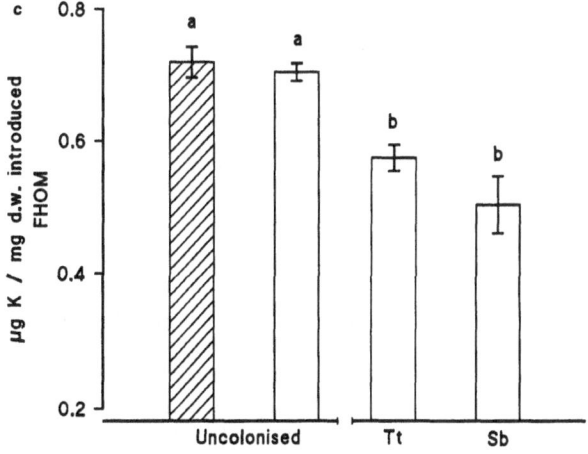

material. Measurements of rates of N mineralisation were made in uncolonised samples of the FH substrate incubated under identical conditions. These revealed that microbial ammonification was insufficient to account for the observed pattern of N depletion, and add support to the view that the ectomycorrhizal fungi are directly involved in exploitation of the organic residues.

However, since the natural substrates employed in these studies contain a normal micro flora and fauna such results cannot elucidate the extent to which the mycorrhizal fungi are involved in mobilisation. A process of facilitated attack by saprotrophs could be occurring, carbon leaked from mycorrhizal fungi being used to sustain detritivores. While some facilitation of this kind may occur, Ingham et al (1991) could find no evidence of consistent increases of bacterial over fungal numbers in soils colonised by ectomycorrhizal mats, and the density of the mycelium formed by the mycorrhizal fungus itself in 'patches' and 'mats' would appear physically to preclude a large increase in biomass of fungal saprotrophs. These observations coupled with the knowledge that some ectomycorrhizal fungi have the ability to produce a suite of enzymes capable of initiating decomposition processes, eg lignase (Griffiths & Caldwell 1992, Haselwandter et al 1990, Trojanowski et al 1984) polyphenol oxidase (Giltrap 1982, Griffiths & Caldwell 1992) peroxidase (Griffiths & Caldwell 1992) and fatty acid esterase (Caldwell et al 1991, Hutchinson 1990a) as well as those proteases phosphatases enabling direct attack upon N & P containing polymers, all suggests that they are directly rather than secondarily involved in mobilisation as well as export of essential nutrients from colonised litter.

The reduction of quality of these substrates, that has been detected most readily as an increase of C:N ratio in patches (Bending & Read 1995) and mats (Griffiths et al 1990, 1991) as exploitation proceeds, may significantly contribute to the slowing in rate of decomposition observed by Gadgil & Gadgil (1971) in litter occupied by mycorrhizal roots of pine.

Colonisation of FH residues by S. bovinus also leads to export of phosphorus. That ectomycorrhizal fungi release phosphatases when exposed to organic sources of P is increasingly clear (Antibus et al 1992). Since the bulk of P in the rooting environment of forest soils is, as with the case of N, in the organic form phosphatase activity is likely to be of great importance. The dynamics of the turnover of the element are also thought to be controlled by the same energy - dependent microbial activities as are those of N (Berg & McClaugherty, 1989). The residues of membranes, mostly phospholipid, and of the nucleic acids DNA & RNA, are likely to be among the primary products of cell turnover in soil. The ability of the non-carbon limited ectomycorrhizal mycelium to express phospho mono- and di-estarase activity in close association with the main sites of turnover in the FH horizon could be of greater biological significance in some circumstances than in their ability to access organic N. In particular in mull humus soils capable of supporting an active population of nitrifying bacteria in the rooting environment, P still largely present in organic form may become the major growth limiting element. Here, and in the wider boreal forest context where N saturation is increasingly occurring as a result of pollutant inputs (Aber et al 1989) phosphatase activities may be key determinants of ecosystem productivity and the need for further study of these enzymes and for improved characterisation of the substrates which they attack has been emphasised (Griffiths & Caldwell 1992, Leake 1995).

Observation of activities of ectomycorrhizal fungi in laboratory microcosms helps to provide a functional basis for selective pattern of distribution of ectomycorrhizal roots, first described by Frank, in the organic horizons between litter and humified layers of the soil. This is the region in which mobilisation of N & P is most active, periods of net immobilisation of the elements associated with freshly fallen litter probably as saprotrophs import them to facilitate exploitation of its carbon (Berg & Söderström 1979), being followed by a phase of mineralisation in the FH layer. The conventional explanation of this process is that its arises when the saprotrophs become carbon limited (Berg & Staaf 1980, Berg & McClaugherty 1989). However, most considerations of the nutrient dynamics of the FH horizon have ignored the fact that, as shown in the microcosms, the FH substrates are selectively colonised by ectomycorrhizal fungi which, because of their attachment to the root, are unlikely to be carbon limited. The possibility is, therefore, that under these circumstances the mycorrhizal fungi are able to express their biochemical capability to mobilise the N & P from the substrates during a particular stage, when, because of their carbon limitation, the remainder of the microbial population is inactive.

The production of an extensive mycelium in order to explore the habitat and exploit its resources is not achieved without cost to the plants. Precise quantification of these costs is difficult to obtain but by direct measurement of mycelial lengths, or by determination of respiratory activity, conversion to biomass and hence some estimate of carbon demand can be achieved. Using the former methods (Read 1992) calculated 200 metres of mycelial length per gram dry peat in microcosms, while, independently, using regression relationships (Finlay & Söderström 1989) arrived at the same value as being representative of a *Lactarius rufus* dominated forest soil in Sweden. Using conversion factors developed by Söderström (1979) this length measurement is equivalent to 3.5 kg live mycelium/ha dry weight of standing crop. Turnover rates are uncertain but Finlay & Söderström assuming a once weekly rate during the 5 months vegetation period obtained an annual production of 70 kg mycelial biomass/ha yr. When the fungal content in the mycorrhizal sheath, c 730 kg/ha, and that of fruit bodies, c 30 kg/ha, are added, a total fungal biomass of 830 kg/ha year was obtained (Finlay & Söderström 1992). Assuming then 40% of this biomass is in the form of carbon, and that the conversion efficiency is 60%, the carbon demand of the ectomycorrhizal fungi in this forest will be 830 kg C/ha/year. Since estimates of photosynthetic fixation of carbon for the same forest are known to be of the order of 5,800 kg C/ha/year, it become evident that the ectomycorrhizal fungi consume 14-15% of the assimilated carbon. This figure is comparable with those obtained by Vogt *et al* (1982) and Fogel & Hunt (1979) for ectomycorrhizal forests in North America. It is somewhat lower than the 30% of total assimilate production observed by direct measurement of mycelial respiration in microcosms (Söderström & Read 1989) but this is to be expected, in view of the fact that the former system consists of mature trees which themselves have a large component of maintenance respiration, and the latter upon seedlings with little storage capability. Söderström & Read (1989) showed by measurement of respiratory output of the mycelium of a number of fungal species before and after their connection with the plant was cut, that activity of the fungi was almost completely dependent upon the supply of current assimilate from the host plant.

The ectomycorrhizal system thus appears to involve allocation of considerable amounts of a non-growth limiting resource, carbon, to secure the capture of those resources, nitrogen and phosphorus, which do limit growth in many of the natural ecosystems domi-nated by plants with this type of symbiosis. While the importance of such carbon investment is here emphasised largely in terms of nutrition of the tree, its role in sustaining the activities of the whole of the soil ecosystem should not be overlooked.

ECOLOGICAL CONSIDERATIONS - A SYNTHESIS OF LABORATORY AND FIELD OBSERVATIONS

Analyses of the structural and functional attributes of selected fungi such as *Suillus bovinus*, growing on natural substrates, while useful in clarifying likely nutritional roles, still provide greatly oversimplified views of the situations prevailing in nature where mixed communities of plant and fungal species co-exist and interact. In order to understand their interrelationships there is no alternative but to depend upon careful field observation of the kind traditionally made by those adopting the more holistic alternative path in mycorrhizal research. Access to the powerful tools of molecular biology now, however, promises to provide a much greater rigour in studies which attempt to unravel the complicated web of these microbial interactions. So far these tools have allowed us only to plot the distribution of distinctive genomes. In future they may be developed to permit a more dynamic understanding.

Since few, if any, ectomycorrhizal fungi have the ability to develop normally in the absence of an autotrophic partner, the prerequisite for expression of any of the biochemical attributes described earlier, is the 'capture' of the root of a compatible host plant which can then serve as a carbon donor. In practice, in undisturbed natural communities, such capture will probably by facilitated by the foraging activities of mycelia growing from a pre-established host, and inter-specific linkages between host plants will be facilitated by the low host specificity shown by the majority of ectomycorrhizal fungi (Molina *et al* 1992). It is important if we are to determine the ecological significance of mycorrhizal foraging patterns, and the inter-linking at the intra and interspecific levels which it can facilitate, to know more about the nature of specificity phenomena.

Understanding of these relationships has progressed furthest in the Douglas-fir - Pine forests of the Pacific north west of America. Trappe (1977) originally calculated that up to 2000 fungal species were capable of forming ectomycorrhiza with Douglas fir, and a similar number are now thought to be associated with pine. However, knowledge of the extent to which these fungi are capable of forming mycorrhizal associations on both hosts is necessary for an understanding of the function of the symbiosis at the ecosystem level. It has been estimated (Molina *et al* 1992) that of the approximately 2000 species compatible with each genus around 1800, or 72%, may be able to form mycorrhiza with both. Thus, although they each have fungi, notably in the genera *Rhizopogon* and *Suillus,* which are specific to them, they still have the potential to share the great majority of the potential symbionts that may be growing from one host towards the other. Since low host specificity increases the potential number of food bases available to the fungus in a community of mixed host species, selection would be expected to favour this trait rather than one involving narrow host range. The same general argument can be applied to the host if it is dependent upon rapid infection for establishment and nutrient acquisition.

Examples of situations at the ecosystem level in which there are possible advantages to the plant of showing broad receptivity are beginning to emerge. Significantly better post-disturbance regeneration (Amaranthus *et* 1990) and early growth (Amaranthus & Perry 1989) of conifers has been observed close to plants such as *Arbutus* or *Arctostaphylos* which act as sources of appropriate fungal inoculum, as compared with that obtained in grassland, open areas or shrubs with VA colonisation. Using vegetation surveys and seedling survival assays Horton & Parker (1994) showed the *Pseudotsuga* preferentially established in *Arctostaphylos* dominated patches and hypothesised that the latter was a source of mycorrhizal inoculum for the conifer. By means of PCR amplification of nuclear ribosomal DNA extracted either from *Arctostaphylos* or *Pseudotsuga* mycorrhiza, as well as from fungal fruit bodies, they were able to show that several of the fungi, including species of *Russula* and *Lactarius* were found on both hosts.

Applications of molecular techniques of this kind suggest that species of mycorrhizal fungi normally regarded as being exclusively involved in ecto- assoications may also form or be associated with monotropoid and orchid mycorrhiza (Bruns personal communication). These observations are of great interest not least because they suggest a possible source of, and pathway for, carbon transfer from an autotroph to the heterotrophic or partially heterotrophic, monotropoid and orchid plant. If these observations are confirmed they will provide great incentive for experimental analysis of the physiological relationship between both types of plant and their fungal symbionts.

Since the demonstration, using simplified laboratory microcosms, of the rapid formation by ectomycorrhizal fungi of linkages at both the intra- and inter specific level (Brownlee *et al* 1983, Read 1984, Read *et al* 1985) there has been much discussion of the possible significance of the phenomenon in the field (Read *et al* 1985, Newman 1988, Perry *et al* 1989, Read 1990). While such linkages provide the potential for flow of nutrients between plants, it is, as Newman points out, difficult to obtain unequivocal evidence of net transfer by this route. In fact as Read *et al* (1985) point out, it is likely that the main benefits of such linkage will accrue at the establishment phase and be associated with improved access to soil resoucres. As the emerging radicle enters the soil at the time of germination, its seed reserves already depleting, it encounters an environment in which most of the resources are already being exploited by established plants. Rapid incorporation into the pre-formed mycorrhizal mycelial network, favoured by low specificity, will enable the seedling at little energy cost to itself potentially to be in contact with the large catchment provided by the foraging activities of the heterotroph. Indirect evidence for the occurrence of such benefits in the field has been provided by the observation that all Douglas-fir seedlings found under the canopy of maturing 60-75 year old stand of the same species were associated with mycelial mats formed by ectomycorrhizal fungi (Griffiths *et al* 1991). The experiments of Amaranthus and colleagues reported earlier are also strongly suggestive of these effects.

The advantages of low host specificity to both host and fungus appear to be so great that the occurrence of the converse phenomenon, strong specificity, albeit only at the genus level in both hosts and fungi may, at first sight, be a puzzle. The answer, however, in part may lie in the fact that most genera that are hosts of ectomycorrhizal fungi produce litter of a distinctive quality which can be clearly defined in terms of its nutrient or C:N ratio (Read 1991). There is thus a substrate induced selection which would in turn be expected to lead to preferential compatibility between hosts and those fungi that were particularly efficient at mobilising key nutrients from their own residues. The inter-generic heterogeneity of litter quality produced by ectomycorrhizal hosts and the parallel diversity of physiological attributes seen in their fungal associates contrasts markedly with the homogeneity and high resource of quality of most plant litters produced by hosts of vesicular-arbuscular (VA) fungi and the parallel lack of biochemical specialisation or specificity shown by the fungi forming the VA type of mycorrhiza.

If, as suggested by this view, selection favouring strong specificity has been influenced strongly by aspects of resource quality this should be reflected in nature by the appearance, on a given species, of increasing proportions of host specific fungi as its characteristic organic residues accumulate with time. This type of successional pattern was reported by Last *et al* (1987) who observed fungal generalists of low host specificity to predominate in the pioneering stages of development of a *Betula* plantation on an agricultural soil, numbers of species of narrow host range increasing with age of the stand, in effect as organic residues accumulated. However, Molina *et al* (1992)observed that in the case of Douglas fir a high proportion of genus specific *Rhizopogon* species was found throughout the life of the plant. In terms of litter quality, that of Douglas fir along with Larch, is amongst the poorest and it may be that in both of these genera the requirement, from an early stage, for fungi effective at mobilising nutrients from peculiarly recalcitrant substrates was a driving force selecting in favour of narrow specialists. This pattern of selection can be termed 'resource selectivity' and there is a need for further investigation of its role in defining the ecological specificity widely revealed in nature. Understanding of resource selectivity will be fundamental if we are to progress further in attempts to use mycorrhizal inoculation as a management tool.

PRACTICAL APPLICATIONS

What emerges from any study of the ectomycorrhizal relationship as it occurs in natural ecosystems is a complex picture in which many fungal symbionts successfully co-exist with the equally diverse population of microbial saprotrophs, on the one hand, and with a generally less diverse range of host plants on the other. The factors that determine success in these fungi are thus, compatibility with the host (host-specificity), with the substrate (resource-specificity) and with the wider environment containing other microorganisms that may be competing for these substrates (ecological specificity).

It would appear that in the relatively simple environment of the forest nursery in which infection of forest trees often first arises in managed systems, processes of cultivation, fertilisation and probably also fungicide application, combine to select fungi of low host and resource specificity. There is evidence that in these situations generalists such as *Laccaria laccata, Thelephora terrestris* (LeTacon *et al* 1992) and *Pisolithus tinctorius* (Marx *et al* 1992) are natural dominants. Laboratory experiments such as those described earlier suggest that these will have a low ability to mobilise some or all of the essential nutrients from complex organic resources, though inoculation experiments indicate that they can be strongly beneficial to seedlings in skeletal or mineral soils.

The problem arises that in the diverse and spatially heterogeneous environment of the outplanting site such 'generalists' may not be at all appropriate to the needs of the host. Here, especially if as often happens in replant sites, organic residues of the previous crop of the same species are to be encountered, fungi with a combination of host and resource specificity may be essential. It has been shown in a number of outplanting trials with Douglas-fir and pine that inoculation with *Rhizopogon* species of narrow host range gives better response than that achieved with the generalists described above (Bledsoe *et al* 1982, Molina *et al* 1992). Further studies of resource specificity are required but it is hypothesised that growth and survivorship of generalists will be adversely affected in environments which contain a large proportion of resources from which they are unable to mobilise essential nutrients. Research should focus on whether such effects, if observed, arise directly from mineral starvation of the fungus, or indirectly through impairment of carbon supply from an increasingly N & P starved host. Reduction of vigour of a generalist population, if it occurred for either of these reasons, could be expected to lead to their progressive replacement, by resource specific fungi capable of supporting the photosynthetic activity of their hosts.

Faced with the complexity of the natural environment, with the uncertainty about the dynamics of microbial communities in general, and with little awareness of the stability and longevity of mycorrhizal populations in particular, those wishing to manipulate mycorrhizal fungi in the field would be well advised to adopt a pragmatic approach.

When introducing a host with a pre-inoculated fungus into a soil environment which has not hitherto supported or does not now contain a population of mycorrhizal fungi, experience suggests (Marx 1992, Grove & LeTacon 1993) that there is a good chance of persistence of the association.

In any naturally occurring mycorrhizal community, in contrast, selection operating throughout the development of the ecosystem, will have produced a population of fungal symbionts that is in equilibrium with local conditions. This is particularly so if resource specificity is as important as was suggested earlier. Under these circumstances it is unlikely that an alien organism will compete successfuly or even survive in completion with the established community. There may now be the potential to genetically engineer or select particularly vigorous races of a fungal symbiont, even to proudce those with desired physiological attributes. However, it is the diversity - physical, chemical and biological - which characterises the soil environment of the ectomycorrhizal community, and it is doubtful, even if it was desirable, that an organism selected to express one or two attributes would function effectively in face of such complexity. It is appropriate to caution strongly against introduction of any 'vigorous' ectomycorrhizal symbiont, produced by whatever means, if it is to challenge the diversity of the extant population. Biological diversity has its own intrinsic merits not least because it can provide a buffer against changes, abrupt or gradual with which any ecosystem can inevitably be challenged. The pragmatist wishing to manipulate the population of fungal symbionts should therefore do so with sensitivity and where possible select from the spectrum of organisms naturally occurring in the site to be managed.

CONCLUSION

Research on ectomycorrhiza has followed somewhat independent paths since the time of Frank. One, laboratory based, has progressed towards providing us with some understanding of the nutritional role of mycorrhizal fungi, and in particular with knowledge of the activities of their mycelia in natural substrates. The other, field based, has described the communities of organisms which occur in nature and highlighted the complexities of their popultion dynamics.

These paths should now come together so that questions concerning the interactive behaviour of these fungi in nature can be addressed. Access to molecular techniques for identification of mycelia promises to tell us much more about the structure and composition of mycorrhizal populations but the information which these approaches provide will only be of value if we are able to interpret species composition in terms of the function of the identified organisms in the ecosystem. This necessitates close collaboration, indeed, a series of symbiotic interactions, between field ecologists, molecular biologists, and physiologists designed to elucidate inter-specific, spatial and temporal aspects of the acitivities of ectomycorrhizal fungi in nature. Only by combining our efforts in this way will clearer understanding of the role of the ectomycorrhizal symbiosis in the dynamics of natural ecosystems be achieved.

REFERENCES

Aber, J D , Nadelhoffer, K J , Steudler, P, and Melillo, J M , 1989, Nitrogen saturation in northern forest systems, *Bio Science* 39 378-386

Abuzinadah, R A , and Read, D J , 1986a, The role of proteins in the nitrogen nutrition of ectomycorrhizal plants I Utilization of peptides and proteins by ectomycorrhizal fungi, *New Phytol* 103 481-493

Abuzinadah, R A , and Read, D J , 1986b, The role of proteins in the nitrogen nutrition of ectomycorrhizal plants III Protein utilization by *Betula, Picea,* and *Pinus* in mycorrhizal association with *Hebeloma crustuliniforme, New Phytol* 103 507-514

Abuzinadah, R A , and Read, D J , 1989, The role of proteins in the nitrogen nutrition of ectomycorrhizal plants V Nitrogen transfer in birch (*Betula pendula*) grown in association with mycorrhizal and non-mycorrhizal fungi, *New Phytol* 112 61-68

Agerer, R., 1987, *Colour Atlas of Ectomycorrhizae*, Einhorn-Verlag, Schwabisch Gmund

Agerer, R , 1991, Characterization of ectomycorrhizae, Methods in Microbiology 23 25-73, Academic Press, London

Agerer, R , 1992, Ectomycorrhizal rhizomorphs - organs of contact In· *Mycorrhizas in Ecosystems,* D J Read, D H Lewis, A H. Fitter and I J Alexander (eds), CAB, Wallinford, UK, pps 84-90

Alexander, I. J , and Hogberg, P , 1986, Ectomycorrhizas of tropical angiospermous trees, *New Phytol* 102: 541-549

Amaranthus, M P , and Perry, D A , 1989, Interaction effects of vegetation type and Pacific madrone soil inocula on survival, growth, and mycorrhiza formation of Douglas-fir, *Can J For Res* 19: 550-556

Amaranthus, M. P., Molina, R., and Perry, D. A., 1990, Soil organisms, root growth and forest regeneration, *Proceedings of the Society of American Foresters*, pp. 89-93, National Convention, Spokane, Washington, September, 1989.

Antibus, R. K., Sinsabaugh, R. L., and Linkins, A. E., 1992, Phosphatase activities and phosphorus uptake from inositol phosphate by ectomycorrhizal fungi, *Can. J. Bot.* 70: 794-801.

Bending, G. D., and Read, D. J., 1995, The structure and function of the vegetative mycelium of ectomycorrhizal plants. V. The foraging behaviour of ectomycorrhizal mycelium nd the translocation of nutrients from exploited organic matter, *New Phytol.* (In press).

Berg, B., and McClaugherty, C. A., 1989, Nitrogen and phosphorus release from decomposing litter in relation to the disappearance of lignin, *Can. J. Bot.* 67: 1148-1156.

Berg, B., and Söderström, B., 1979, Fungal biomass and nitrogen in decomposing Scots pine needle litter, *Soil Biol Biochem* 11: 339-341.

Berg, B., and Staaf, H., 1981, Leaching, accumulation and release of nitrogen in decomposing forest litter. In: Clark, F. E., Rosswall, T. (eds), *Terrestrial nitrogen cycles.* (Ecological Bulletins 33) Swedish Natural Science Research Council, Stockholm.

Blasius, D., Feil, W., Kottke, I., and Oberwinkler, F., 1986, Hartig net structure and formation in fully ensheathed ectomycorrhizas, *Nord. J. Bot.* 6.

Bledsoe, C. S., Tennyson, K., and Lopushinsky, W., 1982, Survival and growth of outplanted Douglas-fir seedlings inoculated with mycorrhizal fungi, *Can. J. For. Res.* 12: 720-723.

Borchers, S. L., and Perry, D. A., 1990, Growth and ectomycorrhiza formation of Douglas-fir seedlings grown in soils collected at different distances from pioneering hardwoods in southwest Oregon clear-cuts, *Can. J. For. Res.* 20: 717-721.

Brownlee, C., Duddridge, J. A., Malibari, A., and Read, D. J., 1983, The structure and function of mycelial systems of ecto-mycorrhizal roots with special reference to their role in forming inter-plant connections and providing pathways for assimilate and water transport, *Plant Soil* 71: 433-443.

Carlton, T. J., and Read, D. J., 1991, Ectomycorrhizas and nutrient transfer in conifer-feathermoss ecosystems, *Can. J. Bot.* 69: 778-785.

Chilvers, G. A., and Pryor, L. D., 1965, *Aust. J. Bot.* 69: 245-261.

Clement, A. J., Garbaye, A. J., LeTacon, I., 1977, Importance des ectomycorrhizas dans la résistance au calcaire du Pin noir (*Pinus nigra* Arn. ssp. *nigricans* Host). *Oecol Plant* 12: 111-132.

Coutts, M. P., and Nicoll, B. C., 1990, Growth and survival of shoots, roots and mycorrhizal mycelium in clonal Sitka spruce during the first growing season after planting, *Can. J. For. Res.* 20: 861-868.

Dominik, T., 1969, Key to ectotrophic mycorrhizae, *Folia For. Pol. Ser.* 15: 309-321.

Duddridge, J. A., and Read, D. J., 1984a, The development and ultra-structure of ectomycorrhizas. I. Ectomycorrhizal development on pine in the field, *New Phytol.* 96: 565-573.

Duddridge, J. A., and Read, D. J., 1984b, The development and ultra-structure of ectomycorrhizas. II. Ectomycorrhizal development on pine in vitro, *New Phytol.* 96: 575-582.

Egli, S., and Kälin, I., 1990, The root window - a technique for the *in vivo* observation of mycorrhiza in the field, *Agri. Ecosy. Env.* 28: 107-111.

Entry, J.A., Rose, C. L., Cromack, K. Jr., 1991a, Litter decomposition and nutrient release in ectomycorrhizal mat soils of a Douglas fir ecosystem, *Soil Biol. Biochem.* 23: 285-290.

Finlay, R. D., and Read, D. J., 1986a, The structure and function of the vegetative mycelium of ectomycorrhizal plants. I. Translocation of ^{14}C-labelled carbon between plants interconnected by a common mycelium, *New Phytol.* 103: 143-156.

Finlay, R. D., and Read, D. J., 1986b, The structure and function of the vegetative mycelium of ectomycorrhizal plants. II. The uptake and distribution of phosphorus by mycelial strands inter-connection host plants, *New Phytol.* 103: 157-165.

Finlay, R. D., Ek, H., Odham, G., and Söderström, B., 1988, Mycelial uptake, translocation and assimilation of nitrogen from ^{15}N-labelled ammonium by *Pinus sylvestris* plants infected with four different ectomycorrhizal fungi, *New Phytol.* 110: 59-66.

Finlay, R. D., and Söderström, B., 1989, Mycorrhizal mycelia and their role in soil and plant communities. In: *Developments in Plant and Soil Sciences, Vol 39, Ecology of Arable Lane, Perspectives and Challenges* (Ed. by M. Clarholm & L. Bergström), pp. 139-148. Kluwer Academic Publishers, Dordrecht/London.

Finlay, R. D., and Söderström, B., 1992, Mycorrhiza and carbon flow to the soil. In: *Mycorrhiza Functioning*, M. Allen ed, Chapman & Hall, London, pp. 134-160.

Fogel, R., and Hunt, G., 1979, Fungal and arboreal biomass in a western Oregon Douglas fir ecosystem: Distribution patterns and turnover, *Can. J. For. Res.* 9: 245-256.

Fontana, A., 1963, Micorrize ectotrofiche in una Ciperacea: *Kobresia bellardii, Giorn. Bot. Ital.* 70: 639-641.

Frank, A B , 1894, Die Bedeutung der Mykorrhizapilze fur die gemeine Kiefer, *Forstwiss Centralb* 16 1852-1890

Gadgil, R L , and Gadgil, P D , 1971, Mycorrhiza and litter decomposition, *Nature* 233 133

Giltrap, N J , 1982, Production of polyphenol oxidases by ectomycorrhizal fungi with special reference to *Lactarius* spp , *Trans Brit mycol Soc* 78 75-81

Griffiths, R P , Galdwell, B A , Cromack, K Jr, and Morita, R Y , 1990, Douglas-fir forest soils colonized by ectomycorrhizal mats I Seasonal variation in nitrogen chemistry and nitrogen cycle transformation rates, *Can J For Res* 20 211-218

Griffiths, R P , Caldwell, B A , Ingham, E R , Castellano, M A , and Cromack, K Jr , 1991a, Comparison of microbial activity in ectomycorrhizal mat communities in Oregon and California, *Biol Fert Soils* 11 196-202

Griffiths, R P , Castellano, M A , and Caldwell, B A , 1991b, Ectomycorrhizal mats formed by *Gautieria monticola* and *Hysterangium setchellii* and their association with Douglas-fir seedlings, a case study, *Plant and Soil* 134 255-259

Griffiths, R P , and Caldwell, B A , 1992, Mycorrhizal mat communities in forest soils In *Mycorrhizas in Ecosystems* Eds D J Read, D H Lewis, A H Fitter and I J Alexander CAB International, Cambridge

Grove, T S , and LeTacon, F , 1993, Mycorrhiza in plantation forestry, *Adv Plant Pathol* 9 191-227

Harley, J L , and Smith, S E , 1983, *Mycorrhizal Symbiosis*, Academic Press, London

Harvey, A E , Larsen, M J , and Jurgensen, M F , 1976, Distribution of ectomycorrhizae in a Douglas fir/Larch forest system in Western Montana, *For Sci* 22 393-398

Haselwandter, K , Read, D J , 1980, Fungal associations of roots of dominant and sub-dominant plants in high-alpine vegetation systems with special reference to mycorrhiza, *Oecologia* (Berl) 45 57-62

Haselwandter, K , Bobleter, O , and Read, D J , 1990, Degradation of ^{14}C-labelled lignin and dehydropolymer of coniferyl alcohol by ericoid and ectomycorrhizal fungi, *Archiv Microbiol* 153 352-354

Hesselman, H , 1900, Om mykorrhizabildingar hos arktiska vaxter Bih Svenska Vetensakad Handl 26 1-46

Hogberg, P , 1992, Root symbioses of trees in African dry tropical forests, *J Veg Sci* 3 393-400

Horton, T , and Parker, V T , 1995, Mycorrhizal facilitated succession is a viable hypothesis to explain the establishment of *Pseudotsuga menziesii* in *Arctostaphylos* dominated chapparal, (In press)

Hutchinson, L , 1990a, Studies on the systematics of ectomycorrhizal fungi in axenic culture II The enzymatic degradation of selected carbon and nitrogen compounds, *Can J Bot* 68 1522-1530

Hutchison, L , 1990b, Studies on the systematics of ectomycorrhizal fungi in axenic culture III Patterns of polyphenol oxidase activity, *Mycologia* 82 424-435

Ingham, E R , Griffiths, R P , Cromack, K Jr, and Entry, J A , 1991, Comparison of direct versus fumigation-flush microbial biomass estimates from ectomycorrhizal mat and non-mat soils, *Soil Biol Biochem* 23 465-71

Kottke, I , and Oberwinkler, F , 1986, Mycorrhiza of forest trees, *Trees* 1 1-24

Kottke, I , Oberwinkler, F , 1986, Root fungus interactions observed on initial stages of mantle formation and Hartig net establishment in mycorrhizas of *Amanita muscaria* (L ex Fr) Hooker on *Picea abies* (L) Karst in pure culture, *Can J Bot* 64 2348-2354

Kottke, I , Oberwinkler, F , 1987, Cellular structure and function of the Hartig net coenocytic and transfer cell-like organization, *Nord J Bot* 7 85-95

Lapeyrie, F F , and Chilvers, G A , 1985, An endomycorrhiza - ectomycorrhiza succession associated with enhanced growth of *Eucalyptus dumosa* seedlings planted in a calcareous soil, *New Phytol* 100 93-104

Lapeyrie, F F , 1990, The role of ectomycorrhizal fungi in calcareous soil tolerance by "Symbiocalcicole" woody plants, *Ann Sci For* 21 579-589

Last, F T , Dighton, J and Mason, P A , 1987, Successions of sheathing mycorrhizal fungi, *Trends Ecol and Evol* 2 157-161

LeTacon, F , Alvarez, I F , Bouchard, D , Henrion, B , Jackson, R M , Luff, S , Parlade, J I , Pera, J , Stenstrom, E , Villeneuve, N , and Walker, C , 1992, Variations in field responses of forest trees to nursery ectomycorrhizal inoculation in Europe In *Mycorrhizas in Ecosystems*, D J Read, D H Lewis, A Fitter, I J Alexander eds (1992), CAB International, Wallinford, UK, pps 119-134

Leake, J R , 1995, Nutrient mobilisation from organic matter by ericoid and ectomycorrhizal fungi some recent advances *Proc 4th European Symposium in Mycorrhiza*, (J Barea *et al* eds) in press

Lindeberg, G , 1944, Uber di Physiologie Ligninabbauender Boden/hymenomyzeten, *Symb botanicae Upsal* 8(2), 1-183

Lundeberg, G , 1970, Utilisation of various nitrogen sources, in particular bound nitrogen, by mycorrhizal fungi, *Stud For Sue* 79 1-95

Max, D H , Ruehle, J L , and Cordell, C E , 1992, Methods for studying nursery and field response of trees to specific ectomycorrhiza, *Methods in Microbiology* 23, Academic Press, London

Massicotte, H B , Peterson, R L , Ackerley, C A , Piche, Y , 1986, Structure and ontogeny of *Alnus crispa-Alpova diplophloeus* ectomycorrhizae, *Can J Bot* 64 177-192

Melin, E , 1925, Untersuchungen uber die Bedeutung der Bedeutung der Baummykorrhiza Eine okologisch physiologische Studie, *Jena Gustav Fischer* pp 151

Melin, E , and Nilsson, H , 1950, Transfer of radioactive phosphorus to pine seedlings by means of mycorrhizal hyphae, *Physiol Plant* 3 88-92

Melin, E , & Nilsson, H , 1952, Transfer of labelled nitrogen from an ammonium source to pine seedlings through mycorrhizal mycelium, *Svensk Bot Tids* 46 281-5

Melin, E and Nilsson, H , 1953, Transfer of labelled nitrogen from glutamic acid to pine seedlings through the mycelium of *Boletus variegatus* (S W) Fr , *Nature*, London 171 434

Meyer, F H , 1973, Distribution of ectomycorrhizae in native and man-made forests In Marks, G C and Kozlowski, T T (eds), *Ectomycorrhizae their ecology and physiology*, Academic Press, New York, pp 79-105

Miller, C S , 1974, In *Biology of Plant Litter Decomposition*, C H Dickinson and G F J Pugh Eds, Vol I, Academic Press, London, pp 105-129

Molina, R , Massicotte, H , and Trappe, J M , 1992, Specificity phenomena in mycorrhizal symbiosis Community ecological consequences and practical applications pp 357-423 in *Mycorrhizal Functioning* (Ed M F Allen) Chapman & Hall, New York

Mosca, A M L , and Fontana, A , 1974, Sull utilazione dell'azoto proteico da parte del micelio di *Boletus luteus* L , *Allionia* 20 47-55

Moser, M , 1967, Die ectotrophe Ernahrungsweise an der Waldgrenze, *Mitt Forstl Bundes-Versuchsanst Wien* 75 357-380

Newman, E I , 1988, Mycorrhizal links between plants Functioning and ecological significance, *Adv Ecol Res* 18 243-270

Newman, E I , and Reddell, P , 1987, The distribution of mycorrhizas among families of vascular plants, *New Phytol* 106 745-752

Norkrans, B , 1950, Studies in growth and cellulytic enzymes of *Tricholoma*, *Symb Bot Upsal* 11, 1-126

Perry, D A , Amaranthus, M P , Borchers, J G , Borchers, S L , and Brainerd, R E , 1989, Bootstrapping in ecosystems, *Bio Science* 39 230-237

Peyronel, B , 1922, Nuovi casi di rapporti micorizici tra Basidiomiceti e Fanerogame arboree, *Boll Soc Bot Ital* , 1 3-14

Post, J M , Pastor, J , Sinke, P J , and Stangenberger, A G , 1985, Global patterns of nitrogen storage, *Nature* 317 613-616

Read, D J , 1982, In support of Franks organic nitrogen theory, *Ang Bot* 61 25-37

Read, D J , 1984, The structure and function of the vegetative mycelium of mycorrhizal roots In *The Ecology and Physiology of the fungal Mycelium* (Ed by D H Jennings and A D M Rayner, pp 215-240 Cambridge University Press, Cambridge

Read, D J , 1990, Ecological integration by mycorrhizal fungi, *Endocytobiology IV*, INRA, Paris, pp 97-107

Read, D J , 1991, Mycorrhizas in ecosystems, *Experientia* 47 476-391

Read, D J , 1992, The mycorrhizal mycelium In Allen, M (ed) *Mycorrhizal functioning An integrative plant fungal process* Chapman & Hall, London, pp 102-103

Read, D J , Francis, R , and Finlay, R D , 1985, Mycorrhizal mycelia and nutrient cycling in plant communities In *Ecological Interactions in Soil Plants, Microbes and Animals* (Ed by A H Fitter, D Atkinson, D J Read, and M B Usher) pp 193-217 British Ecological Society Special Publication 4 Blackwell Scientific Publications, Oxford

Read, D J , Leake, J R , and Langdale, A R , 1989, The nitrogen of mycorrhizal fungi and their host plants In *Nitrogen, Phosphorus and Sulphur Utilization by Fungi* (Ed by L Boddy, R Marchant, and D J Read), pp 181-204 Cambridge University Press, Cambridge

Soderstrom, B E , 1979, Seasonal fluctuations of active fungal biomass in horizons of a podzolized pine-forest soil in central Sweden, *Soil Biol Biochem* 11 149-154

Soderstrom, B , and Read, D J , 1987, Respiratory activity of intact and excised ectomycorrhizal mycelial systems growing in unsterilized soils, *Soil Biol and Biochem* 19 231-236

Staaf, H , and Berg, B , 1977, Mobilisation of plant nutrients in a Scots pine forest mor in central Sweden, *Silva Fenn* 11 210-217

Trappe, J M , 1977, Selection of fungi for ectomycorrhizal inoculation in nurseries, *Ann Rev Phytopath* 15 203-22

Trappe, J. M., 1987, Phylogenetic and ecological aspects of mycotrophy in the angiosperms from an evolutionary standpoint. In Safir, G. R. (ed) *Ecophysiology of VA mycorrhizal plants*, CRC, Boca Raton, pp. 2-25

Trappe, J. M., 1988, Lessons from alpine fungi, *Mycologia* 80: 1-10.

Trojanowski, J., Haider, K., and Hüttermann, A., 1984, Decomposition of ^{14}C-labelled lignin, holocellulose and lignocellulose by mycorrhizal fungi, *Arch. Microbiol* 139: 202-206.

Unestam, T., 1991, Water replellency, mat formation, and leaf-stimulated growth of some ectomycorrhizal fungi, *Mycorrhiza* 1: 13-20.

Visser, S., and Parkinson, D., 1975, Fungal succession on aspen poplar leaf litter, *Can. J. Bot.* 53: 1640-1651.

Vogt, K. A., Grier, C. C., Meier, C. E., and Edmonds, R. L., 1982, Mycorrhizal role in net primary production and nutrient cycling in *Abies amabilis* ecosystems in western Washington, *Ecology* 63: 370-380.

2

TAXONOMY OF ECTOMYCORRHIZAL FUNGI

A Starting Point for Their Biotechnology

James M. Trappe and Ari Jumpponen

Department of Forest Science
Oregon State University
Corvallis, Oregon 97331-7501

ABSTRACT

Early mycorrhiza researchers paid little heed to differences between fungal taxa or genotypes in host response. The common difficulty of duplicating experimental results was probably due in part to differences between the fungi involved. Now the importance of identifying fungi in experiments is widely recognized. Taxonomy thus is an integral part of mycorrhiza research. Taxonomy is a process of hypothesis testing. A Latin fungal name is a brief statement of hypotheses: 1) the species differs inherently from all other species previously described, and 2) it is more closely related to other species in the genus to which it is assigned than to species in other genera. Fungal taxonomy was originally based on macroscopic features. Microscopic and physiological features were subsequently added as important characters. Morphological, anatomical and physiological characters still provide the starting point for erecting and testing taxonomic hypotheses, but these have limitations for separating closely related taxa or revealing phylogenetic relationships. Now, DNA analyses provide powerful tools for more precise testing of taxonomic hypotheses.

WHY KNOW THE FUNGI?

In the 77 years between the coining of the term "mycorrhiza" by B. Frank (1885) and the review of the fungi that form ectomycorrhizae by Trappe (1962), about 2,300 papers had been published on mycorrhizae or mycorrhizal fungi (Trappe and Castellano, 1992). About 400 of these papers listed specific fungus-host ectomycorrhizal associations, about 1,500 combinations in all proposed or demonstrated (Trappe, 1962). The phenomenon of host specificity of many ectomycorrhizal fungi had been well established, but otherwise only 25 of the 2,300 papers dealt with other physiological or ecological differences between ectomycorrhizal fungal species, and only a dozen reported comparisons of host response to different fungi (Trappe, 1962).

Biotechnology of Ectomycorrhizae, Edited by Vilberto Stocchi et al.
Plenum Press, New York, 1995

During that era, then, the possibility that different ectomycorrhizal fungi might affect hosts in different ways was subordinated to broader examination of the significance of mycorrhizae. Major experiments on effects of mycorrhiza formation on hosts were designed with little or no attention paid to the fungi involved. Uncertainties and controversies about effects of mycorrhizae and conditions for their formation abounded, and experiments under one set of conditions often produced different results and conclusions from those under other conditions (Harley, 1965).

In the ensuing 15 years, evidence mounted that the particular fungus or fungi involved in an ectomycorrhizal association could profoundly affect host response under given conditions (Trappe, 1977), and modern research reflects a general understanding of this fact (Read *et al.*, 1992). We suspect that many early controversies about effects of mycorrhiza formation on hosts and difficulties in duplicating results from one experiment to another stemmed from the now well documented fact that fungal species or even genotypes within species generally vary in their effects on hosts. Indeed, the success of ectomycorrhizal trees as dominants over much of the world probably is due in good part to this fungal diversity: decline of ectomycorrhizal forests has been related in part to decline of ectomycorrhizal fungal diversity (Arnolds, 1991).

Knowledge of the fungal participants is clearly essential to sound experimentation on mycorrhizal symbioses and to interpreting mycorrhizal-ecological phenomena. Molecular approaches to studying mycorrhizae represent yet another area in which knowledge of fungal identities is critical, as witnessed by other papers in these proceedings. Molecular biologists need dependable identifications of the fungi for studies of species, populations and mycorrhizal associations and for accumulating DNA sequence data bases. Taxonomists equally need molecular tools for resolving taxonomic puzzles not solvable by traditional morphological approaches or mating compatibility experiments and for testing hypotheses on phylogenies and evolution of the fungi. The two disciplines form a natural research symbiosis: molecular biologists and fungal taxonomists can not be satisfyingly effective without each others' collaboration.

SCIENTIFIC METHOD IN TAXONOMY

Contrary to common misperception, taxonomy is not analogous to file clerking, establishing categories and putting objects in the proper one. Taxonomists, as other biologists, deal with organisms, their relationships and their interactions. The scientific method is basic to the taxonomic process: hypotheses are erected and tested by the best means available. Original hypotheses often hold up under prolonged testing. In other cases, hypotheses that withstand original testing fail as new information or better techniques become available. Then the original hypotheses must be rejected and exchanged for new ones that appear to more accurately explain phenomena observed.

When a new species is proposed and named, it is hypothesized to be inherently different from all other species named up to that time (hypothesis #1). Its assignment to a genus hypothesizes that it is more closely related to other taxa in that genus than to taxa in other genera (hypothesis #2). Or, if a new genus is erected for a species, the species is hypothesized to differ inherently from those in previously described genera and requires a new genus (alternative hypothesis #2). In either event, the genus and species expressed as a Latin binomial provide a shorthand statement of the two hypotheses. Similar hypotheses are raised by assignment of a genus to a family, the family to an order, etc.

These hypotheses are then tested over time. If the new species proves to be the same as or a phenotypic variant of a species described earlier, it is reduced to synonymy with the earlier described species. Hypothesis #1 is thus rejected. If new information reveals that the species should be assigned to a different genus than originally hypothesized, it is transferred and assumes the other generic name while preserving the species name. This constitutes rejection of hypothesis #2, which is replaced by a new generic hypothesis, but retains hypothesis #1.

It occasionally happens that a common fungus is widely known by a Latin name that must be changed in keeping with the International Code of Botanical Nomenclature (Voss *et al.,* 1983). The Code was established by botanists to bring clarity and consistency to the scientific process of describing and naming plants and fungi, *i.e.* to the testing of hypotheses about relationships. A major goal of the Code is to bring stability to names of organisms. An example is the commercially valuable ectomycorrhizal American matsutake mushroom.

This species was described and named by Peck as *Agaricus ponderosus* Peck (see Redhead, 1984, for the detailed nomenclatural history). Later, improved understanding of fungal genera led to its transfer to the genus *Armillaria.* Hypothesis #2 was thereby refuted and replaced with the new hypothesis expressed by the Latin binomial *Armillaria ponderosa* (Peck) Saccardo. Additional information accumulated to suggest that this species belonged in the genus *Tricholoma.* With this further transfer, the species became designated as *Tricholoma ponderosum* (Peck) Singer; the second version of hypothesis #2 was repudiated and replaced with the third version.

Then, in study of Peck's original collections, Redhead determined that Peck had named the same species twice. Of the two names Peck applied to the single species, *Agaricus magnivelaris* Peck had been published earlier than *A. ponderosus.* By the International Code of Botanical Nomenclature, the earlier species name has priority and should be used. Hypothesis #1 was refuted: the species epithet *ponderosus* was not the first name applied to the species. The new combination *Tricholoma magnivelare* (Peck) Redhead was established as the name currently in use. Thus earlier hypotheses #1 and #2 are replaced. The current hypothesis #1 is that *T. magnivelare* differs inherently from all other species named up to that time; hypothesis #2 is that *T. magnivelare* is more closely related to other species in *Tricholoma* than to species in other genera. If these hypotheses withstand future testing, the name *T. magnivelare* will be stable.

The changes in names discussed above may not seem to exemplify stability to the casual observer. However, as new knowledge emerges and old hypotheses are refuted, change is inevitable, the old must be replaced by the new in the quest for reality. Stability in organism names, which are statements of hypotheses, will be the ultimate result, as true relationships are established. In all branches of science, old hypotheses are replaced when new evidence so indicates; taxonomic scientists must follow the same course.

Hypotheses erected at higher taxonomic levels are similarly tested. For example, the hypogeous Basidiomycete genus *Rhizopogon* was long hypothesized as related to the Gasteromycetes (puffballs) due to its rounded shape and the enclosure of its spore-bearing tissues within a peridium. Clements and Shear (1931) placed it in the family Hymenogastraceae, order Lycoperdales. Subsequent anatomical and ecological study of *Rhizopogon,* however, led to rejection of those hypotheses. Heim (1971) concluded that a tie of *Rhizopogon* to the Boletales was "virtually certain." Some mycologists placed *Rhizopogon* in its own family, Rhizopogonaceae, either in the order Hymenogastrales (Smith, 1973) or Boletales (Jülich, 1981). Miller (1982) hypothesized that *Rhizopogon* belonged in the Boletaceae, order Boletales. However, the limitations of morphologic, anatomic and ecologic data could not overcome uncertainty about its proper family and order. Molecular approaches have recently clarified these relationships. Miller's (1982) hypotheses that it belongs in the family Boletaceae, order Boletales are supported by DNA analyses (Bruns *et al.,* 1989; Mehmann *et al.* 1994).

TRADITIONAL TOOLS OF FUNGAL TAXONOMY

Early mycologists relied on macroscopic features of fungal sporocarps to erect hypotheses about species and their relationships. In the 19th century microscopic characters such as spores and spore-bearing cells were added as tools to test those hypotheses. Subsequent researchers added anatomy of various parts of sporocarps. Ecophysiologic considerations also proved useful, such as type of substrate utilized or, in the case of symbiotic fungi, the range of hosts involved. Physiological characters such as response of tissues to various reagents and protein and pigment chemistry have been used more recently when morphologic/anatomic study has proven inadequate for testing taxonomic hypotheses. Mating compatibility has proven useful for sorting out morphologically similar species (Mueller, 1992).

These methods have proven extremely useful in erecting and testing hypotheses about fungal taxa; morphology/anatomy remains the foundation, with the science greatly strengthened by physiological and experimental approaches. All, however, have limitations that prevent adequate testing of many hypotheses. Addition of DNA methods to the taxonomists' arsenal now greatly enhances the precision of delimiting relationships within the Kingdom Fungi, as is abundantly clear from other papers in these proceedings. In addition, studies on population ecology of ectomycorrhizal fungi have become increasingly important in assessing effects of forest practices, pollution, and climate change. Ectomycorrhizal fungal fruiting does not necessarily reflect populations of fungi on root systems, so identification of the fungi actually occupying roots is needed. Much can be done by morphological/anatomical study of ectomycorrhizae, especially for fungi which produce distinctive mycorrhiza morphologies on the rootlets (Agerer, 1986; Trappe, 1967). However, many fungi produce similar types of ectomycorrhizal anatomies. Even mycorrhizae that have a distinctive anatomy may change appearance as they progress through developmental stages. DNA approaches offer particular value for resolving such ambiguities.

DNA ANALYSIS IN FUNGAL TAXONOMY AND IDENTIFICATION

Recent advances in identification of mycorrhizal fungi from sporocarps or mycorrhizae are based on analyzing the genetic information of target organisms. Rapid, simple methods of tissue and DNA preparation for such analysis have been developed (e.g. Cenis, 1992; Graham et al. 1994; Lecellier and Silar, 1994; Rogers et al., 1989). Extraction of DNA may be a problem, especially when pigments or interfering compounds such as phenols or tannins are present (Erland et al., 1994; Jumpponen, unpublished). These problems can usually be overcome by using the freshest material possible, diluting the DNA sample , or completing additional steps in the extraction process, such as running the extract through agarose gel or using 'gene clean' (Bruns et al., 1990; Erland et al., 1994; Rygiewicz and Armstrong, 1991).

Extraction of DNA even from very old tissues has succeeded (Bruns et al., 1990; Pääbo et al., 1988; Pääbo and Wilson, 1988). Thus mycologists can use herbarium specimens for molecular taxonomy and phylogeny. However, quality of the DNA extract can be problematical (Bruns et al., 1990). This is true especially for fungi, since fungal DNA seems to degrade during storage (Bruns et al., 1990) or sometimes rapidly during the extraction procedure (Jumpponen, unpublished). This may result in a low quality initial product which, depending on choice of the molecular tool, can affect reliability of the results.

Identification and systematics of fungi by molecular tools was initiated with DNA-DNA hybridization. This technique is laborious and involves technical problems. Relatedness between fungal species is difficult to determine because the extent of DNA reassociation falls rapidly between fungal species, even in closely related species (Horgen et al., 1984; Vilgalys, 1988). Though these problems were known to exist, DNA-DNA hybridizations have been used for fungal systematics (Kohn, 1992).

Analysis of genetic inter- and intraspecific variation of ectomycorrhizal fungi has become more accessible with restriction fragment length polymorphism (RFLP) analysis. RFLPs have been used for taxonomic, phylogenetic, and population studies (Anderson et al., 1989; Bruns and Palmer, 1989; Egger et al., 1991; Vilgalys and Gonzales, 1990). Inter- and intraspecific variations in RFLPs of ribosomal DNA (rDNA) or mitochondrial DNA (mtDNA) have been reported for several taxa of mycorrhizal fungi: *Cenococcum geophilum* (LoBuglio et al., 1992), E-strain fungi (Egger and Fortin, 1990; Egger et al., 1991), *Hebeloma* spp. (Marmaisse et al., 1992), and *Laccaria* spp. (Armstrong et al., 1989; Gardes et al., 1990; Martin et al., 1991). RFLP analysis seems to provide the resolution needed for separation of species and, in some cases, even strains (Egger et al., 1991; Marmaisse et al., 1992; Martin et al., 1991). On the other hand, cost, labor intensity and the use of radioactive isotopes prevent widespread use of RFLP in identifying mycorrhizal fungi. In addition, analysis of the restriction patterns on autoradiographs can be difficult and sometimes questionable (Gardes et al., 1990).

Early limitations due to tissue availability for DNA analysis have been overcome by the polymerase chain reaction (PCR), in which target sequences of genomic DNA are rapidly amplified in an enzymatically directed procedure (Saiki et al.,1988). PCR enables increasing the number of copies of the genomic region of interest without need to culture the target organism. In addition, selective primers can be chosen to amplify target DNA from a mixture of DNAs from several origins (Gardes and Bruns, 1993; Simon et al., 1992). With selection of the target region in the genomic DNA, levels of resolution can be chosen according to the objectives of a given study (Bruns et al., 1991).

Amplification and sequencing of highly conserved regions of the genome have been used to confirm hypotheses of evolutionary biology and higher level phylogenetic framework. Wainwright et al. (1993), working at the kingdom level, suggested that fungi and animals share "a unique evolutionary history". Swann and Taylor (1993) studied higher taxa of Basiodiomycetes and proposed three major monophyletic lineages: Ustilaginales, simple septate and hymenomycete lineages. The conserved sequences also have been applied to resolve questions on families and genera in Basidiomycetes (Bruns et al., 1989; Hibbett and Vilgalys, 1993). Usually higher level phylogenetic studies face limitations in adequate availability of target sequences (Swann and Taylor, 1993) or the calculation capacity of computers. Furthermore, additional data— morphological, protein, or other DNA sequences—are considered necessary to confirm results of studies of higher level phylogeny (Baldauf and Palmer, 1993).

Noncoding regions of rRNA genes such as internal transcribed spacer (ITS) or intergenic spacer (IGS) provide an often highly variable target between morphologically distinct mycorrhizal species (Erland et al., 1994; Gardes et al., 1991). The intraspecific variation in the ITS region has been reported to be low (Gardes et al., 1991). The ITS region has additional advantages: 1) the region has a convenient size (usually ranging between 600 and 800 base pairs) for amplification and can be readily amplified with 'universal primers' (White et al., 1990), 2) rDNA, with the ITS region included, is present in the genome as multiple copies, permitting amplification from very small and diluted samples (Bruns et al., 1990). Because analysis based on the ITS region is relatively easy, it is of increasing interest in identification of ectomycorrhizal fungi colonizing rootlets.

Though some length polymorphism exists in the ITS region, and tentative identification in some cases can be based only on length of the amplified fragment (Feibelman et al., 1994; Gardes et al., 1991; Henrion et al., 1992, 1994; Jumpponen and Trappe, 1994), additional characters are needed, *e.g.* , from restriction enzyme digests of the amplified sequence. The restriction digest of PCR-amplified target sequences has been successfully applied to identification of fungal symbionts from mycorrhizae at least to the genus level and even to the species level (Erland *et al.*, 1994; Gardes *et al.*, 1991; Henrion *et al.*, 1994). Concentration on highly variable regions of the genome provides a means to increase resolution and the likelihood of species identification.

In addition to problems of resolution, the ratio of successful amplifications may be low. Erland *et al.* (1994) reported only 12 % of the amplifications from *Picea sitchensis/Tylospora fibrillosa* as successful. Studies which require the use of herbarium specimens often yield high proportions of poor amplifications by PCR (Camacho and Jumpponen, unpublished). For example, Bruns *et al.*. (1990) had no success in amplifying DNA from a 50-year-old specimen of *Boletus edulis*. Unfortunately, such data are not usually reported when results are published.

Identification or recognition of an individual, genet or strain can be the focus, especially in population studies. A PCR-based method that uses arbitrary oligonucleotide primers has been developed to identify strains of bacteria (Welsh and McClelland, 1990). Williams *et al.* (1990) described a similar method that uses autoradiography of ethidium bromide rather than radioactivity in visualizing the amplification results. Due to the ease of this latter procedure, it has been used more frequently than that by Welsh and McClelland (1990). The method is now known as random amplified polymorphic DNA (RAPD). RAPD procedure differs from previously discussed PCR-techniques in two important features: 1) only one primer is needed for the amplification, which occurs between two identical priming sites, and 2) the priming sites within the target genome are unknown.

Few studies of mycorrhizal fungi have taken advantage of these new tools of high resolution. Jacobson *et al.* (1993) concluded that RAPD is superior to more traditional tests of vegetative compatibility, which have been applied, for example, to populations of *Suillus bovinus* (Dahlberg and Stenlid, 1990). RAPD markers have been shown to work well in identification of strains or isolates of nonsporulating root endophytes (Jumpponen and Trappe, 1994) and in recognition of races within *Fusarium solani* species (Crowhurst *et al.*, 1991).

The RAPD method provides a tool of great intraspecies resolution. It is reasonably cost efficient and rapid and can be applied in any laboratory equipped for PCR-techniques. Moreover, no previous information on the target genome is needed for successful amplification, as opposed to amplification of a known target region in the PCR procedure. Unfortunately, RAPD has a reputation of problems with reproducibility of the characters (Welsh and McClelland, 1990). It seems necessary to control the reproducibility of the characters and replicate each sample with more than one primer. Additionally, the same thermocyclers and batch of chemicals must be used from one amplification to another (Meunier and Grimont, 1993).

CONCLUSIONS

Traditional taxonomic methods provide the starting point for research on taxonomy and ecology of mycorrhizal fungi. Through these methods hypotheses can be formulated and tested. Frequently, however, traditional methods are limited in scope and precision or extremely time consuming. Many such barriers can be broken by use of DNA analyses, the most powerful tools developed for taxonomy since initiation of morphologic approaches. Molecular approaches, too, have limitations, but new methodologies and applications are in constant development. Use of the methods discussed in this paper in various combinations as appropriate will provide answers to questions that until now have evaded taxonomists and mycorrhizologists alike.

Success in this arena demands collaboration between specialists, because few, if any, scientists can master all of the specialties involved. All specialists must learn enough about the fields of their collaborators to speak a common language and jointly interpret results of analyses and experiments. Thus fungal taxonomists and mycorrhizologists must be trained well enough about molecular approaches to work effectively with molecular biologists, and molecular biologists working with ectomycorrhizal fungi, in turn, must acquire an adequate understanding of fungal taxonomy, ecology and mycorrhizal relationships.

Institutions of higher learning need to pay better attention to these needs in training young scientists. An unfortunate trend in recent years in universities around the world has been to reduce or eliminate taxonomic mycology from their programs. This is leading to a crisis situation: where will molecular scientists in future years find the needed taxonomic expertise? The diminished population of taxonomic mycologists is already overburdened by demands for their services. In recognition of this looming problem, the U. S. National Science Foundation (1994) has initiated a special program in support of taxonomic research and the training of the next generation of taxonomists.

ACKNOWLEDGMENTS

We thank Dr. Vilberto Stocchi and the organizers of the International Symposium on Biotechnology of Ectomycorrhizae: Molecular Approaches for the opportunity to participate. This paper was prepared with support from National Science Foundation Grant DEB-9310006 and the U. S. Forest Service, Pacific Northwest Research Station.

REFERENCES

Agerer, R., 1986, Studies on ectomycorrhizae II. Introducing remarks on characterization and identification, *Mycotaxon* 26: 473-492.

Armstrong, J.L., Fowles, N.L., and Rygiewicz, P.T., 1989, Restriction length polymorphisms distinguish ectomycorrhizal fungi. *Plant and Soil* 116: 1-7.

Anderson, J.B., Bailey, S.S., and Pukkila, P.J., 1989, Variation in ribosomal DNA among biological species of *Armillaria*, a genus of root-infecting fungi, *Evolution* 43: 1652-1662.

Arnolds, E., 1991, Decline of ectomycorrhizal fungi in Europe, *Agric., Ecosyst. Environ.* 35: 209-244.

Baldauf, S.L., and Palmer, J.D., 1993, Animals and fungi are each other's closest relatives: Congruent evidence from multiple proteins, *Proc. Nat. Acad. Sci.* 90: 11558-11562.

Bruns, T. D., Fogel, R., and Taylor, J. W., 1990, Amplification and sequencing of DNA from fungal herbarium specimens, *Mycologia* 82 (2): 175-184.

Bruns, T. D., Fogel, R., White, T. J., and Palmer, J. D., 1989. Accelerated evolution of a false-truffle from a mushroom ancestor, *Nature* 339: 140-142.

Bruns, T.D., and Palmer, J.D., 1989, Evolution of mushroom mitochondrial DNA: *Suillus* and related genera, *J. Mol. Evol.* 28: 349-362.

Bruns, T. D., White, T. J., and Taylor, J. W., 1991, Fungal molecular systematics, *Ann. Rev. Ecol. Syst.* 22: 525-564.

Cenis, J.L., 1992, Rapid extraction of fungal DNA for PCR amplification, *Nucleic Acids Res.* 20: 2380.

Clements, F. E., and Shear, C. L., 1931, The Genera of Fungi, H. W. Wilson Co., New York. 496 pp.

Crowhurst, R.N., Hawthorne B.T., Rikkerink, E.H.A., and Templeton, M.D., 1991, Differentiation of *Fusarium solani* f. sp. *cucritae* races 1 and 2 by random amplification of polymorphic DNA, *Curr. Gen.* 20, 391-396.

Dahlberg, A., and Stenlid, J., 1990, Population structure and dynamics in *Suillus bovinus* as indicated by spatial distribution of fungal clones, *New Phytol.* 115: 487-493.

Egger, K.N., Danielson, R.M., and Fortin, J.A., 1991, Taxonomy and population structure of E-strain mycorrhizal fungi inferred from ribosomal and mitochondrial DNA polymorphisms, *Mycol. Res.* 95: 866-872.

Egger, K.N., and Fortin, J.A., 1990, Identification of taxa of E-strain mycorrhizal fungi by restriction fragment analysis, *Can. J. For. Res.* 68: 1482-1488.

Erland, S., Henrion, B., Martin, F. Glover, L.A., and Alexander, I.J., 1994, Identification of the ectomycorrhizal Basidiomycete *Tylospora fibrillosa* Donk by RFLP analysis of the PCR-amplified ITS and IGS regions of ribosomal DNA, *New Phytol.* 126: 525-532.

Feibelman, T., Bayman, P., and Cibula, W.G., 1994, Length variation in the internal transcribed spacer of ribosomal DNA in chantarelles, *Mycol. Res.* 98: 614-618.

Frank, B., 1885, Über die auf Wurzelsymbiose beruhende Ernährung gewisser Bäume durch unterirdische Pilze, *Ber. Deut. Bot. Gesell.* 3: 128-145.

Gardes, M., and Bruns, T.D., 1993, ITS primers with enhanced specificity for Basidiomycetes- application to the identification of mycorrhizae and rusts, *Mol. Ecol.* 2: 113-118.

Gardes, M., Fortin, J.A., Mueller, G.M., and Kropp, B.R., 1990, Restriction fragment length polymorphisms in the nuclear ribosomal DNA of four *Laccaria* spp.: *L. bicolor, L. laccata, L. proxima* and *L. amethystina, Phytopathol.* 80: 1312-1317.

Gardes, M., White, T. J., Fortin, J. A., Bruns, T. D., and Taylor, J. W., 1991, Identification of indigenous and introduced symbiotic fungi in ectomycorrhizae by amplification of nuclear and mitochondrial ribosomal DNA, *Can. J. Bot.* 69: 180-190.

Graham, G.C., Mayers, P., and Henry, R.J., 1994, A simplified method for the preparation of fungal genomic DNA for PCR and RAPD analysis, *Biotechniques* 16: 48-50.

Heim, R., 1971, The interrelationships between the Agaricales and Gasteromycetes, *In* Petersen, R. H. (ed.), Evolution in the Higher Basidiomycetes, Univ. Tennessee Press, Knoxville. pp. 505-534.

Henrion, B., Chevalier, G., and Martin, F., 1994 Typing truffle species by PCR amplification of the ribosomal DNA spacers, *Mycol. Res.* 98: 37-43.

Henrion, B., Le Tacon, F., and Martin, F., 1992, Rapid identification of genetic variation of ectomycorrhizal fungi by amplification of ribosomal RNA genes, *New Phytol.* 122: 289-298.

Hibbett, D.S., and Vilgalys, R., 1993, Phylogenetic relationships of *Lentinus* (Basidiomycotina) inferred from molecular and morphological characters, *Syst. Bot.* 18: 409-433.

Horgen, P.A., Arthur, R., Davy, O., Moum, A., Herr, F., Straus, N., and Anderson, J., 1984, The nucleotide sequence homologies of unique DNAs of some cultivated and wild mushrooms, *Can. J. Microbiol.* 30: 587-593.

Jacobson, K.M., Miller, O.K., Jr., and Turner, B.J., 1993, Randomly amplified polymorphic DNA markers are superior to somatic incompatibility tests for discriminating genotypes in natural populations of the ectomycorrhizal fungus *Suillus granulatus, Proc. Nat. Acad. Sci.* 90: 8159-9163.

Jülich, W., 1981, Higher Taxa of Basidiomycetes, J. Cramer, Vaduz. 484 pp.

Jumpponen, A.M., and Trappe, J.M., 1994, Population structure of *Phialocephala fortinii* root endophyte on a glacial forefront, *Proc. Fourth Eur. Symp. on Mycorrhizas*, Granada (in press).

Kohn, L.M., 1992, Developing new characteristics for fungal systematics: an experimental approach for determining the rank of resolution, *Mycologia* 84: 139-153.

Lecellier, G., and Silar, P. 1994, Rapid methods for nucleic acids extraction from Petri dish-grown mycelia, *Curr. Gen.* 25: 122-123.

Lobuglio, K.F., Rogers, S.O., and Wang, C.J.K., 1992, Variation in ribosomal DNA among isolates of the mycorrhizal fungus *Cenococcum geophilum*, Canadian Journal of Botany 69: 2331-2343.

Marmaisse, J.C., Debaud, J.C., and Casselton, L.A., 1992, DNA probes for species and strain identification in the ectomycorrhizal fungus *Hebeloma, Mycol. Res.* 96: 161-165.

Martin, F., Zaiou, M., Le Tacon, F., and Rygiewicz, P., 1991, Strain specific differences in ribosomal DNA from the ectomycorrhizal fungi *Laccaria bicolor* (Maire) Orton and *Laccaria laccata* (Scop ex Fr) Br., *Ann. For.* 48: 297-305.

Mehmann, B., Brunner, I., and Braus, G. H., 1994, Nucleotide sequence variation of chitin synthase genes among ectomycorrhizal fungi and its potential use in taxonomy, *App. Environ. Microbiol.* 60: 3105-3111.

Meunier, J.-R., and Grimont, P.A.D., 1993, Factors affecting reproducibility of random amplified polymorphic DA fingerprinting, *Res. Microbiol.* 144: 373-379.

Miller, O. K., Jr., 1982, Ectomycorrhizae in the Agaricales and Gasteromycetes. Can. J. Bot. 61: 909-916.

Mueller, G. M., 1992, Systematics of *Laccaria* (Agaricales) in the continental United States and Canada, with discussion on extralimital taxa and descriptions of extant types, *Fieldiana n.s.* 30: 1-158.

Pääbo, S., Gifford, J.A., and Wilson, A.C., 1988, Mitochondrial DNA sequences from a 7000-year-old brain, *Nucleic Acids Res.* 16: 9775-9787.

Pääbo, S., and Wilson, A.C., 1988, Polymerase chain reaction reveals cloning artifacts, *Nature* 334: 387-388.

Read, D. J., Lewis, D. H., Fitter, A. H., and Alexander, I. J., 1991, Mycorrhizas in Ecosystems, C.A.B. International, Wallingford, U.K. 419 pp.

Redhead, S. A., 1984, Mycological observations 13-14: on *Hypsizygus* and *Tricholoma, Trans. Mycol. Soc. Japan* 25: 1-9.

Rogers, S. O., Rehner, S., Bledsoe, C., Mueller, G. J., and Ammirati, J. F., 1989, Extraction of DNA from Basidiomycetes for ribosomal DNA hybridizations, *Can. J. Bot.* 67: 1235-1243.

Rygiewicz, P.T., And Armstrong, J.L., 1991, Ectomycorrhizal DNA: isolation, RFLPs and probe hybridization. *In* Norris, J.R., Read, D.J., and Varma, A.K., (eds.), Methods in Microbiology, vol. 23, London, Academic Press. pp. 253-281.

Saiki, R. K. , Gelfand, D. H., Stoffel, S., Scharf, S. J., Higuchi, R., Horn, G. T., Mullins, K. B., and Erlich, H. A., 1988, Primer-directed enzymatic amplification of DNA with a thermostable DNA polymerase, *Science* 239: 487-491.

Simon, L., Lalonde, M., and Bruns, T. D., 1992, Specific amplification of 18S fungal ribosomal genes from vesicular-arbuscular endomycorrhizal fungi colonizing roots, *Applied Environ. Microbiol.* 58: 291-295.

Smith, A. H., 1973, Agaricales and related secotioid Gasteromycetes, *In* Ainsworth, G. C., Sparrow, F. K. ,and Sussman, A. S. (eds.), The Fungi–an Advanced Treatise, Vol. IVB, Academic Press, New York. pp. 421-450.

Swann, E.C., and Taylor, J.W., 1993, Higher taxa of Basidiomycetes: an 18S rRNA gene perspective, *Mycologia* 85: 923-936.

Trappe, J. M., 1962., Fungus associates of ectotrophic mycorrhizae, *Bot. Rev.* 28: 538-606.

Trappe, J. M., 1967, Principles of classifying ectotrophic mycorrhizae for identification of fungal symbionts, *14th Congr. Internat. Un. For. Res. Org. Proc. Sec.* 24: 46-59.

Trappe, J. M., 1977, Selection of fungi for ectomycorrhizal inoculation in nurseries, *Ann. Rev. Phytopathol.* 15: 203-222.

Trappe, J. M., and M. A. Castellano, 1991. Mycolit: a Mycorrhiza Bibliography, 1758-1991, Mycologue Publications, Waterloo, Ontario. 550 pp.

U. S. National Science Foundation, 1994, Special Competition in Systematic Biology: Partnerships for Enhancing Expertise in Taxonomy (PEET). Washington, D. C. 4 pp.

Vilgalys, R., 1988, Genetic relatedness among anastomosis groups in *Rhizoctonia* as measured by DNA/DNA hybridization, *Phytopathol.* 78: 698-702.

Vilgalys, R., and Gonzales, D., 1990, Ribosomal DNA restriction fragment length polymorphisms in *Rhizoctonia solani, Phytopathol.* 80: 151-158.

Voss, E. G., Burdet, H. M., Chaloner, W. G., Demoulin, V., Hiepko, P., McNeill, J., Meikle, R. D., Nicolson, D. H., Rollins, R. C., Silva, P. C., and Greuter, W., 1983, International Code of Botanical Nomenclature, Bohn, Scheltema & Holkema, Utrecht. 472 pp.

Wainwright, P.O., Hinkle, G., Sogin, M.L., and Stickel, S.K., 1993, Monophyletic origins of the metazoa: an evolutionary link with fungi, *Science* 260: 340-342.

Welsh, J., and McClelland, M., 1990, Fingerprinting genomes using PCR with arbitrary primers, *Nucleic Acids Res.* 18: 7213-7218.

White, T.J., Bruns, T.D., Lee, S., and Taylor, J., 1990, Amplification and direct sequencing of fungal ribosomal RNA genes for phylogenetics, *In* Innis, M.A., Gelfand, D.H., Sninsky, J.J., and White, T.J. (eds.), CR Protocols: a Guide to Methods and Applications, Academic Press, New York. pp. 315-322,

Williams, J.G. K., Kubelik, A.R., Livak, K.J., Rafalski, J.A., and Tingey, S.V., 1990, DNA polymorphisms amplified by arbitrary primers are useful as genetic markers, *Nucleic Acids Res.* 18: 6531-6535.

3

MOLECULAR APPROACH TO THE IDENTIFICATION OF MYCORRHIZAL FUNGI

L. Simon

Université Laval
Sainte-Foy
Québec G1K 7P4
Canada

Molecular techniques that can be used for ecological studies of arbuscular mycorrhizal fungi will be presented, such as protocols for the identification of AMF by using family-specific primers and single strand conformation polymorphism analysis. To the extent that taxonomic and phylogenetic considerations are important in the design of molecular identification tools, we will also present our results based on the study of rDNA genes from 13 species of arbuscular endomycorrhizal fungi. The recently revised morphologically based classification of arbuscular mycorrhizal fungi into 3 families, the Glomaceae, Acaulosporaceae, and Gigasporaceae, was supported by rDNA sequence data. Also, approximate dates were obtained for the divergence of major branches on the phylogenetic tree. These include an estimate of 353 - 462 Myr ago for the origin of VAM-like fungi that is consistent with the hypothesis that VAM were instrumental in the colonization of land by ancient plants.

INTRODUCTION

The arbuscular fungi living in symbiosis within the roots of the majority of the vascular plants have recently been classified in a new order, the Glomales (7). Within this order, three families have been recognized, the Glomaceae, Acaulosporaceae, and Gigasporaceae. The family Acaulosporaceae includes the genera Acaulospora and Entrophospora, while the family Gigasporaceae includes the genera Gigaspora and Scutellospora. The taxonomic asset of this important group of fungi has been proposed on the basis of careful examination of the available morphological characters, mainly those observed on the large soil borne spores produced in the vicinity of the colonised roots. Also, it was assumed that the unifying characteristic of these fungi, namely their ability to colonize roots and establish what is known as the arbuscular endomycorrhizal symbiosis, was the reflection of the uniqueness of this group of organisms, and sufficiently important to justify their classification in a separate order.

Biotechnology of Ectomycorrhizae, Edited by Vilberto Stocchi et al.
Plenum Press, New York, 1995

As more progress is being made in the understanding of the mechanisms of the arbuscular endomycorrhizal symbiosis, it has become apparent that proper identification of the fungal symbionts is becoming more and more necessary. Simply to be able to keep up with the sheer amount of data reported in the literature (6), and to grasp meaningful relationships in the results coming from the many laboratories that all over the world testify to both the fundamental importance and practical applications of this symbiosis, it is essential that the organisms studied be adequately identified.

The recent availability of powerful molecular techniques that allow the characterization of nucleic acids amplified from minute amounts of biological material has made possible the study of otherwise nearly intractable problems. The development of the Polymerase Chain Reaction and the advances in DNA sequencing technologies have brought spectacular progress in many fields of biology and medicine, and are opening whole new avenues of research in many disciplines (1, 5, 13).

PCR amplification of nuclear encoded ribosomal genes, namely the 18S gene, followed by direct sequencing or single-strand conformation analysis, were used to characterize arbuscular fungi, to develop a phylogenetic framework and to provide molecular tools for identification of the fungal symbionts, even in colonized roots.

PHYLOGENY OF THE GLOMALES

The first DNA sequence data to be obtained from Glomalean fungi were the two 18S ribosomal genes reported for *Glomus intraradices* and *Gigaspora margarita* (11). Axenic spores obtained from *in vitro* colonized root organ culture were used as starting material, from which the 18S gene could be amplified by PCR. The universal ribosomal primers could be used confidently in this case, and the sequences obtained proved to be different from all previously known fungal 18S sequences. A Glomales-specific primer was then designed, VANS1, and demonstrated to be useful for the specific amplification of the 18S gene of Glomalean fungi, even in the presence of extraneous DNA from plants or other fungi. This VANS1 primer could then be used to obtain 18S gene sequences from more species within the Glomales (Table 1). The analysis of the 18S gene sequences obtained from 13 species representing the genera Glomus, Acalauspora, Entrophospora, Gigaspora and Scutellospora has shown that the Glomales forms a coherent monophyletic group of fungi (10). Recent evidence presented by K. O'Donnell indicates that the Glomales are, within the Zygomycetes, the closest group to the derived classes Ascomycotina and Basidiomycotina. A molecular clock analysis of our 18S dataset yielded an estimate for the origin of the Glomales between 353 to 462 Myr ago (Fig.1). This date has been confirmed by another group also using a molecular clock analysis, but with a different dataset comprising only fungal 18S sequences and a different set of calibration events (3). Furthermore, new evidence presented by T.N. Taylor on the Devonian Aglaophyton from the Rhynie Cherts demonstrates the presence of arbuscules in some 400 Myr old fossils.

All these data underscore the uniqueness of this group of fungi, and reinforces the notion that their classification in a separate order is justified.

IDENTIFICATION OF GLOMALEAN FUNGI

There are numerous advantages in being able to work with rDNA genes for the identification of Glomalean fungi. Because of their high copy numbers, extraction, amplification and analysis can be carried out rather easily, using simplified protocols. Since the spores contain between hundreds and thousands of nuclei (2, 4), depending on the species

Table 1. List of Glomales species whose SSU sequence is available

Species	Accession#	Strain used
G. etunicatum	Z14008	UT316A [b]
G. mossae	Z14007	FL156B [b]
G. intraradices	X58725	197198 [c]
G. vesiculiferum	L20824	nd
E. columbiana	Z14006	FL356 [b]
E. contigua [a]	Z14011	WV201 [b]
A. rugosa	Z14005	WV949 [b]
A. spinosa	Z14004	WV860 [b]
G. margarita	X58726	194757 [c]
G. gigantea	Z14009	WV932 [d]
G. albida	Z14009	FL927 [b]
S. pellucida	Z14012	WV873 [b]
S. heterogama [a]	Z14013	WV858B [b]

[a]Revised identification (J. Morton, pers. comm.), differs from original publication.
[b]INVAM Collection, West-Virginia University.
[c]DOAM Collection, Ottawa.
[d]Strain lost, no longer available at INVAM.

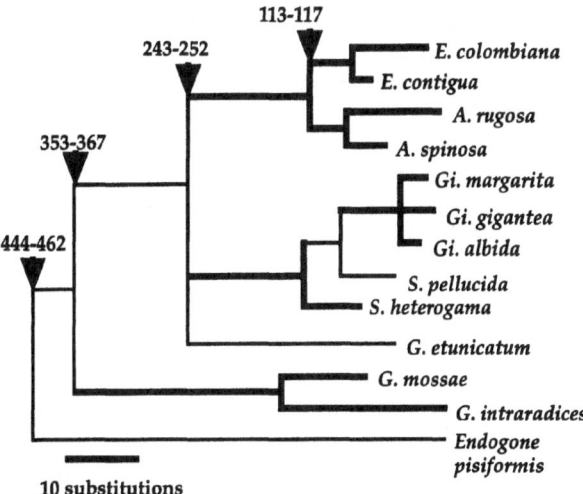

Figure 1. Phylogenetic tree of the *Glomales*. The 13 available SSU sequences from different species of arbuscular fungi (Glomales) were aligned, and 450 characters whose alignment was questionable were removed. Parsimony analysis was conducted on the remaining 1459 characters. This consensus tree was obtained after 600 bootstraps using the Branch and Bound option of PAUP v. 3.0s. Thick lines represent the branches that were present in more than 90% of the bootstraps. The shortest tree obtained had 232 steps, excluding the outgroup branch, CI: 0.746. The misplacement of the *Glomus etunicatum* branch indicates the large distance separating it from the other two *Glomus* species analysed. *Endogone pisiformis*, the type species of Endogonales, was selected as an outgroup. A similar topology was obtained by subjecting the derived distance matrix to the neighbour-joining method of tree construction (data not shown). Divergence times were estimated for the nodes indicated by the bold arrows, and are reported in Myr.

analysed, it should be possible to succesfully amplify rDNA genes consistently from single spores. Since the 18S rDNA gene is a useful gene for phylogenetic reconstructions, many sequences have been obtained and there are now many hundreds available in international databases. The exploration of this large dataset, looking for signature sequences, allows for the rational design of primers or probes. The term "rational design" means that the specificity of a primer or probe targeted to the 18S rDNA gene can be predicted, and the primer or probe can effectively be designed using the prior knowledge afforded by this large 18S dataset. For instance, the VANS1 primer was designed to recognize all arbuscular fungi, and this group-level specificity has now been validated experimentally many times in different laboratories. The VANS1 primer allows the amplification of the 18S rDNA gene of glomalean fungi, even in the presence of extraneous DNA from plants, other fungi or bacteria.

Based on sequence differences between species in the 18S rDNA genes, identification schemes can be devised to detect these sequence polymorphisms. The first approach is to design primers that are specific to a given group of organisms, members of the same family, genera or even species. After an appropriate validation, an assay will be scored as positive or negative, indicating either the presence or absence of the targeted DNA in the sample analysed. An example of this type of assay is illustrated by the different family-specific primers that can be used to identify arbuscular fungi (12).

A second approach is to characterize a region of the 18S rDNA gene that is known to be highly polymorphic, and thus informative. Instead of designing primers that will directly target the informative sites, the primers can be used to amplify a small informative region for further analysis. An example of this type of assay is illustrated by the SSCP analysis of the VANS22-VANS32 fragments used to identify arbuscular fungi. These two primers were designed to amplify a fragment of approx. 150 bp from any fungi, and might even be used on plant or animal DNA. That fragment was shown to be polymorphic among the Glomales, often allowing species-level identification(12). This way, a single primer pair can be used to characterize many specimens, even some that are uncharacterized, and whose 18S sequence is not available. In the example shown in Fig. 2, the primer VANS22 was labeled with the fluorochrome FAM, while the VANS32 primer was labeled with the fluorochrome JOE. This allows the two strands of the amplified fragment to be distinguished by the laser induced fluorescence detector of an ABI 373 automated sequencer. When the VANS22-VANS32 fragments from four different species of *Glomus* were analysed by SSCP, migration differences were observed. It can be noted that the ability to distinguish between the differently colored strands allowed us to clearly demonstrate that the VANS32 strand exhibited more conformation polymorphism than the VANS22 strand, and for the 4 species compared, the migration differences could easily be scored. This simple analysis could be performed in less than two days, and could unequivocally distinguish between the four species analysed, *Glomus etunicatum*, *G. mosseae*, *G. vesiculiferum* and *G. manihot*. This procedure is also effective in identifying species belonging to other families and genera of glomalean fungi, and if needed could be made more discriminating by simultaneously analysing more polymorphic fragments.

CONCLUSION

The Glomales, a group of obligatory biotrophic fungi forming arbuscular endomy-corrhizae, were shown by analyses of their 18S rDNA genes sequences to form a unique and ancient clade among the true fungi. By applying a molecular clock analysis, estimates of 353-462 Myr were calculated for their origin, in agreement with recently uncovered evidence of arbuscular structures in 400 Myr old *Aglaophyton* from the Rhynie Cherts (9). These lines

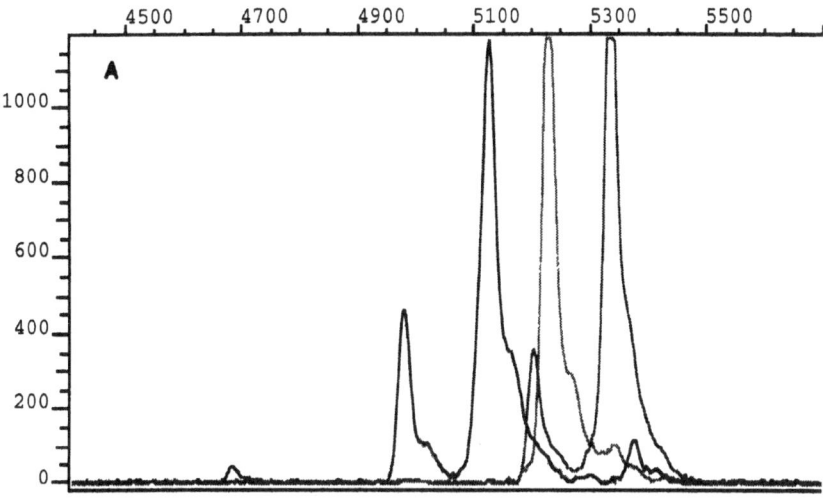

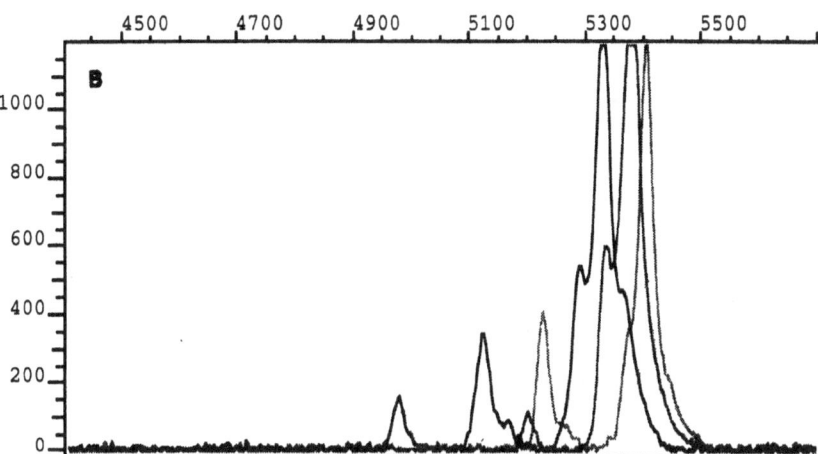

Figure 2. SSCP analysis of 4 *Glomus* species. A fragment of the 18S gene of *G. etunicatum, G. mosseae, G. vesiculiferum* and *G. manihot* was amplified using primer VANS22 and VANS32. The primers used had been labeled with different fluorochromes, VANS22-FAM and VANS32-JOE, to enable color discrimination of the two strands using the laser induced fluorescence detector of the ABI 373 sequencer. Following amplification, a 2 μL aliquot of each reaction was mixed with 35 μL of deionized formamide, 2 μL of ROX-labeled GeneScan 2500 size standards, and 5 μL of loading buffer. This mixture was heated at 90°C for 3 minutes, cooled on ice, and 4 μL were loaded per well on a non-denaturing 6% acrylamide gel. Electrophoresis was performed on an ABI 373 sequencer, at 600 V for 10 hours. The results were analysed with ABI GeneScan software. A) Electropherograms representing the green channel, specific for JOE fluorescence, of 4 adjacent lanes loaded with the fragments amplified from the different *Glomus* species. Scan numbers 4500 to 5700, corresponding to the data collected between 5 and 6.5 hour are depicted. The fastest migrating band (scan number 4990) belongs to *G. etunicatum*, followed by *G. manihot* (scan number 5125), *G. mosseae* (scan number 5225) and *G. vesiculiferum* (scan number 5330). B) Similarly, the electropherograms representing data collected in the blue channel correspond to the FAM fluorescence. The fastest migrating band (scan number 5325) belongs to *G. manihot* followed by *G. vesiculiferum* (scan number 5380), *G. mosseae* (scan number 5410, stronger signal) and *G. etunicatum* (scan number 5410, lower signal). The red channel was monitored to certify that the migration of the Rox-labeled bands (GeneScan 2500 size standards) was uniform in the four adjacent lanes compared (data not shown).

of evidence strengthen the hypothesis that the association with ancestral fungi was instrumental in the colonization of land by plants (8).

The analysis of 18S rDNA genes is not only useful for its phylogenetic interpretation
or the corroboration of current taxonomical hypotheses, but the observation of sequence
similarities among conserved genes analysed in Glomalean species can also be used to devise
molecular tools useful for their identification.

There are already published protocols and primers that allow the identification of
Glomalean fungi using molecular approaches, and these can be applied to facilitate quality
control in spore production facilities. In both field and laboratory experiments, these methods
could be used to confirm the identity of arbuscular fungi, and researchers should be made
aware of the usefulness of this certification and of the importance of conserving voucher
specimens that could be used for future reference.

As the work of characterizing the 18S rDNA genes from more Glomalean species
progresses, it will become possible to refine the identification process, and ultimately devise
protocols that will easily yield species-level resolution.

REFERENCES

1 Arnheim N, White T, Rainey WE. Application of PCR: organismal and population biology. Polymerase
chain reaction can produce large quantities of specific DNA from small, degraded, and impure samples.
1990;40(3):174-182.

2 Becard G, Pfeffer PE. Status of Nuclear Division in Arbuscular Mycorrhizal Fungi During In vitro
Development. Protoplasma 1993;174(1-2):62-68.

3 Berbee ML, Taylor JW. Dating the Evolutionary Radiations of the True Fungi. Can J Bot 1993;71(8):1114-
1127.

4 Bianciotto V, Bonfante P. Quantification of the Nuclear DNA Content of 2 Arbuscular Mycorrhizal Fungi.
Mycol Res 1992;96(Part 12):1071-1076.

5 Innis MA, Gelfand DH, Sninsky JJ, White TJ. PCR protocols, a guide to Methods and applications . San
Diego: Academic Press, 1990: 482.

6 Klironomos JN, Kendrick WB. Research on mycorrhizas: trends in the past 40 years as expressed in the
"MYCOLIT" database. New Phytol 1993;125:595-600.

7 Morton JB, Benny GL. Revised classification of arbuscular mycorrhizal fungi (Zygomycetes): A new
order, Glomales, two new suborders, Glominae and Gigasporinae, and two new families, Acaulosporaceae
and Gigasporaceae, with an emendation of Glomaceae. Mycotaxon 1990;37:471-491.

8 Pirozynski KA, Malloch DW. The origin of land plants: a matter of mycotrophism. BioSystems
1975;6:153-164.

9 Remy W., Taylor T.N., Hass H., Kerp H. 4-hundred-million-year-old vesicular arbuscular mycorrhizae.
PNAS (USA) 1994; 91: 11841-11843.

10 Simon L, Bousquet J, Lévesque RC, Lalonde M. Origin and diversification of endomycorrhizal fungi and
coincidence with vascular land plants. Nature 1993;363(6 may):67-69.

11 Simon L, Lalonde M, Bruns TD. Specific Amplification of 18S Fungal Ribosomal Genes from Vesicular-Arbuscular Endomycorrhizal Fungi Colonizing Roots. Appl Environ Microbiol 1992;58(1):291-295.

12 Simon L, Lévesque RC, Lalonde M. Identification of endomycorrhizal fungi colonizing roots using
fluorescent SSCP-PCR. AEM 1993;59(12):4211-4215.

13 White TJ, Arnheim N, Erlich HA. The Polymerase Chain Reaction. TIG 1989;5(6):185-189.

4

COINCIDENCE BETWEEN MOLECULARLY OR MORPHOLOGICALLY CLASSIFIED ECTOMYCORRHIZAL MORPHOTYPES AND FRUITBODIES IN A SPRUCE FOREST

B. Mehmann,[1] S. Egli,[1] G. H. Braus,[2] and I. Brunner[1]

[1] Swiss Federal Institute for Forest, Snow and Landscape Research (WSL)
Zuercherstrasse 111
CH-8903 Birmensdorf, Switzerland
[2] Institute of Microbiology, Biochemistry and Genetics
Friedrich-Alexander-University
Staudtstrasse 5
D-91058 Erlangen, Germany

SUMMARY

In a study of ectomycorrhizal diversity in a spruce stand in Switzerland the macroscopically classified ectomycorrhizal morphotypes were characterised molecularly by Restriction Fragment Length Polymorphism (RFLP)-patterns of the internal transcribed spacer (ITS) region. In four different plots characterised by different levels of species diversity a total of eighteen ectomycorrhizal morphotypes were macroscopically classified. The evaluation of similarities between the different plot types revealed a more or less homogenous distribution of morphotypes which stands in contrast to the different grades of fruitbody diversity in these plots. Characterisation by molecular tools resulted in a minimum of twenty-three RFLP-types. Seven pattern-types could be referred to fungal species found as fruitbodies in the spruce stand which represents about one third of all mapped fungal species. Different types of correlation between morphotypes and fungal species were observed: (i) one morphotype represents one species, (ii) one morphotype represents several species, (iii) several morphotypes represent one species and (iv) a combination of (ii) and (iii). A low coincidence between spatial distribution and abundance of molecularly identified morphotypes and mapped fruitbodies could be found. The problems associated with classification of ectomycorrhizas and the significance of above-ground fruitbody pattern for below-ground mycorrhizal pattern is discussed.

Biotechnology of Ectomycorrhizae, Edited by Vilberto Stocchi et al.
Plenum Press, New York, 1995

INTRODUCTION

Monitoring and classification of ectomycorrhizal populations are indispensable to assess and describe their diversity or dynamics. Most of the information regarding fungal communities has been based on a record of fruitbody development. Recent studies have shown that fruitbody appearance may not be a good indicator of mycorrhiza composition (Gibson and Deacon, 1988; Danielson and Pruden, 1989; Dahlberg, 1991). The value of counting fruitbodies is limited as the appearance of fruitbodies does not necessarily reflect the below-ground colonisation of root tips (Read, 1984; Taylor and Alexander, 1990). Most current knowledge concerning the community structures of ectomycorrhizal fungi and their dynamics (Dighton et al., 1986) is mainly based on the occurrence of reproductive structures in nature, whereas fruitbody production is often influenced by environmental conditions such as precipitation, temperature or humidity.

Morphotyping of mycorrhizas is another way to describe mycorrhizal patterns. However only a few ectomycorrhizas can be unambiguously macroscopically identified by morphological characteristic features. It is possible to determine some species or at least to tentatively assign them to a particular genus, e.g. ectomycorrhizas formed by fungi of the genera *Tuber, Hebeloma, Laccaria* or *Inocybe* (Voiry, 1981; Ingleby et al., 1990).

Other methods used to identify ectomycorrhizas trace mycelial connections between ectomycorrhizas and fruitbodies, as described by Agerer (1991). This can provide reliable results but is limited to species forming rhizomorphs or conspicuously emanating mycelia, and it is only applicable during the fungus fruiting season. Other methods, such as comparison by thin layer chromatography (Thoen, 1977) are restricted to species with a pigmented mycelium or macrochemical reactions. Melin's Erlenmeyer flask technique for *in vitro* synthesis of ectomycorrhizas has been considered a possible tool to identify fungal partners of naturally occurring ectomycorrhizas through morphological comparison (Zak, 1973). Recent studies using seedlings of *Picea abies* and *Larix leptolepis* and the fungal partners *Amanita muscaria, Hebeloma crustuliniforme* and *Suillus grevillei* confirmed that natural and *in vitro* synthesised mycorrhizas of these species showed no distinct morphological and anatomical differences (Kottke, 1986; Brunner et al., 1991). However, *in vitro* syntheses do not seem to be appropriate for routine identification of natural ectomycorrhizas obtained by large-scale root sampling (Brunner et al., 1992).

Since reliable identification of ectomycorrhizas is restricted to certain species, scientists have to resort to an alternative classification of mycorrhizal morphotypes based on the description of morphological and anatomical features. Molecular methods based on the polymerase chain reaction (PCR) offer several rapid and sensitive means of identifying ectomycorrhizal fungi (Gardes et al., 1991). The PCR method allows the amplification of target sequences and is capable of detecting a single molecule of target DNA (Saiki et al., 1988) and is therefore also useful for identification of the fungal symbiont in the ectomycorrhiza. The use of RFLP comparison of the ITS region of ribosomal DNA is a particularly fast and easy way to confirm species-level identification when known DNA patterns are available for comparison (Gardes et. al., 1991; Gardes and Bruns, 1993a; 1994). Other DNA methods used to differentiate ectomycorrhizal fungal species and/or isolates are the (i) RFLP-typing of the intergenic spacer (IGS) region of ribosomal DNA, or of (ii) a partial region of coding sequences of chitin synthase genes, (iii) random amplified polymorphic DNA (RAPD) or (iv) the recently developed technique of directed amplification of microsatellite-region DNA (DAMD) (Henrion et al., 1994; Mehmann et al., 1994; Wyss and Bonfante, 1993; Potenza et al., 1994a; Costa and Martin, 1994). The amplification of DNA with short primers (RAPD) or microsatellite primers allows a high degree of differentiation using fungal probes as templates. Investigating mycorrhizas, the interference of the host genome renders more difficult or even prevents the identification of the fungal symbiont (Lanfranco et al., 1994; Potenza et al., 1994b).

MATERIAL AND METHODS

Study Area

Investigations were carried out in a 40 year-old pure spruce (*Picea abies*) stand ('Gaesi'), a reforestation of a former alluvial stand (midlands, 425 m a.s.l.). A complete site description is available in Egli (1992). A full inventory and mapping of fruitbodies was made at weekly intervals during the fungus fruiting season over a period of seven years. Fruitbodies were collected, determined (nomenclature following Moser, 1983) and dried. Based on species distribution four different types of plots (0.25 m^2, two per type) have been evaluated (Fig. 1):

- A1, A2: high species diversity: 9 species: *Cortinarius infractus, C. odorifer, C. salor, C. (Telamonia) sp.* 1, *C. varius, Lactarius deterrimus, Russula erythropoda, R. nauseosa, Inocybe fastigata*
- B1, B2: no species ("fungus desert")
- E1, E2: low species diversity: one species *Russula nauseosa*
- F1, F2: low species diversity: one species *Cortinarius odorifer*

Root Sampling and Morphological Classification of Mycorrhizas

Root material was collected in autumn 1993. Ten root samples in each plot were taken by a soil coring cylinder of 6.4 cm diameter to a depth of 10 cm. Soil cores were saturated in water and roots were washed out carefully and stored in cold water pending examination within a few days. Examination was done under a dissecting microscope with a maximum magnification of 50x. Distinctive features for macroscopic classification into morphotypes were type of ramification, colour, structure of mantle surface, and presence and nature of emanating hyphae and rhizomorphs (Agerer, 1987). From each morphotype one to six probes were taken and dried for molecular characterization. (The mycorrhiza of *Cenococcum geophilum* is an identified type and therefore not considered in this study).

Similarity Assessment

Morphotypes present in different plots were compared by using a simple coefficient of similarity calculation $S = 2c/(a+b)$, where **a** is the number of morphotypes in one plot, **b** is the number of the other, and **c** is the number of morphotypes in common (Christensen, 1981). The coefficient is presented as a percentage. The distribution between the different types is calculated by summing the type distribution of plot X1 and plot X2 in one plot X (presence in X means presence in X1 and/or X2) and comparing plot X to plot Y.

Molecular Characterisation

Chromosomal DNA was extracted from dried mycorrhizas, sampled fruitbodies (dried and fresh) and fungal cultures according to Gardes and Bruns (1993a). Using the universal primer ITS4 in combination with the fungus-specific primer ITS1-F preferentially fungal ITS sequences were amplified from infected plant material by the PCR reaction (Gardes and Bruns, 1994). The reaction mixture consisted of a template solution of genomic

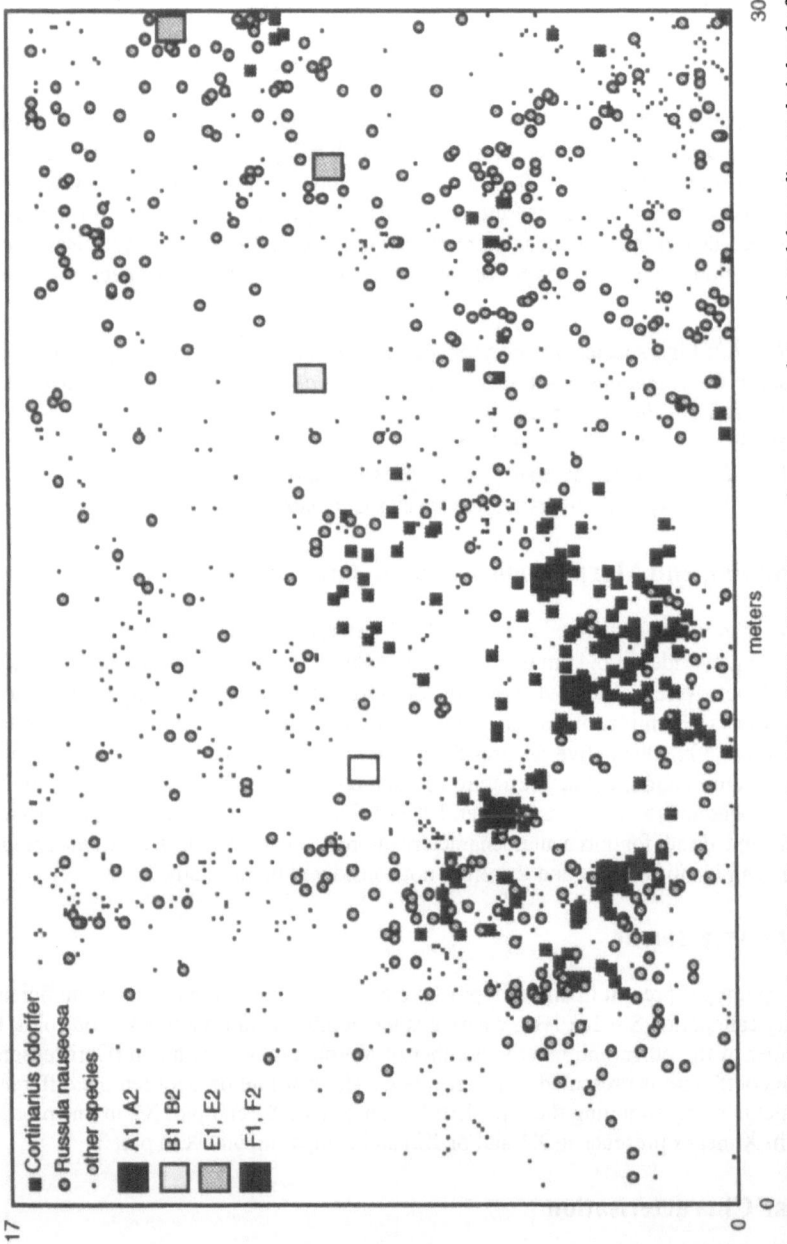

Figure 1. Distribution of mapped fruitbodies in the 'Gaesi' spruce forest from 1986 to 1993. Rectangles represent plots evaluated depending on their level of species diversity for root sampling. Blank rectangles represent no species ('fungus deserts') (B-plots), black rectangles represent high species diversity (A-plots), shadowed rectangles represent one single species plot with *Russula nauseosa* (E-plots) or *Cortinarius odorifer* (F-plots).

DNA and a master mix (4 nmol of each nucleotide, the appropriate amount of 10x Super *Taq* reaction buffer: 100 mM Tris-HCl [pH 9.0], 500 mM KCl, 0.1% (wt/vol.) gelatine, 15 mM $MgCl_2$, 1% Triton X-100) containing 0.5 units of Super *Taq* DNA polymerase (Staehelin, Switzerland) and 0.25 µM of each of the two primers in a 20 µl total reaction volume, overlaid with mineral oil. Temperature cycling according to Gardes and Bruns (1993a) was carried out with a thermocycler (Biometra, Germany). Restriction digestions of the amplified fragments were performed on unpurified PCR products (5-10 µl) in a total volume of 15 µl containing the appropriate amount of restriction enzyme buffer and 5 units of *Alu*I, *Cfo*I, *Hin*fI, *Rsa*I and *Taq*I (Fermentas, Switzerland; Gibco, Switzerland). The reactions were incubated for 2h at suitable temperature for each enzyme. PCR products and their restriction digests were electrophoresed for 1-4h on 2% Agarose gels (Biomol Feinchemikalien, Germany) in TAE buffer. Gels were stained with ethidium bromide and photographed under ultraviolet light.

Molecular Identification

The number and size of restriction fragments from the amplified ITS region from mycorrhizas were compared to patterns derived from fungal species in the reference library (Table 1). This reference library consisted of fungal species found as fruitbodies in the investigated spruce stand (Table 1a) and other specimens (fruitbodies or fungal cultures) of potential spruce ectomycorrhizal fungi (Table 1b). To confirm pattern identity, putative identical patterns were run side-by-side on the same gel (Gardes and Bruns, 1994).

RESULTS

Fruitbodies

Fruitbodies of twenty-two different fungal species were recorded over a period of seven years (Table 1a). Their distribution is shown in Figure 1. All representatives of the reference library (Table 1) are characterised by different patterns (data not shown). Most species of the same genus showed the same fragment pattern for at least two restriction enzymes and for additional restrictions similar patterns caused by additional or missing restriction sites.

Morphotypes

On the basis of macroscopic features a total of eighteen morphotypes (M1-M18) were classified (Table 2). In the eight plots nine to fifteen different types were found (Table 3).

Table 1. List of fungal species in the reference library for RFLP-identification

a mapped fruitbodies:		b potential spruce
Boletus luridus	*Hebeloma birrum*	ectomycorrhizal species:
Cortinarius alboviolaceus	*Hygrophorus pustulatus*	*Amphinema byssoides*
C. arquatus	*Inocybe fastigata*	*Cenococcum geophilum*
C. elotus	*I. praetervisa*	*Cortinarius hercynicus*
C. infractus	*Lactarius deterrimus*	*Mycelium radicis atrovirens*
C. odorifer	*Russula adulterina*	*Phialophora dimorphospora*
C. orichalceus	*R. erythropoda*	*Piloderma croceum* (=bicolor)
C. salor	*R. nauseosa*	*P. byssinum*
C. varius	*R. queletii*	*Telephora terrestris*
C. (Telamonia) sp. 1	*Tricholoma sulphureum*	
Geastrum sessile	*Tuber puberulum*	

Repetitive plots exhibit approximately the same diversity of types and therefore are about equal in quantity. Testing the representativity of the repetitive plots regarding composition, the similarity values (S) of these plots were calculated (Table 4). Comparison within the repetition plots (Table 4a) resulted in relatively high similarity values between 70 and 100 percent. Comparison between the different types of plots (Table 4b) gave values varying from 70 to 90 percent. Therefore, the morphotype diversity of all plots has a high similarity, implying that the morphotypes are spread more or less homogeneously over the whole spruce stand.

RFLP-Characterisation and Identification

DNA isolation from dried probes of mycorrhizas, amplification of the ITS region from chromosomal DNA and further restriction analysis resulted in a total of twenty-three different RFLP-patterns (Table 2). Seven RFLP-patterns could be identified by comparison to our references showing identity in all five restriction digests. The residual sixteen patterns remained unidentified. However, five patterns showed partial identity in some enzyme digestions and other similar patterns which could be derived by additional restriction sites. These characterisations were considered as 'genus-like'. Based on our own observations from reference library and those of Gardes and Bruns (1994) they could be identified at the genus level. Therefore, mycorrhizas of only one third of the fungal species found as fruitbodies in the investigated spruce stand were found in the eighteen morphotypes.

Different types of correlation between morphotypes and fungal species were observed. One morphotype represented one species in the case of morphotype M3 formed by *Cortinarius (Telamonia) sp.* 1. Morphotype M1 formed by *C. alboviolaceus* and another *Cortinarius*-like species is an example of the situation where one morphotype represents several species. In the case of *C. odorifer* and morphotypes M12 and M13 several morphotypes represent one species. Even a combination of this correlation could be observed where three morphotypes show common patterns but are themselves characterised by different patterns (Table 2).

Four groups of mycorrhizal probes attached to different morphotypes were characterised by identical RFLP-patterns. *Tuber puberulum* and *Cortinarius odorifer* each seem to form mycorrhizas of two morphotypes (M10/M11 and M12/M13, respectively). Morphotypes M5 and M14 have identical *Cortinarius*-like RFLP-probes, and M6 and M14 have several common *Russula*-like patterns. Probes of types M16, M17, and M18 have a common unidentified RFLP-type R10. In most cases, they were identical in ramification type, mantle surface, and rhizomorph presence. The most common variable characteristic was colour.

DISCUSSION

Molecular Characterisation and Identification

RFLP-typing of the ITS region with at least five different endonucleases allows a division into different species (Gardes and Bruns, 1994). With isolates of the same species but of different geographic origin differences in restriction digest pattern can be found. This problem can be avoided by building up a reference library of fruitbodies derived from the observed spruce stand and other indigenous specimens and testing several examples of a species to confirm RFLP-patterns. Even when not all molecularly characterised morphotypes can be identified, they can be differentiated from each other by their RFLP-pattern, which therefore allows a characterisation at the species level. Due to the identity in the fragment pattern of some endonucleases, an assignment at the genus level is possible (Gardes and

Table 2. Macroscopic and molecular features of the ectomycorrhizal morphotypes of spruce occurring on the 'Gaesi' stand

Type	Macroscopic features				Molecular features	
	Ramification	Colour	Mantel surface	Rhizomorphs	RFLP-pattern	Fungal species
M1	simple	white	wooly	+	G1	*Cortinarius alboviolaceus*
					R1	*Cortinarius*-like
M2	irregular-pinnate	white	wooly to shaggy	+	G8	*Cortinarius salor*
					R8	?
M3	irregular-pinnate	white	cottony	–	G14	*C. (Telamonia) sp. 1*
M4	simple	white	cottony, translucent	+	R7	?
M5	monopodial-pinnate	golden-yellow to brown	smooth	–	R2	*Cortinarius*-like
M6	monopodial-pyramidal	pinkish-white	smooth	–	G25	*Russula nauseosa*
					G34	*R. erythropoda*
					R34	*R. erythropoda*-like
					R6	?
M7	monopodial-pinnate	silver-straw to yellow	cottony, translucent	+	R4	*Russula*-like
M8	monopodial-pyramidal	loamy-grey	smooth	–	R5	?
M9	simple	greenish-grey	smooth	–	R16	?
					R17	?
M10	monopodial-pyramidal	lemon-yellow	smooth, spiny	–	G61	*Tuber puberulum*
M11	monopodial-pyramidal	pale-brown	smooth, spiny	–	G61	*Tuber puberulum*
M12	irregular-pinnate	white	shaggy	+	R1	*Cortinarius*-like
					G5	*Cortinarius odorifer*
M13	irregular-pinnate	white to yellowish	shaggy	+	G5	*C. odorifer*
M14	monopodial-pyramidal	pale-cream to straw	smooth	–	G25	*Russula nauseosa*
					G34	*R. erythropoda*
					R34	*R. erythropoda*-like
					R24	*R. erythropoda*-like
					R15	?
M15	monopodial-pyramidal	fawn yellow	grainy, long-spiny	–	R2	*Cortinarius*-like
M16	monopodial-pyramidal	cigar-brown	grainy	–	R9	?
M17	monopodial-pyramidal	black	grainy	+	R10	?
					R10, R12	?
M18	monopodial-pyramidal	darkbrown to black	grainy	–	R10, R3	?

Table 3. Distribution of morphotypes

Plots	Morphotypes																		Total number	
	M1	M2	M3	M4	M5	M6	M7	M8	M9	M10	M11	M12	M13	M14	M15	M16	M17	M18	Morphotypes	Fruitbody species
A1	+	+	+		+	+		+	+		+	+	+	+	+	+	+	+	15	9
A2	+	+	+	+	+	+		+			+	+	+	+	+	+		+	14	9
B1	+	+	+	+	+	+	+	+			+			+		+	+	+	12	0
B2	+				+	+	+	+	+					+	+	+	+	+	10	0
E1	+	+	+		+	+		+	+	+			+	+		+		+	12	1
E2	+	+	+		+	+		+	+	+	+			+	+	+	+	+	12	1
F1	+	+					+	+	+		+							+	11	1
F2	+		+		+	+	+	+						+		+	+	+	9	1

Table 4. Similarity values

a. within repetition plots	b. between different plots	
A1-A2 S = 83 %	A-B S = 90 %	
B1-B2 S = 73 %	A-E S = 76 %	
E1-E2 S = 100 %	A-F S = 87 %	
F1-F2 S = 70 %	B-E S = 85 %	
	B-F S = 89 %	
	E-F S = 72 %	

Bruns, 1994). The grade of species identification also depends on the composition of the reference library and could be increased by expanding the number of ectomycorrhizal reference species.

With molecular tools we were able to show that several fungal species are able to form mycorrhizas with identical macroscopic characteristics, a fact already observed by Karen et al. (1994). In some cases, RFLP-patterns of these mycorrhizas are similar and can be assigned to the same genus. A classification according to macroscopic features represents a fast and easy method but only allows an incomplete identification at the genus or family level. An unambiguous morphological determination of the species of most ectomycorrhizal fungi has so far only been possible from fruitbodies. In comparison with the large number of morphological features determining a fungal species, only a rough subdivision is expected to be obtained by the few macroscopic characteristics of morphotypes. Mycorrhizas assigned by macroscopic description to the same morphotype could also be distinguished by microscopic features such as the mantle structures (Egli et al., 1993). Mantle structure seems to be a relatively stable feature allowing a differentiation but not necessarily an identification. In some cases an identification of ectomycorrhizas at the species level on the basis of macro- and microscopic features is possible, as the colour atlas of Agerer (1987) shows. However, not all mycorrhizas have species-specific features as shown in a recent study using *in vitro*-synthesised ectomycorrhizas of the two species *Hebeloma crustuliniforme* and *H. cylindrosporum* with spruce (Brunner, 1991). Whereas microscopical characterisation including fixation, dehydration, embedding, sectioning, and staining is time-consuming, the new methods of molecular characterisation are fast and allow a characterisation within one day.

In the present study, mycorrhizas of *Cortinarius odorifer* showed colours varying from white to yellowish and *Tuber puberulum* from lemon-yellow to pale-brown. Each of these two species formed mycorrhizas assigned to two morphotypes and characterised by identical molecular and macroscopic characteristics including ramification, mantle surface, and rhizomorphs. The colour was the only varying character. It is known that macroscopic features, particularly the colour, can be very inconsistent. Colour-changes of ectomycorrhizas during aging have been observed by various authors (see Egli and Kälin, 1991). In addition, ramification characteristics can be influenced by site factors such as different nutritional conditions in different substrates (Palenzona and Fontana, 1970). According to the results of Haug (1987), microscopic features seem to be less influenced by environmental conditions than macroscopic ones.

Morphotyping as well as microscopic classification are in some contexts relatively unreliable as they are based in part on inconsistent features. In addition, the results from different authors can hardly be compared, as classification into types in many cases is subjective and can thus be inaccurate. In this context, DNA is quite stable and therefore an objective characteristic. Using the same methods and conditions, a restriction site on an amplified fragment is present or absent. A disadvantage might be that different groups working with these tools may use different primers, programs or thermocyclers. Their methods to analyse and record the RFLP patterns in their reference libraries are in most cases different due to their technical supplies from the institutes which renders the exchange of data more difficult.

Coincidence between Above-Ground Fruitbody Pattern and Morphotype Pattern

Small plots with different grades of species diversity were chosen to examine the correlation between species diversity and morphotype diversity. For all four plot types, morphotype diversity was more or less equal, regarding quantity and quality. Therefore, morphotypes are probably distributed in a relatively homogenous fashion over the whole spruce stand within single plots. No correlation existed between number of mycorrhizal types and number of ectomycorrhizal species appearing as fruitbodies. Similar observations were made by Jansen and de Vries (1988), Menge and Grand (1978), and Dahlberg and Stenström (1991). Therefore, we conclude that there is no or only a minor correlation between number and diversity of morphotypes and fruitbody composition regarding spatial distribution.

Only one third of RFLP types can be identified by comparison to our reference library. Two thirds of mycorrhizas are formed by species whose fruitbodies were not found during the seven year observation period. These species might either form hypogeous fruitbodies or did not fruit. The small portion of mapped species on the guild structure forming species had already been observed by other authors (Gardes and Bruns, 1993b; Pritsch and Buscot, 1994). Mapped fungal species only represent a small part of mycorrhizas forming species and therefore do not represent below-ground mycorrhizal proportions.

Supposing that an overview of all fruitbodies of one species during the whole observation time would give an impression of the extent of the below-ground mycelium, mycorrhizas of *Russula nauseosa* should be found in all plots, which is the case. This possibility of exclusive presence in potentially mycelium-containing areas could also be confirmed since fruitbodies and mycorrhizas of *Cortinarius odorifer* were present in plots A and F and absent in B and E. It is almost impossible to estimate the mycelium spreading below ground for fungal species producing a small number of fruitbodies. *Cortinarius alboviolaceus* represented less than 1% (3 fruitbodies) of all fruitbodies and was not expected to form mycorrhizas in all types of plots. Additionally, the number of fruitbodies does not necessarily correlate with the mycorrhizal frequency as *C. (Telamonia) sp.* 1, the most frequent fruitbody (29%) forms only 1-5% of mycorrhizas, and *C. alboviolaceus* less than 1% forms up to a fourth of all mycorrhizas. In a few cases, where many fruitbodies were observed, a correlation between fruitbody and mycorrhiza was noticed. An overall look at the different plot types and their surrounding fruitbodies reveals only a low correlation between fruitbody and mycorrhiza presence.

As fungus fruiting is dependent on soil type, climate, and weather conditions even fruiting is not a reliable sign for presence of below-ground mycelium. As many ectomycorrhizal fungal species hardly or never form fruitbodies or only hypogeous fruitbodies, inventories record only a small part of ectomycorrhiza forming species. For classification and monitoring of mycorrhizal guild structures, mapping and inventory of epigeous fruitbodies is incomplete.

Considering the distribution of fruitbody species, morphotypes, and molecularly characterised mycorrhizas, only a low coincidence between molecularly identified ectomycorrhizal morphotypes and fruitbodies was found The above-ground fruitbody composition does not seem to be representative of the below-ground mycorrhizal pattern

ACKNOWLEDGMENT

We thank Mrs Margrit Zollinger and Mrs Bettina Schneider for mapping and collecting fruitbodies and Mrs Biljana Milakovic for preparation of root samples and morphotyping We also thank Heidi Suess for critical reading of the manuscript We are especially grateful to Ralf Hutter for generous support

This work was supported by Swiss National Foundation grant 31-30778 91, by the Swiss Federal Institute of Technology, and by the Swiss Federal Institute for Forest, Snow, and Landscape Research

REFERENCES

Agerer, R , 1987, Colour atlas of ectomycorrhizae Einhorn, Schwabisch Gmund

Agerer, R , 1991, Characterization of ectomycorrhiza In Norris, J R , Read, D J , and Varma, A K (eds) Methods in microbiology, vol 23 Academic Press, London, pp 25-73

Brunner, I , 1991, Comparative studies on ectomycorrhizae synthesized with various *in vitro* techniques using *Picea abies* and two *Hebeloma* species, *Trees* 5 90-94

Brunner, I , Amiet, R , and Schneider, B , 1991, Characterization of naturally grown and *in vitro* synthesized ectomycorrhizas of *Hebeloma crustuliniforme* and *Picea abies Mycol Res* 95 1407-1413

Brunner, I , Amiet, R , Zollinger, M , and Egli, S , 1992, Ectomycorrhizal syntheses with *Picea abies* and three fungal species a case study on the use of an *in vitro* technique to identify naturally occurring ectomycorrhizae, *Mycorrhiza* 2 89-96

Christensen, M , 1981, Species diversity and dominance in fungal communities In Wicklow, D T , and Caroll, G C (eds) The fungal community - its organisation and role in the ecosystems Marcel Dekker Inc , New York, pp 201-232

Costa, G , and Martin, F , 1994, Directed amplification of microsatellite-region DNA in ectomycorrhizal fungi (abstract) Fourth European Symposium on Mycorrhizas, Granada, Spain, p 14

Dahlberg, A , 1991, Ectomycorrhiza in coniferous forest structure and dynamics of populations and communities Doctoral thesis, Swedish University of Agricultural Sciences SLU/Repro, Uppsala

Dahlberg, A , and Stenstrom, E , 1991, Dynamic changes in nursery and indigenous mycorrhiza of *Pinus sylvestris* seedlings planted out in forest and clearcuts, *Plant Soil* 136 73-86

Danielson, R M , and Pruden, M , 1989 The ectomycorrhizal status of urban spruce, *Mycologia* 81 335-341

Dighton, J , Poskitt, J M , and Howard, D M , 1986, Changes in occurrence of basidiomycete fruit bodies during forest stand development with specific reference to mycorrhizal species, *Trans Br Mycol Soc* 87 163-171

Egli, S , 1992, Der Anisklumpfuss, *Cortinarius odorifer* Britz Okologie, Biologie und Ektomykorrhiza, *Mitt Eidg Forschungsanst Wald Schnee Landsch* 67 315-412

Egli, S , and Kalin, I , 1991, The root window - a technique for observing mycorrhizae on living trees, *Agric Ecosyst Environ* 28 107-110

Egli, S , Amiet, R , Zollinger, M , and Schneider, B , 1993, Characterization of *Picea abies* (L) Karst ectomycorrhizas discrepancy between classification according to macroscopic versus microscopic features, *Trees* 7 123-129

Gardes, M , and Bruns, T D , 1993a, ITS primers with enhanced specificity for basidiomycetes - application to the identification of mycorrhizae and rusts, *Mol Ecol* 2 113-118

Gardes, M , and Bruns, T , 1993b, The mycorrhizal guild structure of a bishop pine forest in California (abstract), In Peterson L , Schelkle M (eds) Ninth North American Conference on Mycorrhizae, University of Guelph, Guelph, Ontario, Canada, p 105

Gardes, M , and Bruns, T D , 1994, ITS-RFLP matching for identification of fungi, *Meth Mol Biol* in press

Gardes, M , White, T J , Fortin, J A , Bruns, T D , and Taylor, J W , 1991, Identification of indigenous and introduced symbiotic fungi in ectomycorrhizae by amplification of nuclear and mitochondrial ribosomal DNA, *Can J Bot* 69 180-190

Gibson, F., and Deacon, J.W., 1988, Experimental study of establishment of ectomycorrhizas in different regions of birch root systems, *Trans. Br. Mycol. Soc.* 91:239-251.

Haug, I., 1987, Licht- und elektronenmikroskopische Untersuchungen an Mykorrhizen von Fichtenbeständen im Schwarzwald. Dissertation, Universität Tübingen, Germany.

Henrion, B., Chevalier, G., and Martin, F., 1994, Typing truffle species by PCR amplification of the ribosomal DNA spacers, *Mycol. Res.* 98:37-43.

Ingleby, K., Mason, P.A., Last, F.T., and Fleming, L.V., 1990, Identification of ectomycorrhizas, Institute of Terrestrial Ecology, Res. Publ. 5.

Jansen, A.E., and de Vries, F.W., 1988, Dutch priority programme on acidification. Qualitative and quantitative research on the relation between ectomycorrhiza of *Pseudotsuga menziesii*, vitality of the host and acid rain. Report 25-02. CEC Research contract ENV-896-NL. Comm. no. 373 of the Biological Station, Wijster, The Netherlands.

Karen, O., Nylund, J.-E., Dahlberg, A., Högberg, N., Jonsson, L., and Grip, K., 1994, Influence of drought on mycorrhizal species composition - A comparison of identification by morphotype characters versus the PCR-technique (abstract). Fourth European Symposium on Mycorrhizas, Granada, Spain, p. 19.

Kottke, I., 1986, Charakterisierung und Identifizierung von Mykorrhizen. In: Einsele, G. (ed.) Das landschafts-ökologische Forschungsprojekt Naturpark Schönbuch (Forschungsbericht). Deutsche Forschungsge-meinschaft, Weinheim, pp. 463-485.

Lanfranco, L., Arlorio, M., Matteucci, A., and Bonfante, P., 1994, Truffles: Life cycle and molecular characterization (abstract), International Symposium on Biotechnology of Ectomycorrhizae: Molecu-lar Approaches, Urbino, Italy, pp. 34-35.

Mehmann, B., Brunner, I., and Braus, G.H., 1994, Nucleotide Sequence Variation of Chitin Synthase Genes among Ectomycorrhizal Fungi and its Potential Use in Taxonomy, *Appl. Environ. Microbiol.* 60:3105-3111.

Menge, J.A., and Grand, L.F., 1978, Effect of fertilization on production of epigeous basidiocarps by mycorrhizal fungi in loblolly pine plantations. *Can. J. Bot.* 56:2357-2362.

Moser, M., 1983, Die Röhrlinge und Blätterpilze. Kleine Kryptogamenflora IIb/2. Gustav Fischer Verlag, Stuttgart, Germany.

Palenzona, M., and Fontana, A., 1970, Influenza di tipi di suolo su tre forme micorrhiziche del pino strobo, *Allionia* 16:101-113.

Potenza, L., Amicucci, A., Rossi, I., Palma, F., De Bellis, R., Cardoni, P., and Stocchi, V., 1994a, Identification of *Tuber magnatum* Pico DNA markers by RAPD analysis, *Biotechnology Techniques* 8:93-98.

Potenza, L., Amicucci, A., Rossi, I., Palma, F., De Bellis, R., Cardoni, P., and Stocchi, V., 1994b, Identification of genetic markers for the characterization of *Tuber* species (abstract), International Symposium on Biotechnology of Ectomycorrhizae: Molecular Approaches, Urbino, Italy, pp. 55-56.

Pritsch, K., and Buscot, F., 1994, Biodiversity of ectomycorrhizas - from morphotypes to species, (abstract). Fourth European Symposium on Mycorrhizas, Granada, Spain, p. 10.

Read, D.F., 1984, The structure and function of the vegetative mycelium of mycorrhizal roots. In: Jennings, D.H., Rayner, A.D.M., (eds.) Methods in microbiology, vol. 23. Academic Press, London, UK, pp. 253-281.

Saiki, R.K., Gelfand, D.H., Stoffel, S., Scharf, S.J., Higuchi, R., Horn, G.T., Mullis, K.B., and Ehrlich, H.A., 1988, Primer-directed enzymatic amplification of DNA with a thermostable DNA polymerase, *Science* 239:487-491.

Taylor, A.F.S., and Alexander, I.J., 1990, Demography and population dynamics of ectomycorrhizas of sitka spruce fertilized with N, *Agric. Ecosyst. Environ.* 28:493-496.

Thoen, D., 1977, Identification of ectomycorrhizal fungi by thin layer chromatography, *Bull. Br. Mycol. Soc.* 11:39-43.

Voiry, H., 1981, Classification morphologique des ectomycorhizes du chêne et du hêtre dans le nord-est de la France, *Eur. J. For. Pathol.* 11:284-299.

Wyss, P., and Bonfante, P., 1993, Amplification of genomic DNA of arbuscular-mycorrhizal (AM) fungi by PCR using short arbitrary primers, *Mycol. Res.* 97:1351-1357.

Zak, B., 1973, Classification of ectomycorrhizae. In: Marks G.C., Kozlowski T.T. (eds.) Ectomycorrhizae, Academic Press, New York, pp. 43-78.

5

ECTOMYCORRHIZA MORPHOGENESIS

Insights from Studies of Developmentally Regulated Genes and Proteins

F. Martin, T. Burgess, M. E. Carnero Diaz, D. de Carvalho, P. Laurent,
P. Murphy, U. Nehls, and D. Tagu

Equipe de Microbiologie Forestière
Centre I.N.R.A. de Nancy
54280 Champenoux
France

INTRODUCTION

Ectomycorrhizas are characterized structurally by the presence of a dense mass of fungal hyphæ forming a pseudoparenchymatous tissue ensheathing the root and a Hartig net of intercellular hyphæ characterized by labyrinthine branching. Fungal hyphæ stimulate lateral root formation, radial elongation of epidermal cells of angiosperms, and dichotomy of the apical meristem in conifer species (Massicotte et al. 1987, 1989; Dexheimer & Pargney 1991; Peterson & Bonfante 1993). During the symbiosis development, cell differentiation and tissue patterning lead to a novel spatial organization, changes in form, and the generation of different cell types. Ectomycorrhiza morphogenesis encompasses a series of complex and overlapping ontogenic processes in symbionts: switching of the fungal growth mode, initiation of lateral roots, aggregation of hyphæ, arrest of cell divisions of ensheathed roots, radial elongation of epidermal cells (Fig. 1). Morphological differentiation is accompanied by the onset of novel metabolic organizations in fungal and plant cells leading to the finished functioning symbiotic organ.

Morphological types of ectomycorrhizas are numerous, yet their basic structure (the sequence and spatial arrangements of the different elements) is always the same. This implies the existence of a unique developmental strategy for building an ectomycorrhiza that early on imposes a basic scheme, on top of which subsequent species-specific customizations occur. Therefore, initiation and development of ectomycorrhiza are primarily controlled by genetic factors and secondly by the developmental status of the mycelium and host-plant and environmental factors. What could be the molecular basis of such a progressive, highly organized ontogenic process? What is the role of cell-cell signaling in development? How are patterns established during formation of fungal symbiotic tissues? How many genes control ectomycorrhiza development–as distinct from providing the housekeeping functions

Biotechnology of Ectomycorrhizae, Edited by Vilberto Stocchi et al.
Plenum Press, New York, 1995

53

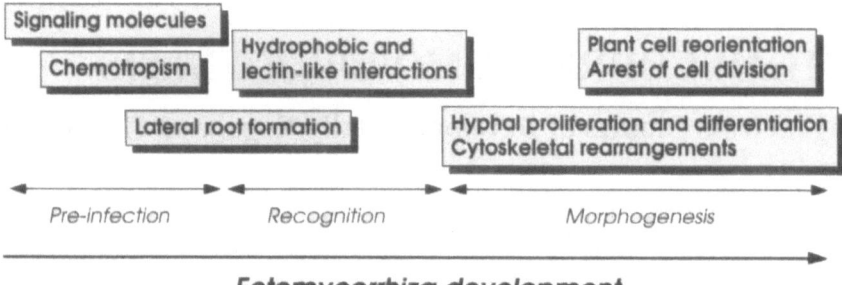

Ectomycorrhiza development

Figure 1. The different interactions between the host root and the ectomycorrhizal fungus and morphogenetic processes observed during ectomycorrhiza development.

of the fungal and plant cells? Unfortunately, experimental data have not yet provided much insight with respect to these questions.

We have speculated that part of ectomycorrhiza development may rely on a precisely choreographed activation of primary regulators (Martin & Hilbert 1991). These genes would encode transcription regulators of specific sets of target genes, which would define the unique developmental pathway of each individual stage (Fig. 2). Target genes may encode

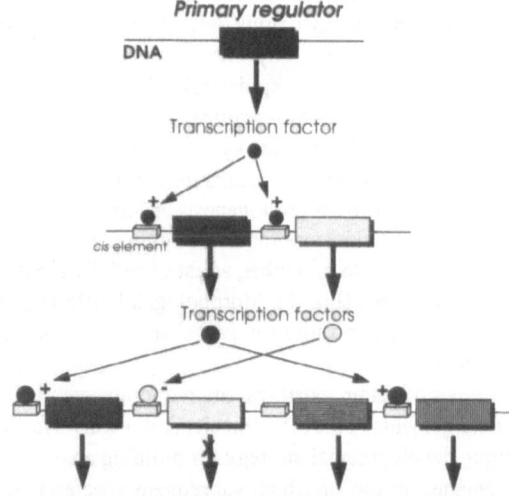

Morphogenetic and symbiosis-regulated proteins

Figure 2. A speculative model showing the basic regulatory mechanisms involving hierarchies of nuclear gene expression which control ectomycorrhiza development. At the top level are a set of primary regulators whose activity as transcription factors would mediate developmental switches (e.g. switch from the indeterminate, apical growth pattern of vegetative mycelium to the highly organized pattern of the ectomycorrhiza). The activity of these master genes will be necessary and sufficient to induce a regulatory cascade that will be partially responsible for ectomycorrhiza morphogenesis. These genes may respond to rhizospheric signals and positional information mediated by 'sensory molecules'. These genes likely activate the mycorrhiza development program, and 'reinforce' decisions during most phases of the ectomycorrhiza morphogenesis. At lower levels in the regulatory hierarchy are genes that encode morphogenetic proteins (e.g. hydrophobins, auxin-regulated protein), which are likely to participate in the symbiotic organ construction.

transcription regulators and, at a lower level, morphogenetic proteins (e.g. cell wall glyco-proteins), membrane transporters, or a specific isozyme which will participate in the structure organization and will provide its specific features and activities. It is widely held that this paradigm may apply to the organization of fungal (Timberlake 1993; Wessels 1993a,b), plant (Coen & Carpenter 1993), and animal (Patel 1994) organs. Recently, molecular approaches have been used with success in several mycorrhizal symbioses, and a number of promising candidates for morphogenetic proteins that may be involved in ectomycorrhiza development have been identified. However, a realistic view of the regulatory gene interactions in ectomycorrhiza morphogenesis has to await identification of the primary regulators control-ling the symbiotic development.

In any case, ectomycorrhiza formation affects both plant and fungal development in a pleiotropic manner. Changes in the morphology, biochemistry and physiology of the plant and fungal cells have been extensively documented (for reviews, see Martin & Hilbert 1991; Bonfante & Perotto 1992; Peterson & Bonfante 1993). In addition, several of the symbiont responses to ectomycorrhiza development are correlated with alterations in gene expression (Tagu et al. 1993; Nehls & Martin, this book). In this review, we examine current knowledge about the nature and function of symbiosis-regulated proteins and genes, and how this can be related to ectomycorrhiza development.

MYCORRHIZA DEVELOPMENT INDUCES CHANGES IN PROTEIN SYNTHESIS

Analysis of natural variants of ectomycorrhizal fungi has clearly shown that ectomy-corrhiza formation results from a cascade of polygenic processes (for a discussion see Martin & Tagu 1995). However, mutational analyses of ectomycorrhizal symbionts have not yet been carried out and the number of genes controlled during the formation of the symbiotic organ is not known. Molecular analysis has therefore been carried out to estimate the number of symbiosis-regulated (SR) genes and proteins. At least three major effects of ectomycor-rhiza development have been reported to be exerted at the gene expression level: (1) the induction of novel abundant polypeptides, dubbed 'ectomycorrhizins' (Hilbert & Martin 1988), (2) the preferential accumulation of fungal polypeptides, and (3) the down-regulation of polypeptide biosynthesis that is present before the plant-fungus interactions (Hilbert & Martin 1988; Hilbert et al. 1991; Guttenberger & Hampp 1992; Burgess et al. 1995; Simoneau et al. 1994). Figure 3 illustrates the effects in polypeptide patterns of ectomycor-rhiza development in the *Eucalyptus grandis–Pisolithus tinctorius* association (Burgess et al. 1995). Among 500 proteins, labelled with [^{35}S]amino acids and analyzed using two-di-mensional electrophoresis, only one polypeptide (E_{32}) appeared to be synthesized exclu-sively in the symbiotic tissues. Protein analyses carried out on other ectomycorrhizal associations (Hilbert & Martin 1988; Hilbert et al. 1991; Simoneau et al. 1993) have also identified symbiosis-specific polypeptides in limited number (5-10). Therefore, regulation of protein patterns in the early stages of ectomycorrhiza formation does not involve activation of an extensive new set of genes. However, two-dimensional electrophoresis only allows the detection of a small set of abundant proteins – 1000 proteins on 10,000-15,000 known to occur in fungal and plant cells – and transcription factors likely to be involved in the control of mycorrhiza morphogenesis remained undetectable.

In contrast, alterations of the biosynthesis of pre-existing polypeptides are massive (Hilbert et al. 1991; Guttenberger & Hampp 1992; Simoneau et al. 1993; Burgess et al. 1995).

In Figure 3, this is most noticeable for acidic 30-32 kDa polypeptides, F_{71}, and F_{95}. These differences in the rate of protein synthesis correlated with the alterations of mRNA

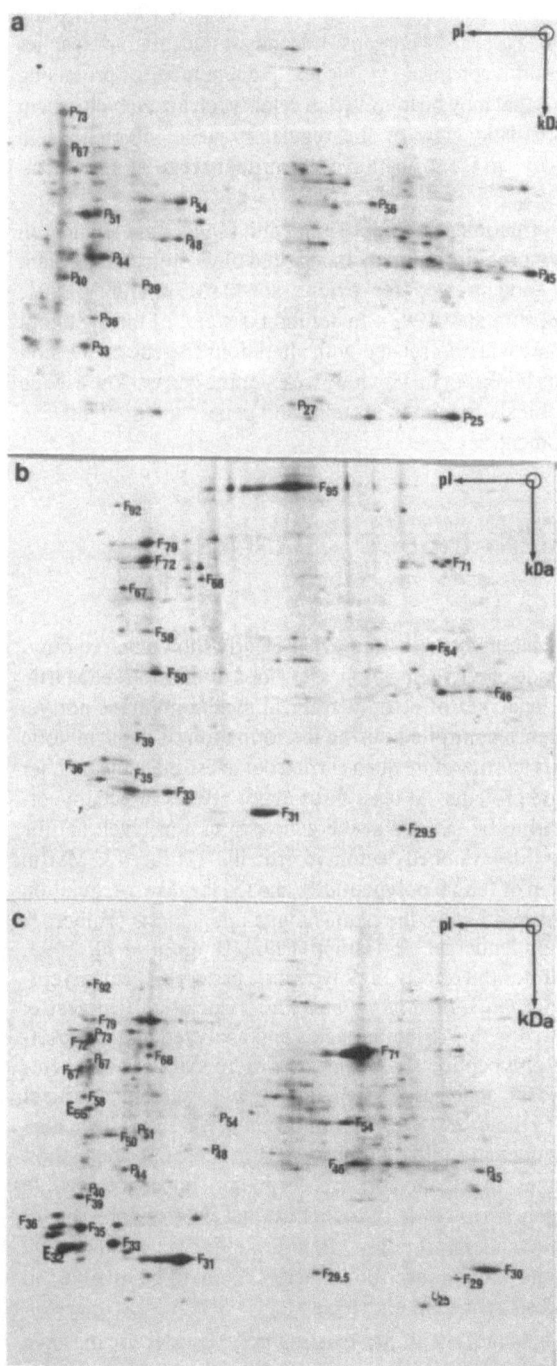

Figure 3. Autoradiograms of two-dimensional gels from a) non-inoculated roots of *E. grandis*, b) free-living mycelium of the isolate H2144 and c) ectomycorrhizas, 4 days post contact. Reference polypeptides of fungal (F) and plant (P) origin are indicated. Symbiosis-specific polypeptides (E), absent in the free-living partners, are present in the symbiosis. Subscripts denote MW (kDa) of polypeptides. The up-regulation of polypeptides F_{31}, E_{32}, F_{36}, and F_{71} is noteworthy. In contrast, the synthesis of F_{95} was down-regulated by ectomycorrhiza development (Burgess et al. 1995).

populations (Tagu et al. 1993; see below). It was deduced from these results that an extensive reprogrammation of the fungal and plant cells is induced by the symbiotic interaction.

Alterations in protein biosynthesis occur in the earliest stages (i.e. 0.5 –1 day postcontact) of root colonization by ectomycorrhizal fungi (Duchesne 1989; Hilbert et al. 1991; Simoneau et al. 1993; Burgess et al. 1995). This is a developmental sequence where dramatic alterations in branching and diameter of hyphæ occur (Jacobs et al. 1989). The latter authors suggested that these early morphogenetic events might be correlated to the compatibility between symbionts. There is now evidence that early changes in protein biosynthesis in *E. grandis* colonized by various isolates of *P. tinctorius* do appear to be correlated with the aggressivity (compatibility) of the fungal isolate (Burgess et al. 1995). Most of the proteins differentially expressed are presumably structural proteins directly related to morphogenesis and enzymes. The regulation of enzymes by symbiosis formation presumably takes place as a consequence of changes in the environment brought about by the developing structure, such as the reorganization of nutrient fluxes.

The induction pattern of SR proteins differs markedly in both temporal, qualitative, and quantitative aspects in different associations (Hilbert et al. 1991; Guttenberger & Hampp 1992; Simoneau et al. 1993; Burgess et al. 1995). Although only a limited number of ectomycorrhiza associations have been yet studied at the protein level, the expression of acidic 32 kDa-ectomycorrhizins appears to be evolutionarily conserved. Although their occurrence has been challenged by Guttenberger & Hampp (1992), these polypeptides have been detected in several different associations: *Eucalyptus globulus–Pisolithus tinctorius* (Hilbert & Martin, 1988; Hilbert et al. 1991), *E. grandis–P. tinctorius* (Burgess et al. 1995), *Betula pendula–Paxillus involutus* (Simoneau et al. 1993), *Pinus sylvestris–Suillus granulatus* (Raudaskoski et al. 1994). Alterations of protein synthesis (down- and up-regulation) have been observed in all of the different endo- and ectomycorrhizal symbioses studied so far, suggesting that it is a general phenomenon. However, the nature and function of most SR proteins is not known, although several of them probably correspond to cell wall proteins (see below) and pathogenesis-related (PR) proteins (e.g. chitinases, peroxidases) (Albrecht et al. 1994 a,b,c; Schwacke & Hager 1992).

BIOSYNTHESIS OF CELL WALL PROTEINS IS ALTERED

It has been suggested that ectomycorrhizal fungi attach to plant hosts by the release of fibrillar adhesive material (Lapeyrie et al. 1989; Lei et al. 1990, 1991; Dexheimer & Pargney 1991), together with amorphous mucilage-like material secreted by the host root (Piché et al. 1988). Following attachment, alterations in cell wall ultrastructure and composition of the apoplastic compartment is taking place (Dexheimer & Pargney 1991). Recently, a series of studies have revived efforts to elucidate the structure and biochemistry of the novel compartment corresponding to the interface (i.e. cell walls and extracellular matrix) between symbionts. Affinity probes (antibodies, enzymes, lectins) have shown that this interfacial compartment contains a complex matrix of polysaccharides and proteins (Piché et al. 1988; Dexheimer & Pargney 1991; Lei et al. 1991; B. Vian & P. Bonfante, personnal communication). This renewed interest was recently sustained by the characterization of several cell wall proteins that are developmentally-regulated during the early stage of ectomycorrhiza development (De Carvalho 1994; Laurent et al. 1995).

Among these are a set of fungal *s*ymbiosis-*r*egulated *a*cidic *p*olypeptides (SRAP), some of which are preferentially accumulated in cell walls during the early stages of the *Eucalyptus–Pisolithus* interaction. Cell fractionation and protein labelling have shown that these acidic 30-32 kDa SRAPs are polypeptides secreted by *P. tinctorius*. They are then either accumulated in cell walls or abundantly secreted in the extracellular medium (De Carvalho

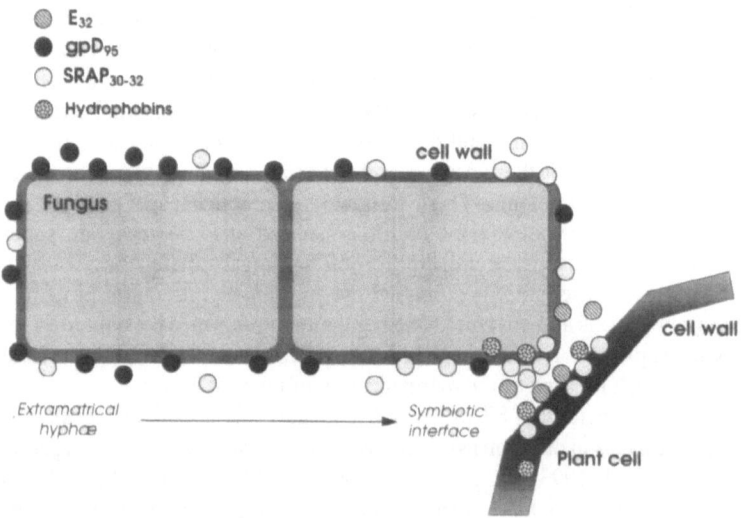

Figure 4. Putative changes in the biosynthesis of fungal cell wall proteins taking place during the recognition phase and hyphal adhesion to the root surface. The cell wall content of the mannoprotein gp95 which is abundant in the free-living mycelium is decreased in the hyphæ in contact with the host root. In contrast, the biosynthesis and secretion of hydrophobins and 30-32 kDa SRAPs is increased. Hydrophobins and 30-32 kDa SRAPs may be components of the fungal microfibrilles bridging the partners.

1994; Laurent & De Carvalho, unpublished results). Some isoforms are constitutively expressed in the non-symbiotic mycelium, whereas other isoforms are preferentially expressed in the fungal symbiotic tissues (Burgess et al. 1995; Laurent, unpublished results). As shown by its high transcript content, another class of fungal secreted cell wall proteins, the so-called hydrophobins (Wessels 1993), is also accumulating when hyphæ colonize the eucalypt root surface (Tagu et al. 1995).

In contrast, the synthesis and concentration of a prominent cell wall mannoprotein (gp95) of *P. tinctorius* are down-regulated in ectomycorrhiza (Laurent et al. 1995). Thus, modifications of cell walls and the extracellular matrix are taking place as fungal cells sense the root surface and differentiate to form symbiotic tissues (e.g. the mantle) and invade host tissues (Fig. 4).

Cell wall proteins are thought to form cross-linked networks with other proteins and polysaccharides in fungal cell walls (Ruiz-Herrera et al. 1994), but the structural properties and the functional significance of such networks are not yet known. Perhaps high levels of up-regulated 30-32 kDa SRAPs and low levels of mannoproteins, such as gp95, function simultaneously to regulate the molecular architecture of protein networks in a manner that allows new developmental fates for both fungal cell aggregation and root colonization by the fungus. Further investigation of the structure and regulation of SR cell wall proteins will provide a more complete picture of the involvement of SRAPs in developing ectomycorrhizal tissues.

PLANT AND FUNGAL CELLS ARE REPROGRAMMED EARLY IN ECTOMYCORRHIZA DEVELOPMENT

Protein analyses suggested that ectomycorrhiza development influences the expression of hundreds of genes. We have therefore developed several molecular approaches (differential hybridisations, shot-gun sequencing, cloning using heterologous genes) to identify genes which are uniquely, or preferentially, expressed in *E. globulus–P. tinctorius* ectomycorrhiza, and thus are targets for developmental regulation (Tagu & Martin 1995; Tagu et al. 1993, 1995; Nehls & Martin, this volume).

Genes Cloned by Differential Hybridisation

A cDNA library from symbiotic tissues sampled during the early stages of the *E. globulus–P. tinctorius* ectomycorrhiza development has been constructed (Tagu et al. 1993). Bacterial colonies carrying mycorrhiza-expressed genes in the λZAPII phage vector have been randomly selected from the library. cDNA clones were amplified by polymerase chain reaction and differentially hybridized with [^{32}P]cDNA probes from either free-living mycelium, non-inoculated roots or mycorrhiza (Fig. 5).

This approach led to the identification of several dozen new fungal and plant loci differentially expressed in mycorrhiza (Tagu et al. 1993; Nehls & Martin, this book). Several of these SR cDNAs have been sequenced and identified by homology search in the National Center for Biocomputing Information (NCBI) database (Altschul et al. 1994). Ectomycorrhiza formation induced both up- and down-regulation of the level of transcripts that are present before the plant-fungus interactions (Tagu et al. 1993; Nehls & Martin, this book) supporting the data obtained from protein patterns.

Among the down-regulated fungal transcripts, those represented by the Mycf102 cDNA showed a high degree of homology (59%) to inducible acid phosphatase genes from yeasts (Murphy & Martin, unpublished results). This SR protein is not directly related to morphogenesis, but its regulation is presumably the result of changes in fluxes of phosphate compounds between fungal cells as a result of the plant-microbe interaction.

Despite the large number of fungal cDNA clones (up to 80%) occurring in the library of 4-day-old eucalypt ectomycorrhiza, plant SR genes with strong homology to diverse plant proteins have also been identified (see Nehls & Martin, this book). Among genes having an up-regulated expression in ectomycorrhiza, sequences homologous to the auxin-regulated gene, *parA*, and α-tubulin genes are highly relevant for the understanding of the symbiosis development. In tobacco, *parA* is induced immediately after the addition of auxin and is expressed upon the initiation of meristematic activity in protoplasts (Takahashi et al. 1989)

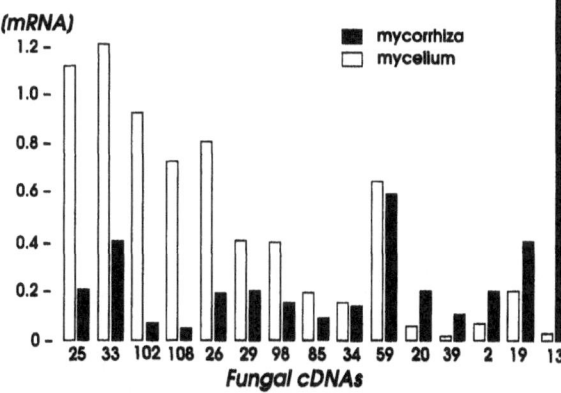

Figure 5. Relative concentration of non-affected, up- and down-regulated fungal transcripts in free-living mycelium (open square) and in ectomycorrhizas (closed square) of *Eucalyptus globulus- Pisolithus tinctorius* (after Tagu *et al.*, 1993).

and during the lateral root formation (Vera et al. 1994). Accumulation of *parA* in eucalypt roots during the colonization by *P. tinctorius* is presumably stimulated by auxins released by hyphæ and correlated to initiation of lateral roots. Increased accumulation of transcripts coding for different plant α-tubulins (Carnero Diaz et al., unpublished results) may be correlated to reorientation of cortical microtubules and cellulose microfibrils in epidermal cells (Peterson & Bonfante 1993).

Transcript Map of Expressed Genes

In addition to differential hybridization approaches, we have recently attempted to generate a transcript map of abundant and moderately expressed genes in eucalypt symbiotic tissues by systematic sequencing of cDNA clones. Extensive analyses of populations of expressed genes, or expressed sequence tags (EST), have been carried out for several plants, *Arabidopsis* (Höfte et al. 1993), rice (Uchimiya et al. 1993), maize (Keith et al. 1993), and *Brassica* (Park et al. 1993).

Bacterial colonies carrying mycorrhiza-expressed genes in λZAPII have been randomly selected from the eucalypt cDNA library (Tagu et al. 1993). In the phage vector, cDNA fragments are inserted unidirectionally into pBluescript plasmid (Short et al. 1988). Clones from the oriented library were sequenced from the 5' end, by using the T3 promoter, to preferentially target the coding region of cDNAs. Sequences were translated in the six open reading frames and compared with NCBI databases (Altschul et al. 1994).

Among these mycorrhiza ESTs, about 25% of the sequences showed a clear similarity with known proteins (Tagu & Martin 1995). The remaining ESTs likely correspond to novel genes. Putative identified genes are categorized according to their expected functions as shown in Table 1 and their likely role discussed below. As stressed in previous random sequencing projects (Höfte et al. 1993), many gene names novel to plant and fungi were identified among the putative genes identified (e.g. cylicin, hemoglobinase). Some of them possess a well-conserved functional domain or are members of superfamily proteins, but several sequence matches (e.g. sphingomyelinase) are difficult to explain by the present understanding of plant or fungal cell metabolism.

Plant Defence Reactions

At least five different types of plant defence (and stress) proteins can be distinguished in the abundantly expressed genes in eucalypt ectomycorrhiza: a proteinase inhibitor II, pathogenesis-related (PR) proteins, a catalase, saposin, and sphingomyelinase (Table 1). Proteinase inhibitors are ubiquitous proteins that inhibit the digestive proteinases of insects and microorganisms (Narvaez-Vasquez et al. 1993). The function of PR1 protein is unknown, whereas saposin is a 10 kDa glycoprotein known to be an activator of sphingomyelinase and other phospholipases. Sphingomyelinases are hemolytic factors able to disrupt cell membranes and they have been involved in animal defence reactions.

Catalase could be involved in the regulation of the production of activated oxygen species elicited by cell wall elicitors of ectomycorrhizal fungi (Schwacke & Hager 1992). Transcript concentration of catalase is down-regulated by 50% in 4-day-old eucalypt ectomycorrhizas (Nehls & Martin, this volume) and this decreased expression might favor the accumulation of the defence molecule H_2O_2. Root colonization by the ectomycorrhizal hyphæ thus induces an entire defence-related gene expression programme that, as an early component, includes the synthesis of an array of diverse plant defence proteins (Sauter & Hager 1992; Albrecht et al. 1994a,b; Simoneau et al. 1994) and activated oxygen species (Schwacke & Hager 1992) and, as a later component, includes the general depression of plant protein biosynthesis, the so-called 'polypeptide cleansing' (Hilbert et al. 1991; Martin

Table 1. Expressed sequenced tags cloned from 4-day-old ectomycorrhizas of *Eucalyptus globulus bicostata–Pisolithus tinctorius* corresponding to known proteins

cDNA clone	Putative protein	Organism	Homology (%)	Similarity (%)
EST 94	Cylicin	*Bos taurus*	20	44
EST 32	Hydrophobin	*Schizophyllum commune*	68	84
EST 141	Hydrophobin	*S. commune*	39	54
EST 167	PR-protein	*Solanum tuberosum*	66	80
EST 149	Proteinase inhibitor	*Glycine max*	86	90
EST 7	Saposin	*Mus musculus*	31	53
EST 57	Sphingo-myelinase	*Clostridium perfringens*	61	78
EST 82	Hemoglobinase	*Schistosoma mansoni*	44	61
EST 60	Proteasome	*Homo sapiens*	63	84
EST 46	Ubiquitin-conjugating enzyme E2	*S. cerevisiae*	42	55
EST 158	Ubiquitin-conjugating enzyme E2	*Arabidopsis thaliana*	48	63

& Hilbert 1991; Burgess et al. 1995). In agreement with the latter 2D PAGE data, a large proportion of genes encoding plant proteins are negatively regulated by mycorrhiza formation (Nehls & Martin, this volume). We speculate that plant defence reactions are activated by the host plant to confine the invading hyphæ to the epidermis. However, elicitation of the plant defence genes might be amplified by the massive fungal colonization of the root resulting from the in vitro system of mycorrhiza synthesis.

Protein Metabolism

Many genes involved in the assembly and turn-over of proteins have been isolated suggesting that they are highly expressed during the early steps of mycorrhiza development. Among these ESTs, two showed strong homologies (55 and 63% matches) with ubiquitin-conjugating enzymes E2, one with the elongation factor 1γ (61%), and another EST with an endopeptidase of the proteasome complex (84%) recently identified in plants (Höfte et al. 1993). Proteasomes are multicatalytic proteinase complexes playing an important role in the degradation of ubiquitinylated proteins in eukaryotes including higher plants (Genschik et al. 1994). Proteasome transcripts accumulate to high levels during cell proliferation (Genschik et al. 1994).

We do not presently know why these genes were so abundant in the mycorrhiza library. However, it is tempting to correlate the expression of these genes involved in protein turn-over with fungal cell proliferation and the large down-regulation of the synthesis of root proteins observed by two-dimensional electrophoresis (Hilbert et al. 1991, Burgess et al. 1995). The latter decrease in protein synthesis in colonized roots is probably correlated to the arrest of the meristematic activity of the ensheathed short roots.

Secreted Fungal Proteins

Two fungal transcripts, corresponding to EST32 and EST141, are very abundant in ectomycorrhiza and aerial hyphæ of *P. tinctorius*. They are not accumulated in hyphæ growing in liquid medium (Tagu et al. 1995). The predicted amino acid sequences of these cDNAs showed features of a novel class of proteins known as hydrophobins and identified in a number of saprophytic and pathogenic fungi (Wessels et al. 1991; Wessels 1993). *Pisolithus* hydrophobins, like their *Schizophyllum commune* homologs, are small cysteine-rich proteins containing an 8 cysteine conserved domain and having strongly hydrophobic domains (Tagu et al. 1995). Some hydrophobins co-polymerize into insoluble rodlet arrays on the surface of conidia and aerial structure conferring hydrophobic properties to the hyphal surface (Stringer et al. 1991; Wösten et al. 1993). Other hydrophobins are morphogenetic proteins that allow hyphae to emerge off the substrate and to adhere to each other during development of aerial multicellular reproductive structures (e.g. conidiophores, basidiocarps) (Wessels 1993). These proteins can also play a role in pathogenic interactions during the appressorium formation in invaded host tissues (St Leger et al. 1992; Talbot et al. 1993; Templeton et al. 1994). The role of hydrophobins in ectomycorrhiza is still unknown, but they could be involved in the aggregation of the dense pseudoparenchymatous mantle ensheathing the colonized root.

CONCLUSION

Ectomycorrhiza development influences both plant and fungus gene expression in a pleiotropic manner. A range of fungal tissues can be distinguished by a combination of anatomical and cytological features (e.g. mantle, Hartig net). On the other hand, root tips proliferate and root cells experience major alteration in their orientation and morphology. Advances of recent years have provided insights on the molecular basis of ectomycorrhiza morphogenesis. With the identification of several developmentally-regulated proteins and genes and a description of their expression and activities, the ground was set for recasting earlier models of symbiosis development in molecular terms. It is apparent from this brief review, however, that there is a vast complexity of genetic programmes with overlapping expression patterns (Fig. 6).

This includes: induction of plant defence reactions, the down-expression of plant protein biosynthesis, the initiation of lateral roots by fungal auxins, the morphogenetic switches of the fungal hyphæ. The coordinated expression of these diverse genetic programmes reflects the pleiotropic alterations imposed on the fungal and plant cells by the

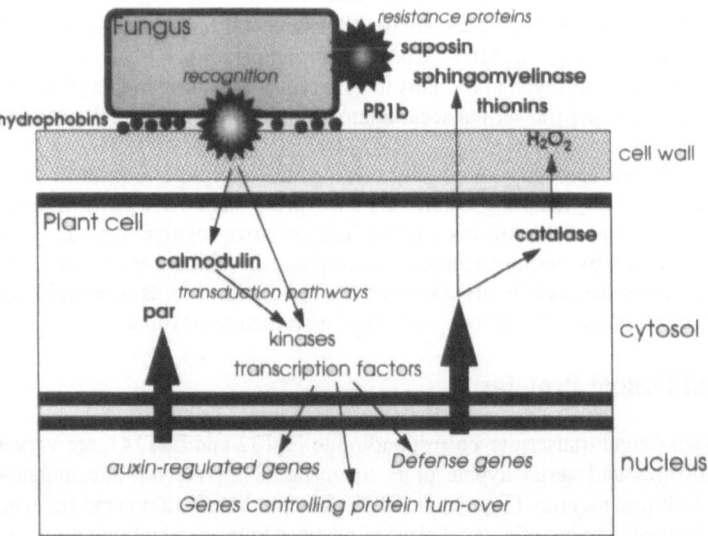

Figure 6. Speculative model for early perception of the ectomycorrhizal fungus by the epidermal root cell and its transduction for induction of mycorrhiza-regulated mechanisms.

construction of symbiotic tissues. Among the many remaining challenges is the elucidation of mechanisms and identification of genes and inducer molecules that integrate the actions of these multiple programmes of gene expression in generating a mature symbiotic organ. Studies in areas such the identification of proteins involved in cell-cell interactions, control of cell expression at the level of signal transduction will be the source for many answers.

A comparative study of gene expression in different types of ectomycorrhizas using the molecular approaches presented here might reveal to what extent similarities and differences in the various types of ectomycorrhizas are the result of variation in the basic mechanisms underlying the respective developmental programmes.

The precise molecular mechanisms responsible for specifying different tissue patterns early in ectomycorrhiza morphogenesis are not known. A major void in our knowledge concerns the identification and expression of master genes and the developmentally-critical genes they control (Fig. 1). In these respects it is crucial to identify pattern mutants altered in the organization of the ectomycorrhiza structure. This would allow the specification events that take place during the early stages of mycorrhiza development to be monitored.

ACKNOWLEDGMENTS

The work from our laboratory was supported by grants from the *Eureka-Eurosilva Cooperation Programme on Tree Physiology* and the I.N.R.A.. TB was a recipient of a Doctoral Fellowship from the INRA and an Australian Postgraduate Scholarship. MECD was supported by a scholarship of the Ministère de la Recherche et de l'Enseignement Supérieur (MRES). PM was an Eureka-Eurosilva postdoctoral fellow and UN was funded by postdoctoral fellowships from the MRES and the Deutsche Forschung Gemeinschaft. Thanks to Prof. B. Dell for a critical reading of the manuscript.

REFERENCES

Albrecht, C , Laurent, P, and Lapeyrie, F 1994a *Eucalyptus* root and shoot chitinases, induced following root colonization by pathogenic versus ectomycorrhizal fungi, compared on one- and two-dimensional activity gels *Plant Sci* **100**. 157-164

Albrecht C , Asselin A , Piche Y , and Lapeyrie F 1994b Chitinase activities are induced in *Eucalyptus globulus* roots by ectomycorrhizal or pathogenic fungi, during early colonization *Physiol Plant* **91**, 104-110

Albrecht C , Burgess T , Dell B , and Lapeyrie F 1994c Chitinase and peroxidase activities are induced in eucalyptus roots according to aggressiveness of Australian ectomycorrhizal strains of *Pisolithus* sp *New Phytol* **127**, 217-222

Altschul S F, Boguski M S , Gish W, and Wootton J C 1994 Issues in searching molecular sequence databases *Nature Genetics* **6**, 119-129

Bell-Pedersen, D , Dunlap, J C , and Loros, J J. 1992. The *Neurospora* circadian clock-controlled gene, *ccg-2*, is allelic to *eas* and encodes a fungal hydrophobin required for formation of the conidial rodlet layer *Genes & Develop* **6**: 2382-2394

Berlin V. and Yanofski C (1985) Isolation and characterization of genes differentially expressed during conidiation of *Neurospora crassa Mol Cel Biol* **5**, 849-855

Bonfante, P and Perotto, S. 1992. Plants and endomycorrhizal fungi: the cellular and molecular basis of their interaction. *In* Molecular signals in plant microbe communications *Edited by* D.P S Verma CRC Press, Boca Raton pp 445-470

Burgess, T., Laurent, P, Dell, B , Malajczuk, N , and Martin, F 1995. Effect of the fungal isolate aggressivity on the biosynthesis of symbiosis-related polypeptides in differentiating eucalypt ectomycorrhiza *Planta* **195**:408-417

Chilvers, G. A., Douglass, P A., and Lapeyrie, F 1986. A paper-sandwich technique for rapid synthesis of ectomycorrhizas *New Phytol* **103** 397-402.

Coen, E S. and Carpenter, R. 1993. The metamorphosis of flowers *Plant Cell* **5** 1175-1181

De Carvalho, D. 1994. Contribution à l'étude des protéines régulées par la symbiose ectomycorhizienne. Caractérisation des protéines membranaires et pariétales de l'ectomycorhize *Eucalyptus globulus–Pisolithus tinctorius*. Ph. D. thesis, Ecole Nationale du Génie Rural, des Eaux et des Forêts, Nancy, France.

Dexheimer, J. and Pargney, J. C. 1991. Comparative anatomy of the host-fungus interface in mycorrhizas. *Experientia* 47: 312-320.

Genschik P., Jamet E., Philipps G, Parmentier Y., Gigot C., and Fleck, J. (1994) Molecular characterization of a β-type proteasome subunit from *Arabidopsis thaliana* co-expressed at a high level with an α-type proteasome subunit early in the cell cycle. *Plant J.* 6: 537-546.

Giovannetti M., Sbrana C., Avio L., Citernesi A.S. and Logi C. 1993. Differential hyphal morphogenesis in arbuscular mycorrhizal fungi during preinfection stages. *New Phytol.* 125, 587-593.

Guttenberger M. and Hampp R. 1992. Ectomycorrhizins - Symbiosis-specific or artifactual polypeptides from ectomycorrhizas? *Planta* 188, 129-136.

Harley, J. L. and Smith, S. E. 1983. Mycorrhizal symbiosis. Academic Press, London.

Hilbert, J. L. and Martin, F. 1988. Regulation of gene expression in ectomycorrhizas. I. Protein changes and the presence of ectomycorrhiza-specific polypeptides in the *Pisolithus-Eucalyptus* symbiosis. *New Phytol.* 110: 339-346.

Hilbert, J. L., Costa, G., and Martin, F. 1991. Regulation of gene expression in ectomycorrhizas. Early ectomycorrhizins and polypeptide cleansing in eucalypt ectomycorrhizas. *Plant Physiol.* 97: 977-984.

Höfte H., Desprez T., Amselem J., Chiapello H., Caboche M., Moisan A., Jourjon M.F., Charpenteau J.L., Berthomieu P., Guerrier D., Giraudat J., Quigley F., Thomas F., Yu D.Y., Mache R., Raynal M., Cooke R., Grellet F., Delseny M., Parmentier Y., De Marcillac G., Gigot C., Fleck J., Phillips G.J., Axelos M., Bardet C., Tremousaygue D., and Lescure B. 1993. An inventory of 1152 expressed sequence tags obtained by partial sequencing of cDNAs from *Arabidopsis thaliana*. *Plant J.* 4, 1051-1061.

Horan, D. P., Chilvers, G. A., and Lapeyrie, F. 1988. Time sequence of the infection process in eucalypt ectomycorrhizas. *New Phytol.* 109: 451-458.

Jacobs, P. F., Peterson, R.L., Massicotte, H.B. (1989) Altered fungal morphogenesis during early stages of ectomycorrhiza formation in *Eucalyptus pilularis*. *Scanning Microscopy* 3:249-255.

Keith C. S., Hoang D. O., Barrett B. M., Feigelman B., Nelson M. C., Thai H., and Baysdorfer C. 1993. Partial sequence analysis of 130 randomly selected maize cDNA clones. *Plant Physiology* 101, 329-332.

Lapeyrie, F., Lei, J., Malajczuk, N., and Dexheimer, J. 1989. Ultrastructural and biochemical changes at the pre-infection stage of mycorrhizal formation by two isolates of *Pisolithus tinctorius*. *Ann. Sci. For.* 46s: 754s-757s.

Laurent P., De Carvalho D., and Martin F. (1995) Major cell wall mannoproteins of *Pisolithus tinctorius* are regulated during development of eucalypt ectomycorrhiza. *Planta*, submitted for publication.

Lei, J., Lapeyrie, F., Malajczuk, N., and Dexheimer, J. 1990. Infectivity of pine and eucalypt isolates of *Pisolithus tinctorius* Pers. Coker & Couch on roots of *Eucalyptus urophylla* S. T. Blake *in vitro*. II. Ultrastructural and biochemical changes at the early stage of mycorrhiza formation. *New Phytol.* 116: 115-122.

Lei, J., Wong, K. K., and Piché, Y. 1991. Extracellular concanavalin-A binding sites during early interactions between *Pinus banksiana* and two closely related genotypes of the ectomycorrhizal basidiomycete *Laccaria bicolor*. *Mycol. Res.* 95: 357-363.

Malajczuk N., Lapeyrie F., and Garbaye J. 1990. Infectivity of pine and eucalypt isolates of *Pisolithus tinctorius* on roots of *Eucalyptus urophylla* in vitro. *New Phytol.* 114: 627-631.

Martin, F. and Hilbert, J. L. 1991. Morphological, biochemical and molecular changes during ectomycorrhiza development. *Experientia* 47: 321-331.

Martin, F. and Tagu, D. 1995. Ectomycorrhiza development: a molecular perspective. *In* Mycorrhiza: Structure, molecular biology and function. *Edited by* B. Hock and A. K. Varma. Springer Verlag, Berlin, Heidelberg, New York, in press.

Martin, F., Lapeyrie, F., and Tagu, D. 1995. Altered Gene Expression during Ectomycorrhiza Development. *In* The Mycota. Vol. VI. Plant Relationships. *Edited by* P. Lemke and G. Caroll. Springer Verlag,, Berlin, Heidelberg, New York, in press.

Massicotte, H. B., Peterson, R. L., and Ashford, A. E. 1987. Ontogeny of *Eucalyptus pilularis-Pisolithus tinctorius* ectomycorrhizae. II. Transmission electron microscopy. *Can. J. Bot.* 65:1940-1947.

Massicotte, H. B., Peterson, R. L., and Melville, L. H. 1989. Ontogeny of *Alnus rubra - Alpova diplophloeus* ectomycorrhizae. I. Light microscopy and scanning electron microscopy. *Can. J. Bot.* 67: 191-200.

Narvaez-Vasquez, J., Franceschi, V. R., and Ryan, C. A.. 1993. Proteinase-inhibitor synthesis in tomato plants: evidence for extracellular deposition in roots through the secretory pathway. *Planta* 189, 257-266.

Nehls, U. and Martin, F. 1995.Gene expression in roots during ectomycorrhiza development. This volume.

Park Y.S., Kwak J.M., Kwon O.Y., Kim Y.S., Lee D.S., Cho M.J., Lee H.H., and Nam H.G. 1993. Generation of expressed sequence tags of random roots cDNA clones of *Brassica napus* by single-run partial sequencing. *Plant Physiol.* **103**, 359-370.

Patel . H. (1994) Developmental evolution: insights from studies of insect segmentation. *Science* **266**, 581-590

Peterson, R. L. and Bonfante, P. 1994. Comparative structure of vesicular-arbuscular mycorrhizas and ectomycorrhizas. *Plant and Soil* **159**, 79-88.

Piché, Y., Peterson, R. L., and Massicotte, H. B. 1988. Host-fungus interactions in ectomycorhizae. *In* Cell to cell signals in plant, animal and microbial symbiosis. NATO ISI Ser. **H17**: 55-71.

Raudakoski, M., Tarkka, M., Timonen S., Niini, S., and Astrom, H. 1994.The role of cytoskeleton in mycorrhizal associations. *Proceedings of the Fifth International Mycological Congress*, August 14-21, 1994, Vancouver, British Columbia, Canada.

Ruiz-Herrera, J., Mormeneo, S., Vanacloca, P., Font-de-Mora, J., Iranzo, M., Puertes, I., and Sentandreu, R. 1994. Structural organization of the components of the cell wall from *Candida albicans*. *Microbiology* **140**: 1513-1523.

Sauter, M. and Hager, A. 1992. The mycorrhizal fungus *Amanita muscaria* induces chitinase activity in roots and in suspension-cultured cells of its host *Picea abies*. *Planta* **179**: 61-66.

Scheres, B., van de Weil, C., Zalensky, A., Horvath, B., Spaink, H., van Eck, H., Zwartkruis, F., Wolters, A.-M., Gloudemans, T., van Kammen, A., and Bisseling, T. 1990. The *ENOD12* gene product is involved in the infection process during the pea-*Rhizobium* interaction. *Cell* **60**: 281-294.

Schwacke, R. and Hager, A. 1992. Fungal elicitors induce a transient release of active oxygen species from cultured spruce cells that is dependent on Ca^{2+} and protein-kinase activity. *Planta* **187**, 136-141.

Short, J.M., Fernandez, J.M., Sorge, J.A. and Huse, W.D. (1988) Lambda ZAP: a bacteriophage lambda expression vector with in vivo excision properties. *Nucl. Acid. Res.* 16, 7583-7600.

Simoneau, P., Viemont, J. D., Moreau, J. C., and Strullu, D. G. 1993. Symbiosis-related polypeptides associated with the early stages of ectomycorrhiza organogenesis in birch *Betula pendula* Roth.. *New Phytol.* **124**: 495-504.

Simoneau, P., Juge, C., Dupuis, J. Y., Viemont, J. D., Moreau, C., and Strullu, D. G. 1994. Protein biosynthesis changes during mycorrhiza formation in roots of micropropagated birch. Acta Bot. Gallica, in press.

St Leger, R. J., Staples, R. C., and Roberts, D. W. 1992. Cloning and regulatory analysis of starvation-stress gene, *ssgA*, encoding a hydrophobin-like protein from the entomopathogenic fungus, *Metarhizium anisopliae*. *Gene* **120**: 119-124.

Stringer, M. A., Dean, R. A., Sewall, T. C., and Timberlake, W. E. 1991. *Rodletless*, a new *Aspergillus* developmental mutant induced by directed gene activation. *Genes Devel.* **5**: 1161-1171.

Tagu, D. and Martin, F. (1995) Expressed sequence tags of randomly selected cDNA clones from *Eucalyptus globulus–Pisolithus tinctorius* ectomycorrhizæ. *Mol. Plant Microbe Interactions*, in press.

Tagu, D., Python, M., Crétin, C., and Martin, F. 1993. Cloning symbiosis-related cDNAs from eucalypt ectomycorrhizas by PCR-assisted differential screening. *New Phytol.* **125**: 339-343.

Tagu, D., Nasse, B., and Martin, F. 1995. Preferential accumulation of transcripts coding for hydrophobins in the *Eucalyptus globulus–Pisolithus tinctorius* ectomycorrhiza. *Gene*, Submitted for publication.

Takahashi, Y., Kusaba, M., Hiraoka, Y., and Nagata, T. 1991. Characterization of the auxin-regulated *par* gene from tobacco mesophyll protoplasts. *Plant J.* **1**: 327-332.

Talbot, N. J., Ebbole, D. J., and Hamer, J. E. 1993. Identification and characterization of MPG1, a gene involved in pathogenicity from the rice blast fungus *Magnaporthe grisea*. Plant Cell **5**: 1575-1590.

Templeton, M. D., Rikkerink, E. H. A., and Beever, R. E. 1994. Small, cysteine-rich proteins and recognition in fungal-plant interactions. *Mol. Plant Microb. Int.* **7**: 320-325.

Timberlake, W. E. (1993) Translational triggering and feedback fixation in the control of fungal development. *Plant Cell* 5, 1453-1460.

Tommerup, I. C. and Malajczuk, N. (1993) Genetics and molecular genetics of mycorrhiza. *Advances in Plant Pathology* 9, 83-102.

Uchimiya H., Kidou S., Shimazaki T., Aotsuka S., Takamatsu S., Nishi R., Hashimoto H.,Matsubayashi Y., Kidou N., Umeda M., and Kato A. 1993. Random sequencing of cDNA libraries reveals a variety of expressed genes in cultured cells of rice (*Oryza sativa* L.). *Plant J.* **2**, 1005-1009.

Vera, P., Lamb, C., and Doerner P. W. 1994. Cell-cycle regulation of hydroxyproline-rich glycoprotein *HRGPnt3* gene expression during the initiation of lateral root meristems. *Plant J.* **6**: 717-727.

Wessels, J. G. H. 1993a. Developmental regulation of fungal cell wall formation. *Annu. Rev. Phytopathol.* **32**: 413-437.

Wessels, J. G. H. 1993b. Wall growth, protein excretion and morphogenesis in fungi. *New Phytol.* **123**: 397-413.

Wessels, J. G. H. 1994. Fruiting in the higher fungi. *Adv. Microb. Physiol.* **34**: 147-202.

Wessels, J. g. h., De Vries, O. m. h., Asgeirsdottirs, A., Schuren, F. h. j. 1991. Hydrophobin genes involved in formation of aerial hyphae and fruit bodies in *Schizophyllum*. *Plant Cell* 3: 793-799.

Wösten, H. A. B., De Vries, O. M. H., and Wessels, J. G. H. 1993. Interfacial self-assembly of a fungal hydrophobin into a hydrophobic rodlet layer. *Plant Cell* 5: 1567-1574.

6

MOLECULES AND GENES INVOLVED IN MYCORRHIZA FUNCTIONING

Silvio Gianinazzi, Vivienne Gianinazzi-Pearson, Philipp Franken,
Eliane Dumas-Gaudot, Diederik van Tuinen, Assem Samra,
Fabrice Martin-Laurent, and Barbara Dassi

Laboratoire de Phytoparasitologie INRA/CNRS
SGAP, INRA, BV 1540
21034 Dijon
France

The positive effect of mycorrhizas on plant growth and health results from a complex, molecular dialogue between the two symbiotic partners. This starts before the physical contact between plant and fungus occurs, then develops and amplifies with their morpho-functional integration and continues all along the life of the plant. Because of this dialogue, symbiosis development depends not only on the genetic make-up of both partners (Gianinazzi-Pearson and Gianinazzi, 1989), but also on environmental conditions which allow optimal coordination in their functioning (Smith and Gianinazzi-Pearson, 1988). Identifying the molecules and genes involved in the symbiotic dialogue is therefore essential not only for improving agroforestry management practices for better plant growth and health, but also for monitoring fungal development and fruit-body production. In this paper we discuss present knowledge about molecules and genes involved in mycorrhiza functioning, taking examples from the most widespread type of root symbiosis, the arbuscular mycorrhiza (Harley and Smith, 1983).

THE ARBUSCULAR MYCORRHIZAL SYMBIOSIS

The fungi involved in this symbiosis all belong to the same order of Glomales (Zygomycetes) and can colonize roots of the majority of terrestrial plants. In spite of the fact that these fungi possess a very large genome (about 10^9 bp, M. Hosny, unpublished data), they are obligate symbionts and, contrary to ectomycorrhizal fungi, they do not form macroscopic fruiting bodies but microscopic asexual spores which are formed isolated or in loose sporocarps (Harley and Smith, 1983). In the soil, host root exudates specifically enhance spore germination and hyphal growth, essential processes leading to root infection and colonization (Giovannetti et al., 1994).

Once inside roots, and under the influence of the host, the fungus forms arbuscules specifically in cells of the parenchyma cortex, and often either intraradical or extraradical vesicles. However, this massive root colonization is not chaotic and plants keep fungal development permanently under control (Gianinazzi, 1991), so that hyphae never invade the root meristem nor the central cylinder. Nevertheless, fungal colonization induces modifications in the host plants at different levels of organisation, but these modifications are not deleterious and they usually improve plant performance. For example, roots infected by arbuscular mycorrhizal fungi show a more dichotomous pattern of development (Atkinson et al., 1994) which, together with the formation of an extraradical mycelium network, are important in determining the plant's relationship with the abiotic and biotic environment, and in particular acquired resistance to different kinds of stress (Rösendahl, 1985) and increased ability in exploiting soil resources (Smith et al., 1994).

A MODEL FOR MOLECULAR EVENTS INVOLVED IN ARBUSCULAR MYCORRHIZA

The arbuscular mycorrhizal symbiosis represents one of the closest interactions between different organisms. This is particularly evident from the features of cortical cells containing arbuscules, which are considered essential structures for reciprocal nutrient exchange between host and fungus (Harley and Smith, 1983). In these cells, the host encloses the fungus with a specialized membrane, the periarbuscular membrane, the properties of which are well adapted to such bidirectional nutrient exchange (Smith et al., 1994). The volume of host cytoplasm and number of organelles (mitochondria and plasts) increase drastically, indicating a higher metabolic activity in the infected cells (Jacquelinet-Jean-mougin et al., 1987). Furthermore, the host nucleus swells, becomes lobed, and shows a decondensed chromatin, reflecting an increased transcriptional activity (Berta et al., 1990).

In order to understand the molecular events involved in this morphofunctional integration between the symbiotic partners in arbuscular mycorrhiza, we propose the working model shown in Figure 1. In this model, molecular interactions between both partners are characterized by the production and perception of signals, which lead to activation of a cascade of genes essential for the metabolic and structural changes which will form the basis of the symbiosis. This exchange of signals and differential gene expression guides the interaction of the two partners from the first step of recognition in the soil to the final development of a highly ordered structure, the intracellular arbuscule.

Plant signals contained in host root exudates (Giovannetti et al., 1994; Gianinazzi-Pearson, 1995) and perceived either through a fungal cell surface or intracellular receptor may activate, via a signal transduction chain, a so-called master gene. This will start a cascade induction of regulator and finally effector genes, so that genes necessary for infection of the plant cell are expressed, leading to the induction of certain structural and metabolic changes important for the symbiotic interaction, such as arbuscule formation (Gianinazzi-Pearson and Gianinazzi, 1989) or synthesis of specific fungal enzymes involved in phosphate metabolism (Gianinazzi-Pearson and Smith, 1993). The overall consequence will be the morphological and physiological changes characterizing arbuscular mycorrhizal fungi during the symbiotic phase of their life cycle.

Colonization of the root tissues could, in turn, stimulate the fungus to produce signal molecules which when perceived by the plant induce an analogous cascade of events leading to changes in the expression of certain host genes (Gianinazzi-Pearson, 1994). This can be seen, for example, in the low expression of plant defence genes (Gianinazzi-Pearson et al., 1992; Gollotte et al., 1993; Harrison and Dixon, 1994), the synthesis of novel proteins or polypeptides (Figure 2) (Garcia-Garrido et al., 1993; Dumas-Gaudot et al., 1994), the enhancement of certain metabolic pathways (Dumas et al., 1990; Ghachtouli El et al., 1994)

Plant   Fungus

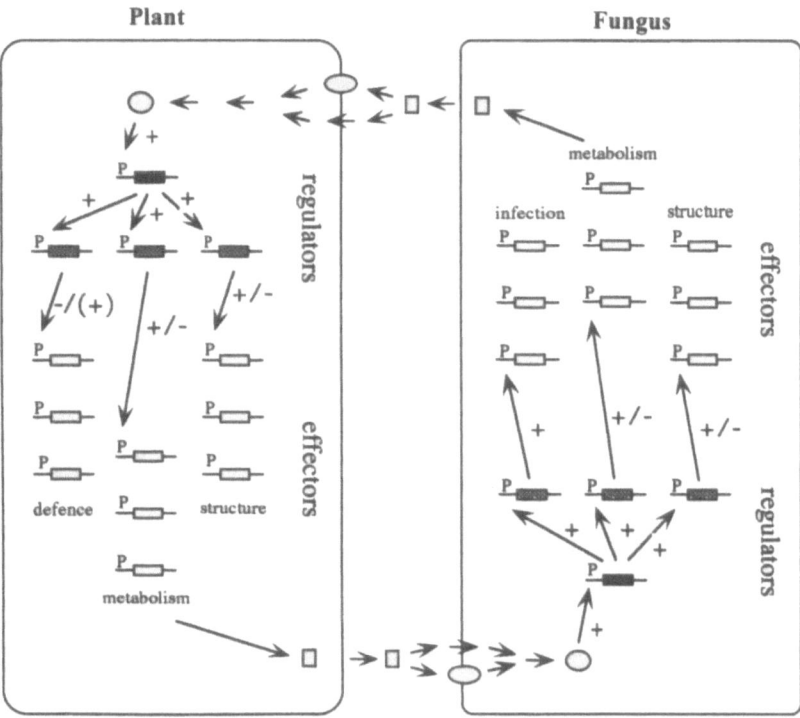

Figure 1. Model for molecular events involved in arbuscular mycorrhizal symbiosis. Plant or fungal signals (open squares) are recognized by membrane or cytoplasmic receptors (open circles) and activate, *via* a signal transduction chain, certain master genes (filled boxes). In this way, a cascade of regulatory (shaded boxes) and effector (open boxes) genes are controlled. Induction is indicated by +, repression by - and changes in both directions by +/-.

and structural changes either at the cellular level (Bonfante, 1994) or that of the whole root system (Atkinson *et al.*, 1994).

GENES INVOLVED IN ARBUSCULAR MYCORRHIZAL SYMBIOSIS

Mutants provide definite proof that a specific gene is involved in a given process. That specific plant genes are essential to arbuscular mycorrhiza establishment was shown following the isolation of two types of mutants, obtained by chemical mutagenesis of pea (Duc *et al.*, 1989) and altered in their ability to form the symbiosis : early mutants (myc^{-1}) that stop fungal development at the root surface after appressorium formation, and late mutants (myc^{-2}) where root colonization occurs, but arbuscule development is defective (Gianinazzi-Pearson *et al.*, 1991). In both cases, the resistance to arbuscular mycorrhizal fungi is found in plants which are unable to complete symbiotic interactions with *Rhizobium* (Duc *et al.*, 1989; Gianinazzi-Pearson *et al.*, 1991). This could reflect links between infection events in nodulation and arbuscular mycorrhiza formation (Duc *et al.*, 1989, Gianinazzi-Pearson *et al.*, 1994). Early mutants are the most frequent and have been most studied ; they are induced by at least four mutated loci and properties of the mycorrhiza-resistant character

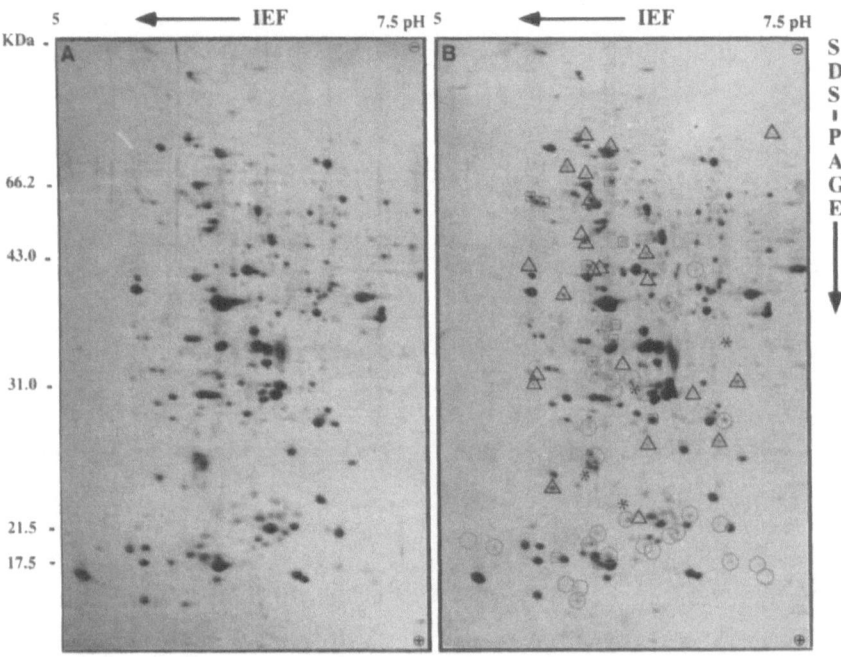

Figure 2. Modifications in root polypeptides with arbuscular mycorrhiza formation Polypeptide patterns in 2D-PAGE from pea roots, uninoculated (A) and one week after inoculation with *Glomus mosseae* (B) polypeptides □ increase, O decrease, ✳ disappear or Δ are new in arbuscular mycorrhizal roots (IEF isoelectric focusing)

can be summarized as follows : monogenic, recessive, genetically stable and indissociable from the nod⁻ character (Duc *et al.*, 1989). Moreover, the mutated genes appear to be symbiosis-specific since they do not affect susceptibility to different pathogens like *Fusarium*, *Agrobacterium* or nematodes (Gianinazzi-Pearson *et al.*, 1994). These mutants represent therefore a very interesting model for studying genes and gene expression essential to arbuscular mycorrhiza functioning (Gollotte *et al.*, 1994).

STRATEGIES FOR IDENTIFYING GENES AND THEIR EXPRESSION IN ARBUSCULAR MYCORRHIZA

Evidence now exists that there must be changes in expression of many plant and fungal genes during the establishment of arbuscular mycorrhiza. Two approaches are possible for identifying specific genes and gene products: blind or orientated.

Two main strategies can be adopted in a **blind approach**. The first one implies direct analysis of polypeptides extracted from mycorrhizal and non mycorrhizal roots (Garcia-Garrido *et al.*, 1993; Dumas-Gaudot *et al.*, 1994). Figure 2 illustrates an example of modifications observed in pea roots one week after inoculation with *Glomus mosseae*. In comparison to nonmycorrhizal roots, an important number of polypeptides of plant origin decrease in amount, some increase and few disappear. However, many new polypeptides can also be detected in the inoculated roots, but it is impossible to distinguish those of fungal origin

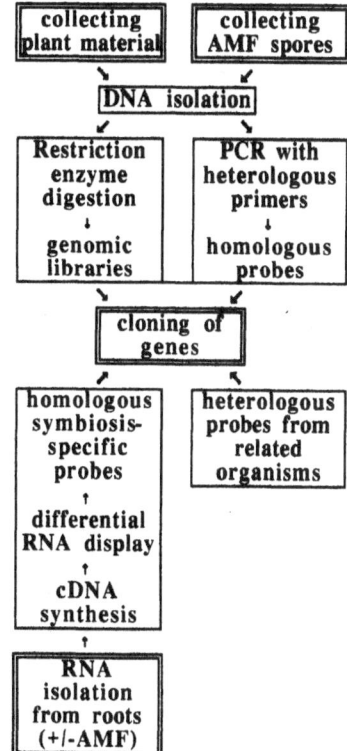

Figure 3. Strategies for cloning arbuscular mycorrhiza-control-led genes. Extracted DNA from fungi and plants is used for restriction enzyme digestion and cloning. Libraries are screened with heterologous and/or with homologous gene probes. Homologous probes can also consist of antibodies or oligonucleotide probes produced after purification and sequencing of target polypeptides. (AMF: arbuscular mycorrhizal fungi).

from those of plant origin. In contrast, when the mycorrhiza-resistant mutant peas are challenged with the same arbuscular mycorrhizal fungus, the major modification is the disappearance of plant polypeptides (Samra *et al.* unpublished results). This could mean that the ability to form arbuscular mycorrhiza is linked to the non-repression of genes that are constitutively expressed in plants. Interesting polypeptides from such analysis could be purified and sequenced, corresponding antibodies or oligoprobes produced and used as homologous probes for screening fungal and plant genomic libraries, in order to clone specific symbiosis genes (Figure 3).

A second strategy consists in extracting mRNA populations from mycorrhizal and non mycorrhizal roots followed either by *in vivo* translation or by cDNA synthesis. In applying these techniques, important modifications occurring in gene expression during arbuscular mycorrhiza formation can be identified (Martin-Laurent *et al.*, 1994). For example, using the differential display reverse transcription-PCR technique (DDRT-PCR) with different combinations of primers, PCR-cDNA fragments that are only present or absent in infected roots have been identified. Figure 4 shows new cDNA fragments representing genes that are only transcribed in arbuscular mycorrhizas, while the fragments that disappear represent genes that are repressed during symbiosis formation. Interesting cDNA fragments have to be cloned and subsequently used as homologous probes against plant and fungal genomic libraries to identify specific arbuscular mycorrhizal genes (Figure 4).

Interestingly, whatever the method used, modifications in gene expression accompanying arbuscular mycorrhiza formation follow the same pattern: many new genes are specifically expressed, whilst a significant number are down or up regulated. The next step is to better characterize these modifications and to discern between those which come from

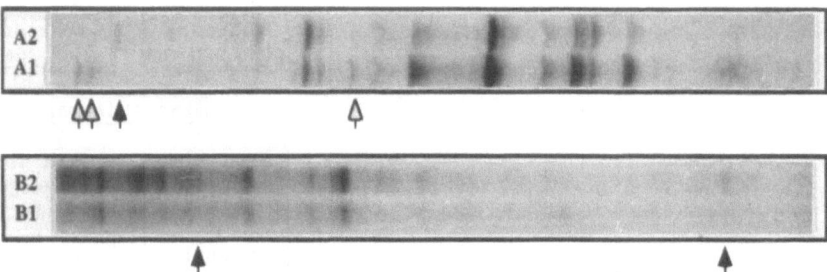

Figure 4. Modifications in DDRT-PCR patterns with arbuscular mycorrhiza formation. Display patterns obtained from RNA samples of pea roots, uninoculated (lane 1) or one week after inoculation with *Glomus mosseae* (lane 2) using a combination of two primers: $(T)_{11}GA$ and GTGATCGCAG (A) or $(T)_{11}CA$ and TCTGTGCTGG (B). (▷) indicate bands specifically repressed during mycorrhization. (▶) indicate bands specifically enhanced during mycorrhization.

the arbuscular mycorrhizal fungus and those which are of plant origin. For this purpose, the development of genomic libraries of arbuscular mycorrhizal fungi (Zézé *et al.*, 1994, and unpublished results) which can be screened with homologous symbiosis-specific probes is essential (Figure 3) (Franken *et al.*, 1994).

In an **orientated approach** molecules of known function, like those involved in fungal phosphate metabolism (alkaline phosphatase), bidirectional nutrient exchanges between the two symbiotic partners (plasmamembrane ATPase) or plant defence reactions (chitinases), are interesting candidates.

Fungal alkaline phosphatase is a potential molecular marker of a functional arbuscular mycorrhiza in terms of phosphate metabolism (Gianinazzi *et al.*, 1992). The activity of this enzyme appears as the fungus colonizes the host root and it is not detected if there is no root colonization as in the case of myc^{-1} plants, indicating that this fungal enzyme is somehow regulated by the host plant (Tisserant *et al.*, 1993). Cytoenzymology and cytochemical techniques have shown that the enzyme molecule and its activity are located inside the same cellular compartment, the fungal vacuole. However, neither the molecule nor the activity of this enzyme could be detected inside the vacuole of germinating hyphae (Gianinazzi-Pearson and Gianinazzi, 1994). This indicates that not only the activity of this fungal enzyme, but probably also its synthesis are regulated by the host plant. In order to understand the regulatory mechanisms involved, it is necessary to clone the fungal gene coding for this enzyme. The strategy for this could be based on the fact that alkaline phosphatase genes are conserved across the animal, fungal and bacterial kingdoms. We have identified conserved regions in *E. coli* and *S. cerevisiae,* PCR primers have been designed and cloning of an amplification product corresponding to a fragment of the alkaline phosphatase gene from arbuscular mycorrhizal fungi is in progress.

Plasma membrane ATPase of both the plant and fungus is activated during arbuscular mycorrhiza formation. This activity is associated with the fungal plasma membrane throughout the intraradicular mycelium, whilst in the host it is specifically induced along the new membrane of cortical cells developing around living arbuscular branches, but not around dead hyphae or around fungal coils in epidermal cells (Gianinazzi-Pearson and Gianinazzi, 1988; Gianinazzi-Pearson *et al.*, 1991). The induction of this host enzyme activity, necessary for generating energy for bidirectional nutrient exchanges between the two symbionts, depends therefore on fungal development. The specificity of host ATPase activation has recently been confirmed by the absence of the active enzyme on the host membrane surrounding the aborted arbuscules in late pea mutants (myc^{-2}) (Gianinazzi-Pearson *et al.*, 1994). Further research is necessary to understand how this host-specific ATPase is regulated

Pea   Tobacco

C   M   P   C   M   P

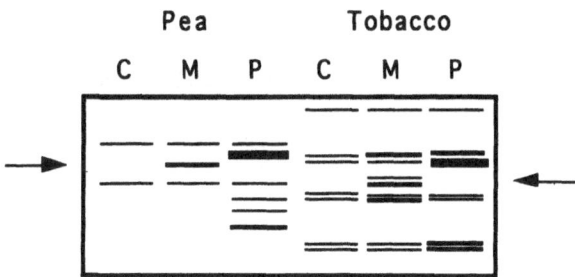

Figure 5. Diagrammatic representation of changes in acidic chitinase isoforms after infection with arbuscular mycorrhizal or pathogenic fungi. Proteins extracted from pea or tobacco, uninoculated (C), or inoculated with *Glomus mosseae* (M) or *Chalara elegans* (P) and separated in the Davis electrophoretic system. (➤) indicate the isoforms specifically induced in mycorrhiza.

by the fungus, but recent cloning from mycorrhizal roots of a cDNA fragment, with high sequence homology with a plant ATPase gene, has shown that higher amounts of corresponding mRNA occur during arbuscular mycorrhiza formation, thus opening the possibility of a spatio-temporal analysis of plant ATPase activation in infected cortical cells (see P.J. Murphy, P. Langridge and S.E. Smith in this book).

Genes coding for *molecules associated with plant defence* are known to be weakly, transiently, very locally or not activated during fungal/root interactions in arbuscular mycorrhizas as compared to pathogen infections (Gianinazzi-Pearson, 1994; Franken and Gnädinger, 1994; Harrison and Dixon, 1994). However, further analysis of acidic chitinase isoforms in pea and tobacco roots has revealed that different isoforms are active depending on whether the roots are infected by a pathogenic fungus (*Aphanomyces euteiches* or *Chalara elegans*) or a symbiotic fungus (*G. mosseae*) (Figure 5). This observation is the first demonstration of the existence of symbiosis-specific plant molecules of a known function (Dumas-Gaudot *et al.*, 1994; unpublished results).

Another demonstration that the relationship between plant defence genes and arbuscular mycorrhiza formation is not simple comes from studies of plants that overexpress defence-related genes. Different *Nicotiana* which have been transformed by introducing foreign genes coding for basic and acid chitinases of different origin (Gianinazzi-Pearson, 1994; Vierhilig *et al.*, 1993), and a *N. debneyi* x *glutinosa* amphidiploid hybrid constitutively expressing different defence-related genes, including chitinase (Gianinazzi and Ahl, 1983; Ahl-Goy *et al.*, 1992), are not altered in their ability to form arbuscular mycorrhizas (Gianinazzi-Pearson, 1994; unpublished results), indicating that arbuscular mycorrhizal fungi can somehow bypass the activity of plant defence genes.

CONCLUSIONS

It is obvious that a complex series of signal exchanges must be involved in interactions between fungal and plant partners of arbuscular mycorrhiza, and there is increasing evidence concerning the nature of molecules, subcellular and cellular structures and genes involved in mycorrhiza functioning. These can be summarized as follows: (i) the arbuscular mycorrhizal association is controlled by specific plant genes, some of which are common to other root symbioses (e.g. *Rhizobium*), (ii) specific gene products occur with mycorrhiza formation, while plant defence genes are only weakly and transiently activated, (iii) an interface, structurally and functionally well adapted to bidirectional nutrient exchanges, develops between the two symbiotic partners, and (iv) transorganismal enzyme activation has been demonstrated

Although there is still a long way to go, these observations clearly indicate that a multidisciplinary approach is the only way by which mycorrhizal research can provide the necessary knowledge for reorientating selection and management toward the optimal use of functional mycorrhizal symbiosis either for plant or fungal production

REFERENCES

Ahl Goy, P, Felix, G, Metraux, J P, Meins, Jr F, 1992, Resistance to disease in the hybrid *Nicotiana glutinosa* x *Nicotiana debneyi* is associated with high constitutive levels of β-1,3-glucanase, chitinase, peroxidase and polyphenoloxidase, *Physiological and Molecular Plant Pathology* 41 11-21

Atkinson, D, Berta, G, and Hooker, J E, 1994, Impact of mycorrhizal colonisation on root architecture, root longevity and the formation of growth regulators In *Impact of Arbuscular Mycorrhizas on Sustainable Agriculture and Natural Ecosystems*, S Gianinazzi and H Schuepp (eds), Birkhauser Verlag, Basel, Switzerland, pps 89-99

Berta, G, Sgorbati, S, Soler, V, Fusconi, A, Trotta, A, Citterio, A, Bottone, M G, Sparvoli, E, and Scannerini, S, 1990, Variations in chromatin structure in host nuclei of a vesicular arbuscular mycorrhiza, *New Phytol* 114 199-205

Bonfante, P, 1994, Ultrastructural analysis reveals the complex interactions between root cells and arbuscular mycorrhizal fungi In *Impact of Arbuscular Mycorrhizas on Sustainable Agriculture and Natural Ecosystems*, S Gianinazzi and H Schuepp (eds), Birkhauser Verlag, Basel, Switzerland, pps 73-87

Duc, G, Trouvelot, A, Gianinazzi-Pearson, V and Gianinazzi S, 1989, First report of non mycorrhizal plant mutants (Myc) obtained in pea (*Pisum sativum* L) and fababean (*Vicia faba* L), *Plant Science* 60, 215-222

Dumas, E, Tahiri-Alaoui, A, Gianinazzi, S, and Gianinazzi-Pearson, V, 1990, Observations on modifications in gene expression with VA endomycorrhiza development in tobacco qualitative and quantitative changes in protein profiles In *Endocytobiology IV*, P Nardon, V Gianinazzi-Pearson, A M Grenier, L Margulis, and D C Smith (eds), INRA, Paris, pps 153-157

Dumas-Gaudot, E, Asselin, A, Gianinazzi-Pearson, V, Gollotte, A, and Gianinazzi, S, 1994, Chitinase isoforms in roots of various pea genotypes infected with arbuscular mycorrhizal fungi, *Plant Science* 99 27-37

Dumas-Gaudot, E, Guillaume, P, Tahiri-Alaoui, A, Gianinazzi-Pearson, V, and Gianinazzi, S, 1994, Changes in polypeptide patterns in tobacco roots colonized by two *Glomus* species, *Mycorrhiza*, 4 215-221

Franken, P, and Gnadinger, F, 1994, Analysis of parsley arbuscular endomycorrhiza infection development and mRNA levels of defense-related genes, *Molecular Plant-Microbe Interactions* 7(5) 612-620

Franken, P, van Tuinen, D, Martin-Laurent, F, and Gianinazzi-Pearson, V, 1994, Molecular characterization and gene analysis of arbuscular mycorrhizal fungi In *4th ESM Proceedings*, J M Barea *et al* (eds) (in press)

Garcia-Garrido, J M, Toro, N, and Ocampo, J, 1993, Presence of specific polypeptides in onion roots colonized by *Glomus mosseae Mycorrhiza* 2 175-177

Ghachtouli El, N, Paynot, M, Morandi, D, Martin-Tanguy, J, and Gianinazzi, S, 1994, Effect of polyamines on endomycorrhizal infection of wild type *Pisum sativum* cv Frisson (nod+myc+) and two mutants (nod-myc+ and nod-myc-), *Mycorrhiza* 5 189-192

Gianinazzi, S, 1991, Vesicular-arbuscular (endo)mycorrhizas Cellular, biochemical and genetic aspects, *Agriculture Ecosystems and Environment* 35 105-119

Gianinazzi, S, and Ahl, P, 1983, The genetic and molecular basis of b-proteins in the genus *Nicotiana Netherland Journal of Plant Pathology* 89 275-281

Gianinazzi, S, Gianinazzi-Pearson, V, Tisserant, B, and Lemoine, M C, 1992, Protein activities as potential markers of functional endomycorrhizas in plants In *Mycorrhizas in Ecosystems*, D J Read, D H Lewis, A H Fitter and D J Alexander (eds), CAB International, Oxon, pps 333-339

Gianinazzi-Pearson, V, 1995, Morphofunctional compatibility in interactions between roots and arbuscular endomycorrhizal fungi molecular mechanisms, genes and gene expression In *Pathogenesis and Host-Parasite Specificity in Plant Diseases* K Kohmoto, R P Singh and U S Singh (eds), Pergamon Press, Elsevier Science, Oxford, vol II, pps 251-263

Gianinazzi-Pearson, V, and Gianinazzi, S , 1988, Morphological integration and functional compatibility between symbionts in vesicular-arbuscular endomycorrhizal associations In *Cell to Cell Signals in Plant, Animal and Microbial Symbiosis*, S Scannerini, D C Smith, P Bonfante-Fasolo and V Gianinazzi-Pearson (eds), NATO ASI, serie H, Cell Biology, Springer Verlag, Berlin, Vol 17 73-84

Gianinazzi-Pearson, V, and Gianinazzi, S , 1989, Cellular and genetical aspects of interactions between hosts and fungal symbionts in mycorrhizae, *Genome* 31 336-341

Gianinazzi-Pearson, V, and Gianinazzi, S , 1994, Proteins and protein activities in endomycorrhizas symbiosis In: *Mycorrhiza Structure, Function, Molecular Biology and Biotechnology*, A Varma and B Hock (eds), Springer Verlag, Heidelberg, pps 251-266

Gianinazzi-Pearson, V, and Smith, S E , 1993, Physiology of mycorrhizal mycelia In *Mycorrhiza synthesis, Advances in Plant Pathology*, D S Ingram, P H Williams and I C Tommerup (eds), Academic Press, London, pps 55-82

Gianinazzi-Pearson, V, Gianinazzi, S , Guillemin, J P, Trouvelot, A , Duc, G , 1991, Genetic and cellular analysis of resistance to vesicular arbuscular (VA) mycorrhizal fungi in pea plants In *Advances in Molecular Genetics of Plant-Microbe Interactions*, Vol I, H Hennecke and D P S Verma (eds), Kluwer, Academic Publishers, Dordrecht, Boston, London, pps 336-342

Gianinazzi-Pearson, V, Gollotte, A , Dumas-Gaudot, E , Franken, P, and Gianinazzi, S , 1994, Gene expression and molecular modifications associated with plant responses to infection by arbuscular mycorrhizal fungi In *Advances in Molecular Genetics of Plant-Microbe Interactions*, M Daniels, J A Downie and A E Osborn (eds), Kluwer Academic Publishers, Boston and London Vol 3, pps 178-186

Gianinazzi-Pearson, V, Gollotte, A , Lherminier, J , Tisserant, B , Franken, P, Dumas-Gaudot, E , Lemoine, M C , van Tuinen, D , and Gianinazzi, S , 1994, Cellular and molecular approaches in the characterization of symbiotic events in functional arbuscular mycorrhizal associations, *Canadian Journal of Botany* (in press)

Gianinazzi-Pearson, V, Smith, S E , Gianinazzi, S , and Smith, F A , 1991, Enzymatic studies on the metabolism of vesicular-arbusuclar mycorrhizas V Is H$^+$-ATPase a component of ATP-hydrolysing enzyme activities in plant-fungus interfaces ?, *New Phytologist* 117 61-76

Gianinazzi-Pearson, V, Tahiri-Alaoui, A , Antoniw, J F , Gianinazzi, S , and Dumas-Gaudot, E , 1992, Weak expression of the pathogenesis related PR-b1 gene and localization of related protein during symbiotic endomycorrhizal interactions in tobacco roots, *Endocytobiosis and Cell Research* 8 177-185

Giovannetti, M , Sbrana, C , Citernesi, A S , Avio, L , Gollotte, A , Gianinazzi-Pearson, V, and Gianinazzi, S , 1994, Recognition and infection process, basis for host specificity of arbuscular mycorrhizal fungi In *Impact of Arbuscular Mycorrhizas on Sustainable Agriculture and Natural Ecosystems*, S Gianinazzi and H Schuepp (eds), Birkhauser Verlag, Basel, Switzerland, pps 61-72

Gollotte, A , Gianinazzi-Pearson, V, Dumas-Gaudot, E , Giovannetti, M , Lherminier, J , and Gianinazzi, S , 1994, Plant mutants as models for studying the cellular and molecular bases of plant fungal interactions In *4th ESM Proceedings*, J M Barea *et al* (eds) (in press)

Gollotte, A , Gianinazzi-Pearson, V, Giovannetti, M , Sbrana, C , Avio, L , and Gianinazzi, S , 1993, Cellular localization and cytochemical probing of resistance reactions to arbuscular mycorrhizal fungi in a 'locus a' myc- mutant of *Pisum sativum* (L), *Planta* 191 112-122

Harley, J L , and Smith, S E , 1983, Mycorrhizal symbiosis, Academic Press, London and New York, 983 pages

Harrison, M , and Dixon, R A , 1994, Spatial patterns of expression of flavonoid/isoflavonoid pathway genes during interactions between roots of *Medicago truncalata* and the mycorrhizal fungus *Glomus versiforme*, *Plant Journal* 6 9-20

Jacquelinet-Jeanmougin, S , Gianinazzi-Pearson, V, and Gianinazzi, S , 1987, Endomycorrhizas in the Gentianaceae II Ultrastructural aspects of symbiont relationships in *Gentiana lutea* L , *Symbiosis* 3 269-286

Martin-Laurent, F A , Dumas-Gaudot, E , Schlichter, U , Franken, P, Antoniw, J A , Gianinazzi-Pearson, V, and Gianinazzi, S , 1994, Differential display reverse transcriptase polymerase chain reaction a new approach to detect symbiosis related genes involved in arbuscular mycorrhiza In *4th ESM Proceedings*, J M Barea *et al* (eds) (in press)

Rosendahl, S , 1985, Interactions between the vesicular arbuscular mycorrhizal fungus *Glomus fasciculatum* and *Aphanomyces euteiches* root rot of peas, *Phytopathologishe Zeitung* 114 31-40

Smith, S E , and Gianinazzi-Pearson, V, 1988, Physiological interactions between symbionts in vesicular-arbuscular mycorrhizal plants, *Annual Review of Plant Physiology and Plant Molecular Biology* 39 221-244

Smith, S E , Gianinazzi-Pearson, V, Koide R , and Cairney, J W G , 1994, Nutrient transport in mycorrhizas structure, physiology and consequences for efficiency of the symbiosis, *Plant and Soil* 159 103-113

Tisserant, B., Gianinazzi-Pearson, V., Gianinazzi, S., and Gollotte, A., 1993, *In planta* histochemical staining of fungal alkaline phosphatase activity for analysis of an efficient arbuscular mycorrhizal infection, *Mycological Research* 97(2): 245-250.

Vierhilig, H., Alt, M., Neuhaus, J.M., Boller, T., and Wiemken, A., 1993, Colonization of transgenic *Nicotiana sylvestris* plants, expressing different forms of *Nicotiana tabacum* chitinase, by the root pathogen *Rhizoctonia solani* and by the mycorrhizal fungus *Glomus mosseae, Molecular Plant-Microbe Interactions* 6: 261.

Zézé, A., Dulieu, H., and Gianinazzi-Pearson, V., 1994, DNA cloning and screening of a partial genomic library from an arbuscular mycorrhizal fungus, *Scutellospora castanea, Mycorrhiza 4: 251-254.*

7

CLONING FUNCTIONAL ENDOMYCORRHIZA GENES

Potential for Use in Plant Breeding

P. J. Murphy,[1] A. Karakousis,[1] S. E. Smith,[1,2] and P. Langridge[1]

[1] Department of Plant Science
[2] Department of Soil Science
Waite Agricultural Research Institute
Glen Osmond
South Australia 5064
Australia

ABSTRACT

Two cDNA clones, BMR6 and BMR78, which detect transcripts of genes that increase in abundance in the roots of *Hordeum vulgare* cv. 'Galleon' during colonisation by the vesicular-arbuscular mycorrhizal fungus *Glomus intraradices* have recently been isolated. Sequence analysis indicated that BMR78 is a partial clone of a H+ - ATPase gene. A putative function for the gene corresponding to BMR6 was not ascertained.

The genomic DNA sequences which hybridise to BMR6 and BMR78 were mapped to barley chromosomes 6H and 2H respectively, by segregation analysis of restriction fragment length polymorphisms (RFLP's) in two doubled haploid mapping populations. The results were confirmed by hybridisation analysis of wheat-barley chromosome addition lines. BMR78 maps to a region of the barley genome which contains a putative nematode resistance gene.

INTRODUCTION

Fungi capable of entering into mycorrhizal associations with higher plants are found in all the taxonomic groups. Genera of the order Glomales (Morton and Benny, 1990) form morphologically distinctive tissues termed vesicular-arbuscular (VA) mycorrhizas. The great majority of terrestial plants, including most agricultural and horticultural species, can enter into symbiotic associations with VA mycorrhizal fungi. As a result of infection, the plant may show greater stress tolerance or disease resistance and enhanced growth, particularly in soils of moderate or low nutrient status. Indeed, some horticulturally important tree species appear to have an almost obligatory requirement for VA mycorrhizal associations in order to attain satisfactory growth (Harley and Smith, 1982).

Biotechnology of Ectomycorrhizae, Edited by Vilberto Stocchi et al.
Plenum Press, New York, 1995

The beneficial consequences associated with VA mycorrhiza formation have stimulated considerable research to determine if these fungi can be manipulated on an agronomic scale. It may be possible to improve plant growth and soil status whilst reducing the use of expensive and environmentally damaging chemical inputs. The major stumbling block in the effective management of these beneficial micro-organisms, both at the experimental and field level, is that they are for all intents and purposes obligate biotrophs. This suggests that specific plant gene products, possibly resulting from chemical signalling between the symbiotic partners, are required to mediate and maintain fungal growth and differentiation (see Smith and Gianinazzi-Pearson, 1988; Koide and Schreiner, 1992).

There is now an increased awareness in the importance of breeding for nutritional traits to improve the productivity of crop plants, particularly in soils with low concentrations of available nutrients such as phosphorus. Baon, Smith and Alston (1994) showed that there is considerable phenotypic variance between several cultivars of barley in their response to VA mycorrhizal colonisation at low phosphorus supply. Agronomic or physiological traits, such as mycorrhizal response and dependency or nutrient aquisition and utilisation, often require phenotypic estimations which are notoriously difficult to incorporate into traditional plant breeding programs. This is primarily because of the considerable time and expense involved in assessing breeding lines in large field trials.

Cost effective DNA-based diagnostics may prove to be particularly useful in breeding programs, particularly those which involve complex multigenic and/or quantitative traits (Rafalski and Tingey, 1993). It is envisiged that using DNA markers to track desirable traits in genetic crosses will greatly accelerate the release of improved crop cultivars. The barley breeding program at the Waite Agricultural Research Institute is increasingly making use of DNA-based technology. One related project involves the production of a saturated genetic map of barley (Langridge *et al.*, unpublished) using a large number of codominant restriction fragment length polymorphisms (RFLP's).

We mapped the mycorrhiza-related cDNA clones, BMR6 and BMR78, both as part of the ongoing DNA-marker mapping program at the Waite Institute and because mapping genes, the expression of which are implicated in mycorrhiza formation, may provide the most suitable markers for monitoring the segregation of nutritional traits of interest to plant breeders (see Smith, Robson and Abbott, 1992).

METHODS AND MATERIALS

Plant Material

Three doubled haploid mapping populations were generated from the crosses of Galleon x Haruna Nijo, Clipper x Sahara and Chebec x Harrington (Langridge *et al.*, unpublished).

Seeds of Chinese Spring wheat, Betzes barley and wheat/barley chromosome disomic addition lines were supplied by Dr. K. Shepherd. Essentially, each addition line comprises a complete Chinese Spring wheat genome as well as one pair of each of the seven chromosomes from Betzes barley (see Islam, Shepherd, and Sparrow, 1981;). Addition line 1 contains both barley chromosomes 1H and 6H.

DNA Extraction

Total genomic DNA was extracted from the adult leaves of all plants as described by Guidet *et al.* (1991). The high molecular weight DNA was then purified and isolated by ultracentrifugation, butanol extraction and dialysis and the DNA concentration determined as described by Maniatis, Fritsch and Sambrook (1982).

Detection of RFLP Markers

Six samples of genomic DNA (10 µg) from each of the six parental lines used to generate the F1 populations were separately restriction digested with each of the following restriction endonucleases in accordance with the manufacturers instructions (Promega); *Eco* R1, *Dra* 1, *Hind* 111, *Bam* H1 and *Eco* RV. The digested samples were then size fractionated on 1% agarose gels and transferred to Hybond N+ (Amersham) hybridisation membranes (Maniatis, Fritsch and Sambrook, 1982). DNA insert from either BMR6 and BMR78 was radiolabelled by random priming (Feinberg and Vogelstein, 1983) with a-^{32}P-dCTP (Amersham) and used to probe the membranes. The autoradiograms were visually assessed to identify RFLP's between the pairs of parental barley cultivars used for the generation of each of the mapping populations.

Mapping

Radio-labelled probe prepared from BMR6 and BMR78 insert (as described above) was hybridised to membranes containing *Bam* HI digested genomic DNA from each line of the doubled haploid population of the Clipper x Sahara cross and *Dra* 1 digested total DNA extracted from each of the Galleon x Haruna nijo doubled haploids.

Linkage Analysis of RFLP Markers

The autoradiograms of the membranes were visually assessed to score the segregation of the RFLP marker bands in each of the F1 progeny. The EXCEL software program was used to order and store the data and the MAPMAKER program (Lander *et al.*, 1987) was used to carry out the linkage analysis. The map distances (in centimorgans) were determined using the Kosambi mapping function.

Addition Line Mapping

Genomic DNA (10 µg) extracted from Chinese Spring wheat, Betzes barley and the wheat-barley chromosome addition lines 1-7 was digested with the restriction endonuclease-*Hind* III in accordance with the manufacturers recommendations (Promega). The DNA samples were size fractionated by 1% agarose gel electrophoresis, transferred to Hybond N+ and hybridised with radio-labelled insert from BMR6 or BMR78. The autoradiograms were then used to localise the genes represented by the cDNA clones to a barley chromosome.

RESULTS

One approach to identifying the types of molecules involved in establishing a functional VA mycorrhiza is to investigate plant gene expression in mycorrhizal roots. We differentially screened a cDNA library prepared from Galleon barley roots colonised by the VA mycorrhizal fungus *Glomus intraradices* Schenk and Smith, and were able to identify four cDNA clones (Murphy *et al.*, in preparation) which detect the mRNA transcripts of genes that appear to be differentially-regulated during the earliest colonisation stages of the symbiosis (see Table 1).

Table 1. Summary of the four barley mycorrhiza-related clones identified by differential screening of a cDNA library prepared from 14 day old *Glomus intraradices* colonised *Hordeum vulgare* cv. 'Galleon' roots

Clone	cDNA size	Transcript Size	Comments
BMR6	615bp	1300b	up-regulated during mycorrhizal colonisation; sequenced but function unknown.
BMR64	480bp	1300b	down-regulated during mycorrhizal colonisation; not sequenced.
BMR78	605bp	3200b	up-regulated during mycorrhizal colonisation; sequence indicates that it is a partial clone of a H+ - ATPase gene.
BMR93	900bp	2000b	down-regulated during mycorrhizal colonisation; not sequenced.

Both clones detected bands of moderate intensity in the genomic DNA blots of all the barley cultivars probed. BMR6 detected RFLP's in several of the restriction enzyme/barley cultivar combinations. Some of these differed only slightly in the restriction fragment length or contained multiple bands and were deemed unsuitable for further use. The *Bam* H1 digestion of Clipper and Sahara genomic DNA produced a single unambiguous RFLP band. BMR78 detected only one RFLP, this was found in the *Dra* 1 digestion of Galleon and Haruna Nijo. On the basis of these results, BMR6 and BMR78 were mapped using *Bam* H1 digestions of the Clipper x Sahara doubled haploid population and *Dra* 1 digestions of the Galleon x Haruna Nijo population.

The gene sequence hybridising to BMR78 was placed in a group containing clones that map (using wheat/barley chromosome addition lines) to chromosome 2H (Figure 1A). Several markers in close proximity to BMR78 have been mapped to the long arm of this chromosome. Another barley chromosome 2H linkage map, generated using the doubled haploid population obtained from the Clipper x Sahara cross, contained several RFLP markers (but not BMR78) which were common to both maps. A comparison of the two maps (not shown) showed that markers common to both maps were in the same linear order and that the genetic distances between many of these loci were in good agreement. BMR78 maps to a region of chromosome 2H which contains another root cDNA marker (AWBMA21) isolated from the mycorrhizal barley-root library and Ha2, a putative marker for the cereal-cyst nematode (CCN) resistance gene, *Cre* (Williams, Fisher and Langridge, 1994).

The chromosomal position obtained for BMR6 was less conclusive than that obtained for BMR78 as only two other probes were in the same linkage group (Figure 1.C). This was primarily because the mapping project is being carried out in chromosome number order (ie. 1H-7H) and at the time of this investigation many more markers had been mapped for chromosomes 1H-3H than chromosmes 4H-7H. Both of the markers linked to BMR6 generate multiple RFLP's which map to various chromosomes but one of the RFLP's generated by each of the probes map to a common region. On this basis the gene corresponding to BMR6 was mapped onto the long arm of chromosome 6H.

BMR6 and BMR78 were also mapped using hybridisation analysis of wheat-barley chromosome addition lines to confirm and clarify the results obtained from the mapping experiments. The strongly hybridising bands observed on the autoradiographs of these hybridisations (Figure 2) confirmed that the sequences which hybridise with BMR6 and BMR78 are located on barley chromosomes 6H and 2H, respectively. Several other hybridising bands of much lower intensity were also observable in the parental wheat and barley lines.

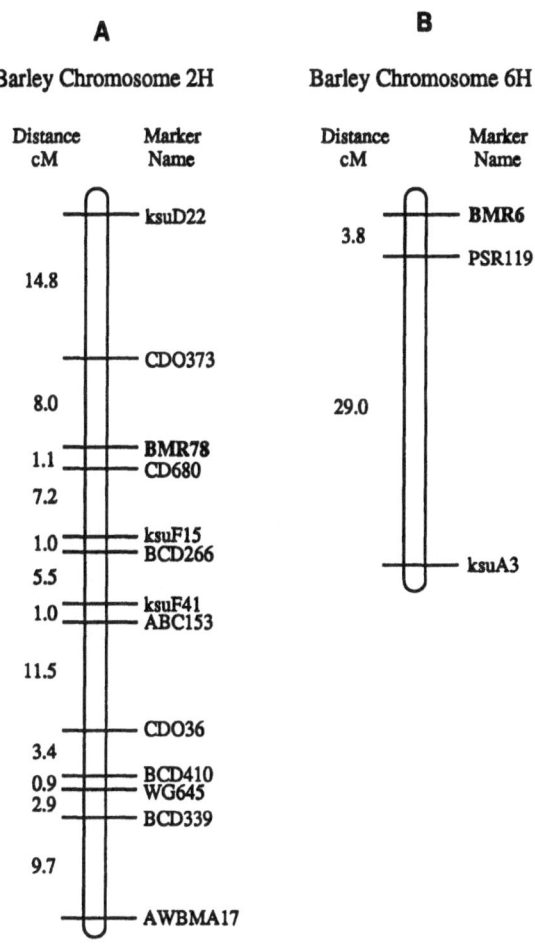

Figure 1. Genetic linkage maps of barley chromosomes 2 and 6 showing the putative positions of the barley mycorrhiza-related cDNA clones BMR78 and BMR6. The maps were generated by linkage analysis (using the 'Mapmaker' program) of doubled haploid populations derived from the crossing of Galleon x Haruna Nijo (for chromosome 2H) and Clipper x Sahara (for chromosome 6H).

DISCUSSION

Two of the clones, BMR6 and BMR78 have been characterised in further detail: It was found that neither of the genes detected by these clones were induced by avirulent or virulent biotypes of the root pathogen *Gaeumannomyces graminis* or water-stress conditions and sequence analysis of BMR78 strongly suggested that it represents the partial sequence of a H+ - ATPase (Murphy *et al.*, in preparation). This lends further support to the proposition that active transport processes are involved in the bi-directional transfer of nutrients between the symbiotic partners (Smith and Smith, 1990; Gianinazzi-Pearson *et al.*, 1991).

BMR78 is almost certainly a partial cDNA of the 3'-OH region of a H+ - ATPase gene. These enzymes are encoded by multigene families that are highly conserved both within and between the genera investigated to date (Boutry *et al.*, 1989; Ewing *et al.*, 1990; Harper, Manney and Sussman, 1994). The weakly hybridising bands observable

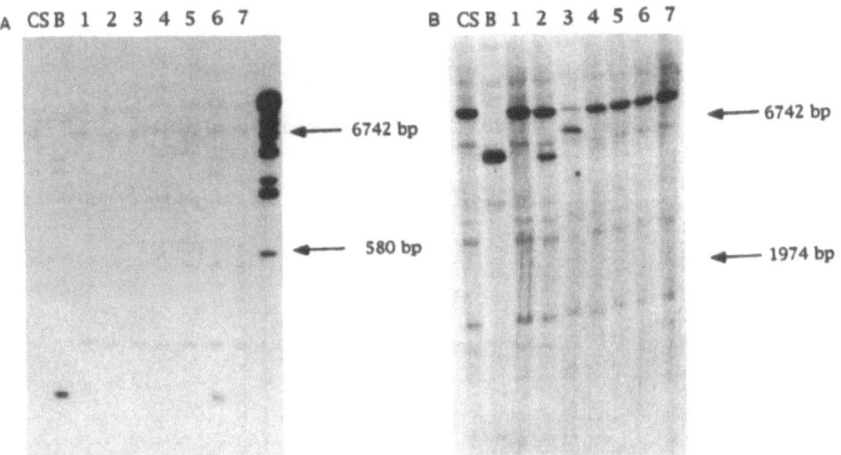

Figure 2. Southern hybridisations of Chinese Spring (CS) wheat, Betzes barley (B) and wheat:barley-chromosome addition lines (1-7) with the barley mycorrhiza-related cDNA clones BMR6 (Figure 2A) and BMR78 (Figure 2B). Plant genomic DNA (10µg) was digested to completion with *Hind* III, separated by gel electrophoresis (1% agarose) and transferred to Hybond N+ (Amersham) hybridisation membranes. The membranes were hybridised overnight with radiolabelled probes generated by random-primer labelling of BMR6 and BMR78. The membranes were washed until no background signal was detectable and autoradiographed for 7 days.

in the genomic DNA blots of both Chinese Spring wheat and Betzes barley when probed with BMR78, indicated that this may also be true for these two important cereal species. Re-probing the original cDNA library with BMR78 detected one group of strongly-hybridising cDNA's and another group of moderately-hybridising cDNA's. Whilst the strongly hybridising clones had identical restriction patterns to BMR78, the second group did not. These results also indicate that at least one other ATPase gene may be present in barley. A more complete molecular investigation of this clone is now planned. It is still not clear if the presence of multiple copies of H+ - ATPases represents the existence of a group of enzymes with distinct regulatory and/or catalytic properties or not (Houlne and Boutry, 1994).

The function of the gene represented by BMR6 is unclear as it has no homology to other reported gene sequences. However, the genomic DNA blots of Chinese Spring wheat and Betzes barley probed with BMR6 suggest that it may be a member of a multigene family.

We have mapped the strongly hybridising sequences detected by BMR6 and BMR78 to the long arms of chromosomes 6H and 2H respectively. The preliminary results indicate that BMR78 (and AWBM21) map in the region of the putative CCN-resistance marker, Ha2. If genetic analysis confirms that this region of chromosome 2H contains a CCN-resistance gene, then BMR78 could prove to be a useful marker in breeding programs and/or a probe for chromosome-walking experiments aimed at cloning the gene.

The high number of polymorphisms (at least 1 for each cultivar tested) detected with BMR6 suggests that it could be used as a common marker for a wide range of cultivars. The accumulation of individual linkage maps with groups of such common markers will eventually lead to a unified map showing regions associated with phenotypic traits.

REFERENCES

Baon J B , Smith S E and Alston A M (1994) Mycorrhizal responses of barley cultivars differing in P efficiency *Plant and Soil* 157 97-105

Boutry M , Michelet B and Goffeau A (1989) Molecular cloning of the family of genes encoding a protein homologous to plasma membrane H+ - translocating ATPases *Biochem Biophys Res Comm* 162 567-574

Ewing N N , Wimmers L E , Meyer D J , Chetelat R T and Bennett A B (1990) Molecular cloning of tomato plasma membrane H+-ATPase *Plant Physiol* 94 1874-1881

Feinberg A P and Vogelstein B (1983) A technique for radiolabelling DNA restriction endonuclease fragments to high specific activity *Anal Biochem* 132 6-13

Giannazzi-Pearson V, Smith S E , Giannazzi S and Smith F A (1991) V Is H+ - ATPase a component of ATP-hydrolysing enzyme activities in plant-fungus interfaces? *New Phytol* 117 61-74

Guidet F , Rogowski P , Taylor C , Weining S and Langridge P (1991) Cloning and characterisation of a new rye-specific repeated sequence *Genome* 34 81-87

Harley J L and Smith S E (1983) Mycorrhizal Symbiosis London, New York Academic Press

Harper J F , Manney L and Sussman M R (1994) The plasma membrane H+ - ATPase gene family in *Arabidopsis* genomic sequence of AHA10 which is primarily expressed in developing seeds *Mol Gen Genet* 244 572-587

Houlne G and Boutry M (1994) Identification of an *Arabidopsis thaliana* gene encoding a plasma membrane H+ - ATPase whose expression is restricted to anther tissues *Plant Journal* 5 311-317

Islam A K M R , Shepherd K W and Sparrow D H B (1981) Isolation and characterisation of euplasmic wheat-barley chromosome addition lines *Heredity* 46 161-174

Koide, R T and Schreiner, R P (1992) Regulation of the vesicular-arbuscular symbiosis *Ann Rev Plant Physiol Mol Biol* 43 557-81

Lander E S, Green J , Abrahamson A , Baarlow M J , and Daly M J (1987) MAPMAKER an interactive package for constructing primary genetic linkage maps of experimental and natural populations *Genomics* 1 174-181

Maniatis T , Fritsch E F and Sambrook J (1982) Molecular Cloning A Laboratory Manual Cold Spring Harbour Laboratory Cold Spring Harbour, New York

Morton J B and Benny G L (1992) Revised classification of arbuscular mycorrhizal fungi (Zygomycetes) A new order, Glomales, two new suborders, Glominae and Gigasporinae, and two new families, Acaulosporaceae and Gigasporaceae *Mycotaxon* 37 471-491

Ralfalski J A and Tingey S V (1993) Genetic diagnostics in plant breeding RAPD's, microsatellites and machines *Trends in Genetics* 9 275-279

Smith S E and Giannazzi-Pearson V (1988) Physiological interactions between symbionts in vesicular-arbuscular mycorrhizal plants *Ann Rev Plant Physiol* 39 221-244

Smith S E and Smith F A (1990) Structure and function of the interfaces in biotrophic symbioses as they relate to nutrient transport *New Phytol* 114 1-38

Smith S E , Robson A D and Abbott K (1992) The involvement of mycorrhizas in assessment of genetically dependant efficiency of nutrient uptake and use *Plant and Soil* 146 169-179

Williams K J , Fisher J M and Langridge P (1994) Identification of RFLP markers linked to the cereal cyst nematode resistance gene (*Cre*) in wheat *Theor Appl Genet* 89 927-930

8

THE SEXUAL CYCLE IN NEUROSPORA

From Fertilization to Ascospore Discharge

Robert L. Metzenberg

Department of Biomolecular Chemistry
University of Wisconsin
1300 University Avenue
Madison, Wisconsin 53706

INTRODUCTION

Species of the genus Neurospora have been much studied as model systems for understanding metabolic pathways and their regulation, the processes and consequences of sexuality, and the natural history and evolution of a representative fungal genus. These subjects have been extensively reviewed in the literature, and there seems to be no urgent need to update them. The initiation and progression of events in fruiting bodies and asci has been the subject of a recent careful and scholarly review, to which the reader is enthusiastically referred (Raju, 1992). By contrast, the subject of nutrition of perithecia of developing asci, and of communication between nuclei as judged by dominance and by success or failure of complementation in the developing perithecial system has received little recent discussion. Much of the knowledge exists as unpublished bits of information, some of it from the laboratory of the author, or as minor observations made in connection with more substantial findings on other subjects. Such information is nonetheless useful in thinking about how development of fruiting bodies occurs. This review will be unabashedly eclectic, inferential, and anecdotal.

MATING TYPE IN *NEUROSPORA CRASSA*, BRIEFLY CONSIDERED

N. crassa is a heterothallic filamentous fungus which mates only when strains of the two mating types, designated A and a, are brought together under suitable conditions. The conspicuous phase of the organism is haploid both in nature and in the laboratory, and the diploid phase produced by caryogamy in the ascogenous hyphae of the perithecium has only a fleeting existence before it proceeds into meiosis. *N. crassa* is monoecious; the wild type of either strain can function as male or as female, in the sense of spermatial donor and donor of the cytoplasm and its elements. The way in which the sex roles are defined is the opposite of that said to have operated in the Garden of Eden: the first arrival on the scene becomes

Biotechnology of Ectomycorrhizae, Edited by Vilberto Stocchi et al.
Plenum Press, New York, 1995

the female. Pheromones are involved in the recognition that leads to fusion (Bistis, 1983). Strains of the two mating types differ only in a single "mating type region", not in having different sex chromosomes; the rest of the chromosome that bears this region can be made isogenic between the two mating types. Cloning and analysis of this region has showed that, between the two mating types, there is no resemblance of the sort that would permit them to be called alleles. (Vollmer & Yanofsky, 1986; Glass et al., 1988; Metzenberg & Glass, 1990). Therefore they have been called "idiomorphs" to indicate their singularity of structure. There are no inactive or archival cassettes of the opposite mating type, hence mating-type switching of the kind seen in *Saccharomyces cerevisiae* is not only not observed — it is not physically possible.

FORMATION OF PROTOPERITHECIA

Protoperithecia are the unfertilized female component in the sexual process, and normally are the source of the cytoplasm of the zygote, and of all of the mitochondria that appear in the progeny of a cross (Mitchell & Mitchell, 1952; Reich & Luck, 1966; see, however, Yang & Griffiths 1993). Either mating type is able to form these female elements, and protoperithecia of either mating type can be fertilized by male elements (microconidia, macroconidia, or hyphae) of the opposite mating type — viz., Neurospora species are monoecious. Protoperithecium formation is an endogenous property , and does not require any inductive signal from the opposite mating type. Indeed, it occurs abundantly in single-mating type cultures. Neither does it require genetic information from the mating type region. The A1 or a1 reading frames can be interrupted by frameshift mutations, resulting in completely sterile mutants, without any obvious effect on the production of protoperithecia. However, these protoperithecia are completely unreactive to attempts to fertilize them with the opposite mating type. In fact, the same result is obtained when the DNA of the entire mating type region is evicted and replaced with unrelated DNA sequences, showing that no genetic information from any part of either idiomorph is required for the formation of this female structure (An, Z., Randall,T. A., & Metzenberg,R. L., unpublished results).

While the formation of protoperithecia seems to ask little or nothing of the mating type locus, the process has rather exacting nutritional and environmental demands. Protoperithecia are not formed on nutritionally-optimal medium, nor in submerged culture, and are formed only sparsely at temperatures above 25C. Little is known about the mechanism behind any of these nutritional and environmental requirements. A frequently-used medium for the induction of protoperithecia, and thus for mating in general, is that described by Westergaard & Mitchell (1947). The prominent difference between this medium and those that do not support formation of protoperithecia is that a moderate concentration of nitrate ions (as opposed to a high concentration of ammonium ions) is the source of nitrogen. The absence of ammonium and of other preferred sources of nitrogen (e.g., glutamine, arginine, alanine) is crucial, not the presence of nitrate, which is used only when preferred sources are exhausted. Other slowly-used sources of nitrogen may substitute for nitrate. This indicates that protoperithecia are induced by a marginal degree of nitrogen starvation. Marginal concentrations of carbon (glucose) can induce mating in cultures that would otherwise not mate because of a very high concentration of nitrate (Westergaard & Mitchell, 1947), and intense deprivation of carbon as well as limitation of nitrogen can induce crossing between strains for which absence of ammonium is not a sufficient stimulus (Ricci et al. 1991). Several workers (e.g., Davis & DeSerres, 1970; Fairfield & Turner,1993) have advocated using a circle of filter paper in place of sucrose or other sugars to enhance either protoperithecium formation or ascospore production. Intense starvation on plain non-nutrient agar can (tardily) induce formation of protoperithecia even at 35C, a temperature at which no sexual cycle is seen on Westergaard-Mitchell medium (Rothschild & Suskind, 1966). Blue light also greatly stimulates the formation of protoperithecia, and thus the entire sexual process (Degli Innocenti et al., 1983; Sommer et al., 1989). The suppression of proto-perithecial formation in submerged or culture with otherwise-appropriate medium is also unexplained. Apparently an air-medium interface is a significant signal for protoperithecium formation, as it is for conidiation. Formation of protoperithecia is also inhibited by elevated temperature in a range that is perfectly compatible with rapid vegetative growth, though the effect of temperature is not as profound as its effect on meiosis and on orderly disjunction (Frost & Greenhill, 1963; McNelly-Ingle & Frost, 1965). At least for the nutritional and temperature limitations on protoperithecium formation and on the subsequent stages in sexual reproduction, the pseudohomothallic species *N. tetrasperma* and the various true homothallic species are less fastidious than is *N. crassa*, with some isolates mating produc-tively on rich medium and/ or at temperatures above 30C. Finally, blue light induction accelerates formation of protoperithecia, exerting its effect through the *bli-7* gene (also known as *ccg-2* and as *eas* — Sommer et al., 1989). The factors favoring protoperithecium formation seem to be additive, at least qualitatively; a high enough fluence of blue light can partially overcome the effects of an abundance of nitrogen, etc. Nutritional control of sexuality shows at least one sharp contrast between *N. crassa* and *S. cerevisiae*: Neurospora must be nutritionally restricted in the haploid state to form courtship and mating structures (protoperithecia) analogous to a "schmoo" in yeast; the latter is formed in rich medium. In yeast, the diploid must be nutritionally restricted to enter into meiosis (Malone, 1990), which in Neurospora requires or at least tolerates abundant external nutrients (see below). The common theme is that the form believed to predominate in nature — haploid in Neurospora, diploid in yeast — must be nutritionally restricted as part of the sexual process.

FORMATION OF PERITHECIA

On fusion of a spermatium (e.g., a conidium) with the trichogyne of a proto-perithecium, the latter becomes a flask-like perithecium. This enlarges greatly and differen-tiates to produce internal structures like ascogenous hyphae, croziers, and paraphyses, and external structures like that of sclera and neck, or beak, and an ostiole. Glass & Lee (1992) have used RIP (Repeat-Induced Point Mutations — Selker et al.1987, 1989) to destroy the functions of a region within the A idiomorph, but outside the region of the A-1 reading frame — regions now designated the A-2 and A-3 reading frames (Glass, personal communication). Such RIP mutants are able to make quite normal-appearing perithecia complete with paraphyses, neck, and ostiole, but formation of croziers and asci is drastically reduced. This indicates that formation of these structures need not occur in lockstep with meiosis and ascus formation. Those few asci that are formed appear normal and complete, as if some critical step governing entry into caryogamy or meiosis has been impaired. If that stage is success-fully traversed, subsequent steps proceed uneventfully.

Mutants at the *peak* locus (abbreviated *pk*) cause abnormalities in a later stage of development. There is a disordering of the usual linear arrangement of eight ascospores to give a thin-walled sac in which the spores are mixed randomly (Srb et al., 1973). The mutations confer a distinctive vegetative phenotype as well, and it seems likely that cell wall composition is affected both in vegetative hyphae and in the ascus. Most though not all *pk* mutations appear to be fully recessive, so it is clear that the product encoded by *pk+* must be sufficient to influence the formation of the entire ascus. The rare dominant mutations (*Pk*) could be formally explained by postulating some form of negative complementation.

Several lines of evidence indicate that the sclera and neck are purely maternal tissue. DeLange & Griffiths (1980) described a mutant called *pen-1* (perithecial neck) which, in heterozygous crosses, forms perithecia lacking any neck or beak and without asci when the mutant allele is contributed by the protoperithecial parent, but is without any abnormal phenotype when the mutant allele is contributed by the conidial or spermatial parent. Similarly, a class of mutations at the *per-1* locus results in a colorless or pale yellow perithecium instead of the normal black color (Howe & Johnson, 1976), but only when *per-1* is the maternal parent. Johnson (1976) used heterocaryons of the constitution *per-1 + per-1⁺* to show that the scleral tissue of most perithecia arises from a single precursor cell, but that perithecia that do arise from more than one cell are of mottled or segmented color pattern. Thus pigment formation is determined locally and autonomously. Raju (1992), working with heterocaryons of *crosswallless-1* and *-2* strains reported that the non-ascogenous material making up the paraphyses is also maternally derived.

As a contrast, something akin to paternally-determined behavior is seen with three non-allelic male-barren mutants, termed *mb-1, mb-2,* and *mb-3.* (Vigfusson & Weijer, 1972). These are almost completely infertile in heterozygous crosses when they are presented as the male parent, but are normal or almost normal as the female parent (See, however, the personal communications by N. B. Raju cited in Perkins et al., 1982). This would seem to indicate that the male-derived nucleus per se has a role either in fertilization or in the progressive development of the ascogenous hyphae that is not duplicated by the female-derived nucleus. The *mb* mutants are recessive, in the sense that if a heterocaryon is prepared between the *mb* mutant and an *mb⁺* partner and that heterocaryon is used as a male parent in a cross, the *mb* nucleus can behave in the cross essentially as if it were wild type. This indicates that something furnished by the *mb⁺* component in the heterocaryon that acts as a male can be used by the *mb* nucleus, whereas this product is either not made by the female *mb⁺* partner, or is made but is unavailable to the *mb* nucleus. It would be extremely interesting if this were due to some necessary pre-fertilization differentiation of the spermatium analogous to the genetic imprinting which occurs during mammalian spermatogenesis (for review, see Hall, 1990), but there is no plausible evidence of this. Another possibility is that both components of a heterocaryon sometimes fertilize the same trichogyne, or different trichogynes of the same perithecium, and that some sort of complementation can occur in a mixed rosette of asci.

Another type of impairment of perithecial development occurs in a mutant called *fmf-1*, which virtually eliminates both female and male fertility in heterozygous crosses. This mutant, known from a single mutant allele, could be thought of in one sense as a perithecium-dominant or ascus-dominant mutant (Johnson, 1979). As in the case of the *mb* mutants, however, the appearance of simplicity disappears when same-mating-type heterocaryons, *fmf-1 + fmf-1⁺*, are studied. When such a heterocaryotic parent, of either mating type, is used as the *fmf-1* parent of a heterozygous cross, the *fmf-1* partner behaves like wild type in the cross. Again, it is as if an *fmf-1⁺* nucleus is able to confer fertility, in this case both as a male and as a female, to an *fmf-1* nucleus with which it is in partnership in a heterocaryon, but the *fmf-1* parent does not receive the complementing substance from its *fmf-1⁺* mate.

A different result — nucleus-autonomous expression — is seen in some other kinds of crosses in which one of the parents is a heterocaryon. Again, a bit of explanation is in order. A cross involving a homocaryon of a poorly-growing mutant to a more robust mate often fails because of the metabolic shortcomings of the disadvantaged partner. However, such a poor mater or non-mater can often be combined with an auxotroph that is disabled in its mating type idiomorph (Griffiths & DeLange, 1978; Perkins, 1984) on minimal medium to give a vigorously-growing, nutritionally-forced heterocaryon. One popular "helper" strain is aᵐˡ ad-3B cyh-1, in which the aᵐˡ represents a mutationally disabled mating type idiomorph of type a, the ad-3B confers an adenine requirement, and cyh-1 confers resistance to cycloheximide. Heterocaryons which include this "helper" usually mate entirely normally because of complementation of the deleterious allele by the "helper", and such a cross produces an abundant crop of ascospores. While the "helper" participates nutritionally and functionally in the cross, it does not seem to do so genetically. The "helper" does not participate in assortment or crossingover with the genomes of other potential parental nuclei, as shown by the absence of cycloheximide-resistant recombinants (Perkins, 1984; note, however, the existence of very rare, exclusively non-recombinant progeny — Griffiths & DeLange, 1978, confirmed by Randall & Metzenberg, unpublished observations). Progeny of the "helper" are absent or rare whether the heterocaryon is presented as the male or the female parent. The simple interpretation is that the mating-type idiomorph of the "helper", in particular the mutant gene within it, is not complementable by the normal mating type idiomorph of the disadvantaged mutant. Yet by definition, the disadvantaged mutant is helped by the "helper", so these two nuclear types must be in intimate communication. Clearly some functions are shared between nuclei while others are not. Why is the disadvantaged mutant not helped by its robust mating partner, as it is by its heterocaryotic "helper"? No simple answer to this question has emerged, and thinking in conventional terms like dominant and recessive is of limited use in such a situation.

Other mutations show various kinds of autonomous expression. Certain mutations at the *per-1* locus can give rise not only to colorless or pale yellow perithecia instead of the normal black coloration if the mutant is the protoperithecial parent (see above), but also to viable colorless or orange ascospores which fail to go into dormancy and fail to become heat-resistant (Johnson, 1976, 1978). The abnormal ascospores are those carrying the *per-1* mutation, showing that the phenotype is cell-autonomous.

A special case that formally could be classified as autonomous expression is that of the *Spore killer* strains, of which two linked mutants, *Sk-2* and *Sk-3*, are the most studied (see Turner & Perkins, 1991, for a review). In a heterozygous cross to most laboratory Neurospora strains, *Sk-2* kills all spores that do not contain it (i.e., those four spores that are sk-2 ^{s}, or sporekiller-2-sensitive). Symmetrically, *Sk-3* in a heterozygous cross kills those ascospores that do not contain it (sk-3 ^{s}). The *sporekiller* traits do not seem to have any comparable effect in the vegetative phase of the life cycle, and heterocaryons between killer and sensitive strains can be established without the sensitive component dying out; the killer effect is restricted to the sexual phase. The initial killer strains from nature were *N. intermedia*, not *N. crassa*, but the genetic region carrying the genes of interest have been introgressed into an otherwise *N. crassa* background for further study. However, the trait seems best thought of as a haplotype of linked genes, not as a single gene with various alleles. This haplotype has resisted fine structure analysis because crossingover in heterozygous crosses is completely suppressed over a chromosomal segment that corresponds to some 30 centimorgans. The physical appearance of the vulnerable sk-2 ^{s} ascospores formed in a heterozygous cross is informative. During the very early stages of ascus growth, before growth and darkening of the spores, the spores that will soon die cannot easily be distinguished from the four killer spores. The killer spores continue to grow and to develop pigment, however, so that after the second post-meiotic mitosis, each ascus is seen to contain four normal-looking black spores and four very small white spores. It is as though the developing killer spores produce something that kills the sensitive spores, or perhaps deprives them of some product essential for their further maturation. However, it it clear that the killer spores do not have to proceed to an advanced stage of development, or even to be viable themselves, to wreak their havoc on the sensitive spores. When *Sk-2* is crossed to *Sk-3* (i.e., *Sk-2 sk-3* ^{s} x *sk-2* ^{s} *Sk-3*) virtually all spores are killed at an early stage of development. This is at least still consistent with a notion of spore autonomy. For example, a nucleus of one killer genotype might be immune to its own biochemical weaponry because it does not contain the specific target for that weaponry. But once again, the simple model runs into trouble when certain kinds of complementation studies are done. A word of explanation is needed about these complementation studies, which are not of the conventional sort. There are several mutants which allow the experimenter to routinely produce ascospores that are heterocaryotic. One such mutant, to be discussed in a different light below, is *Banana* (abbreviated *Ban* — see discussion below). *Ban* is an ascus-dominant mutant in which the entire ascus, including the mitotic descendants of all four meiotic products, is cut out as a single multinucleate, banana-shaped ascospore. These ascospores are capable of germinating and giving rise to heterocaryotic mycelia, from which the constituent homocaryons can be isolated by plating conidia. When a mutant like *Sk-2* is crossed to a sensitive *Ban* strain, one might expect that the nuclei carrying *sk-2*^{s} would be killed, as they would have been in ordinary eight-spored asci, and only *Sk-2* nuclei would survive. But the actual result is very different. Even the sensitive nuclei survive, and homocaryons that contain them can be isolated (Raju, 1979). Alternative approaches, not involving *Ban*, can be used to generate ascospores that are heterocaryotic, i.e., *sk-2*^{s} + *Sk-2*, and the results are the same: the presence of *Sk-2* in the same ascospore as *sk-2*^{s} protects the *sk-2*^{s} from being killed. No completely satisfying model has been presented to explain either the mechanism of killing or the way in which sensitive spores can be protected. The point in introducing the subject of spore-killers here is to emphasize once again how our simplest notions of recessiveness, dominance, and autonomy initially seem to provide a useful conceptual framework, but disappoint us when they are examined more closely.

NUTRITION IN PERITHECIA AND IN ASCOSPORES

If a cross giving rise to a perithecium is homozygous for some auxotrophic marker, the simple increase in mass of the perithecium and the asci within it dictates that the required nutrient must be furnished exogenously. Can a perithecium from a homoallelic cross of an auxotroph take up nutrients rapidly enough to meet this need and allow the formation of viable ascospores? There seems to have been no systematic study of this question, but the answer is clearly "yes" for some auxotrophs and "no" for many others. For example, homoallelic crosses of inositol (*inl*) mutants mate and give normal ascospore production on medium of inositol concentration appropriate for ordinary vegetative growth. By contrast, homoallelic *lys-3* x *lys-3* crosses give only white, inviable spores (Ahmad et al. 1979), as do *lys-1* x *lys-1* (Kinsey, J., personal communication). It is notable that a common phenotype in these homoallelic crosses is a very deficient production of ascospores. This is also common in homozygous crosses of arginine mutants such as *arg-5* (Davis, 1979), *arg-12* (Davis & Thwaites, 1963, and Davis, personal communication), *arg-1* and *arg-10* (Perkins et al., 1982). Often those few that are produced do not turn black and are not capable of germinating. The formation of full-sized perithecia suggests that in many cases, uptake of the required nutrient into the perithecium itself is sufficient, but that the developing ascospore can not transport it from the perithecial milieu. In a similar manner, homozygous crosses of purine mutants like *ad-7* (Perkins et al., 1982) , and *ad-3A* and *ad-3B* (DeSerres, 1956; Griffiths, 1970) give mostly or exclusively pale, inviable spores. Yet homozygous (including homoallelic) *ad-8* x *ad-8* crosses produce viable progeny on adenine-supplemented crossing medium (Ishikawa, 1962), indicating that sufficient adenine does penetrate the ascospores to bring them to maturity. This apparent inconsistency may have a simple answer. *ad-8* is the only known "adenine locus" that is specific to adenine synthesis and is completely uninvolved in guanine synthesis (Perkins et al., 1982, see p. 440). It therefore seems possible, though unproven, that exogenous adenine is freely available to ascospores, but that these spores are unable to convert exogenous adenine to nutritionally necessary guanine derivatives. Homozygous crosses of the polyamine mutant *spe-1* made on ordinary concentrations of spermidine are very infertile, but in a different way. They produce copious ascospores which are mostly black but essentially completely inviable. From crosses on very high concentrations of spermidine, e.g., 5 mM, germinability of the spores is much higher, though in the absolute, still very low (Davis, R. H., personal communication). The perithecium, then, can be regarded as a nutritionally restricted system, but not a completely closed one.

RESIDUAL OUTGROWTH OF AUXOTROPHIC ASCOSPORES ON MINIMAL MEDIUM

Typically, a strain with an auxotrophic marker (say, arg-12) is crossed to one with the wild type allele, arg-12^+. Such a heterozygous cross can be prepared on minimal medium with arg-12^+ as the protoperithecial parent and the resulting progeny germinated and plated onto minimal medium. Under these conditions, the arg-12 sporelings grow out for the first 12 or more hours almost as vigorously as do arg-12^+ sporelings, and it is difficult to distinguish them visually. The auxotrophic sporelings then stop growing, however, showing that their residual growth is not due to "leakiness", but rather to phenotypic lag. The ability of arg-12 spores to germinate and grow out, or indeed even to develop to maturity can be thought of as a form of complementation in which the wild type allele rescues the mutant one. In principle, an arg-12 spore could receive from its prototrophic female parent either arginine, or the missing enzyme to make arginine (ornithine carbamoyl transferase), or an unprocessed proetin precursor of that enzyme, or the mRNA to make that enzyme, or any combination of these. Another arginine locus, arg-5, provides a useful contrast. In this case, auxotrophic spores from a heterozygous cross grow out very little when they are germinated on minimal medium and are almost immediately distinguishable from arg-5^+ sporelings (Metzenberg, R. L., unpublished observations). If a characteristic amount of arginine were routinely included in ascospores, arg-5 and arg-12 sporelings from heterozygous crosses should grow out to the same extent. The fact that they grow out quite different amounts indicates that their wild type partners have endowed them with different amounts of either the wild type enzyme, or of the wild type mRNA. In other words, a capacity to make arginine must, to a considerable extent, be a shared dowry until the time ascospore walls are laid down and the spore becomes impermeable to large molecules. The dowry will be a different size for the products of different genes. In extreme cases, the auxotrophic spores from heterozygous crosses may even fail to mature completely, with the very large majority of auxotrophic ascospores remaining unpigmented and incapable of germinating. cys-3 and lys-5 provide examples of such "spore-autonomous" white-spore traits (Perkins et al., 1982), yet cys mutants and lys mutants at other loci do not show this behavior. This suggests that the dowry provided by the wild type allele is very small or very labile.

The variable dowry that applies to metabolic pathways leading to low molecular weight nutrients can be expected to apply to pathways involving structural and regulatory molecules as well. Where the actions of such a molecule are executed very early (e.g., in the zygote, before the first meiotic division), or where the dowry given to the spores is large compared to the need for that molecule, any mutation for spore development would be seen as recessive in heterozygous crosses. Where the actions occur after formation of cell membranes and walls around the individual spores and where the dowry is small relative to the eventual requirement for the molecule in question, the mutation will be seen as spore-autonomous. There are good examples of both of these, as will be discussed immediately below. A number of genes are also known in which the effect of mutant alleles prevails over those of the wild type alleles — in other words, the mutations are formally ascus dominant. As will be considered after the discussion of recessive and spore-autonomous mutations, these ascus-dominants are not so simply explained.

Recessive mutations that prevent the formation of asci or the ripening of ascospores have been carefully investigated by classical procedures (DeLange & Griffiths, 1980). A simple case of recessiveness that results in morphologically abnormal asci is provided by the behavior of *asd-1* (ascus-development-1 — Nelson & Metzenberg, 1992). These workers used reverse-genetics in the form of subtractive hybridization to enrich for transcripts that were present in mixed-mating-type heterocaryons under conditions appropriate for mating, but were absent in vegetatively-grown cells. A number of the genes corresponding to these transcripts were inserted ectopically into the genome and the resulting transformants were crossed to the opposite mating type to effect pre-meiotic disruption of both the ectopic and endogenous copies by RIP. Homozygous crosses (*asd-1* x *asd-1*) produced no ascospores at all. Asci of approximately wild-type length and diameter were formed, but orderly migration of nuclei into the regions that normally would give rise to spores did not occur, and no ascospores were delineated. Yet crosses heterozygous for *asd-1* were indistinguishable from homozygous wild type crosses. The complete recessiveness of the mutation indicates that its product is efficiently shared by all the emerging meiotic products. Other presumably null RIP disruptants from the Nelson and Metzenberg experiments have been examined, and none has been found to be either spore-autonomous or dominant in crosses. It should be noted, however, that this could be a consequence of the use of RIP to produce the mutations. For example, a spore-autonomous RIP mutation that gave malformed, inviable ascospores would not survive the cross that was intended to generate it.

DEFICIENCY-BEARING STRAINS, INCLUDING DEFICIENCY OF THE NUCLEOLUS ORGANIZER AND rDNA

An instructive case of spore-autonomous lethality is offered by crosses of transloca-tion-bearing strains to those with normal sequence, which Perkins and his colleagues have explored in their monumental studies (Perkins & Barry, 1977, and other widely scattered works). The most conceptually simple case, perhaps, is that in which a terminal segment of a donor chromosome is translocated onto the terminal or non-terminal region of an acceptor chromosome without any essential genes being reciprocally transferred from acceptor to donor. Such a translocation is likely to involve the transfer of many genes and hundreds or even thousands of kilobase pairs of DNA from one chromosome to another. When such a translocation strain is crossed to Normal Sequence, independent assortment of chromatids will result in 1/4 of the progeny being like the Normal Sequence parent, 1/4 like the translocation parent, 1/4 being duplication-bearing progeny with the wild-type form of the donor chromosome and the mutant form of the acceptor chromosome, and 1/4 will be deficiency types carrying the mutant form of the donor chromosome and wild type form of the acceptor chromosome (see Figure 1). These deficiency strains are missing all the genetic material that was transferred away from the donor chromosome in the parental translocation strain, so it is not surprising that they are inviable. It is perhaps more surprising that the deficiency-bearing ascospores develop at all. Usually, though not always, the inviable segregants are easily recognized as white or very lightly pigmented ascospores that fail to mature; often these are smaller than normal spores as well. Viability, then, is "spore-autono-mous". An interesting variant on this, though not quite a unique one, is the situation in which the chromosome arm bearing the sole nucleolus organizer is translocated to another chro-mosome, and with it, all 180-odd copies of the tandemly repeated rDNA. Crosses of the paradigmatic translocation strain [called T(VR → IVR)AR33] produce the expected progeny with duplication or deficiency of all of the rDNA. The deficiency spores are, of course, inviable and fail to germinate. What is astonishing is that they do develop to full size and

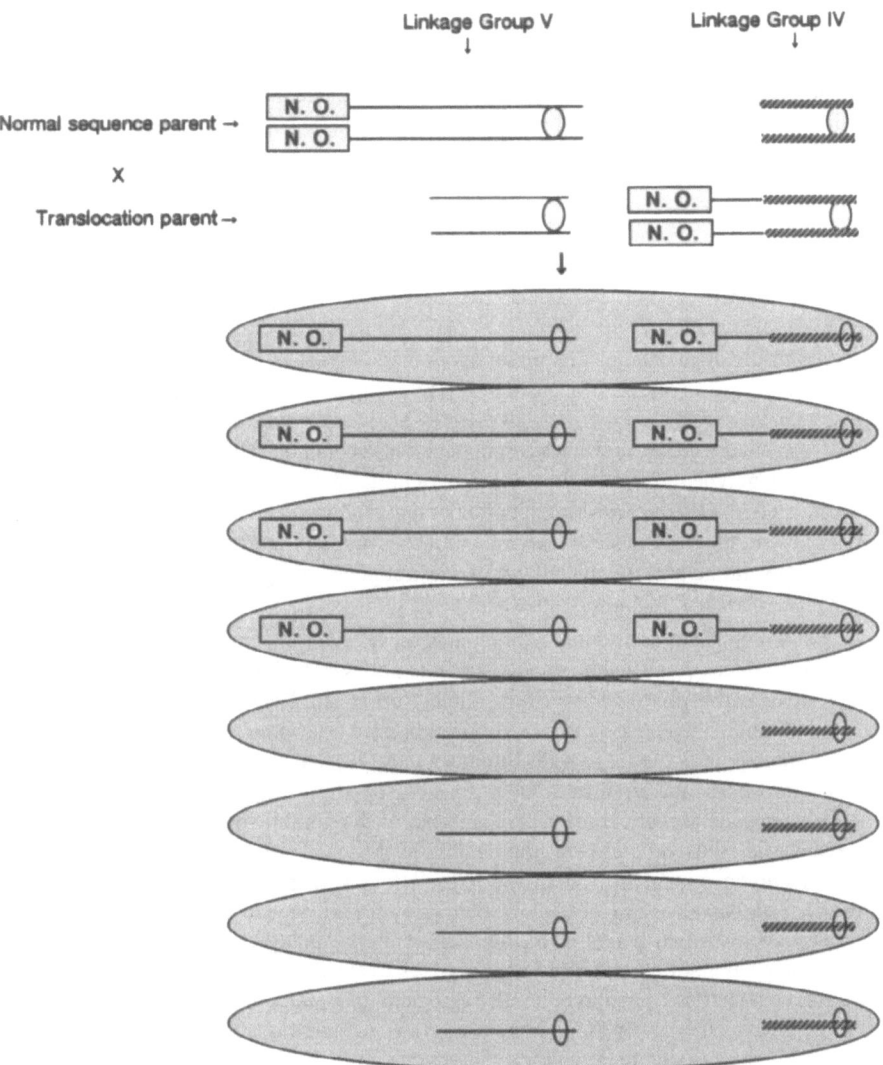

Figure 1. Cross of a Normal sequence parent to a Translocation parent in which the Nucleolus Organizer (N.O.) and some centromere-proximal chromosomal sequences are moved from Linkage Group V to Linkage Group IV. In the absence of crossingover, chromatid assortment will give eight parental-type ascospores in half the asci (not shown) and as shown below, non-parental ditype asci in which half the spores have a duplication of the N.O. and the other half are deficient in it. Centromeres are shown as ovals.

even turn black, sometimes being distinguishable from the other three classes of progeny only by their failure to germinate (Perkins and Barry, 1977; Metzenberg, unpublished). It seems inconceivable that ascospores could develop from naked nuclei in an octad to differentiated ascospores without any involvement of ribosomes. It must be that ribosomes from a common pool, perhaps in the zygote, are given as a dowry to all spores, including those that are unable to code for them. It is almost equally surprising that other genetic elements near the nucleolus organizer, which are also absent in the deficiency spores, are likewise unnecessary for growth and pigmentation of ascospores, or are present in sufficient amounts in the dowry to support this degree of development. The phenomena described here are not peculiar to one translocation strain involving the nucleolus organizer; translocations other than T(VR → IVR)AR33 that generate segregants deficient in the nucleolus organizer also produce inviable, black, normal-sized ascospores.

THE PARADOX OF ASCUS DOMINANT (ZYGOTE DOMINANT) MUTATIONS

A number of mutations alter either ascus or ascospore morphology or development even in heterozygous crosses. Some examples, by no means an exhaustive list, are Banana (Ban), Perforated (Pfr), Indurated ascus , abbreviated Iasc, Round ascospores (abbreviated R), and Asm-1 . As a group, they have certain features in common. They have some abnormal vegetative phenotype, generally slow, semi-colonial growth and tardy conidiation, and all are female-sterile, with the possible exception of Iasc, for which no result has been reported. In each of them, the penetrance of the mutant phenotype falls short of 100%, and for some is much lower. In the asci in which the trait was not penetrant, generally all the spores are phenotypically normal, including those of the mutant genotype.

Pfr, which yields asci somewhat similar to those of Ban except that they have multiple pores in their apices ("saltshaker appearance"), is not even viable unless sheltered in both sexual and vegetative phase by a wild type allele, Pfr+ (Raju, 1987). Iasc , like Ban and Pfr, gives asci in which the entire ascus is a single elongated ascospore, except that in Iasc , the ascus wall functions as the spore wall. But in some respects, the most informative mutant is the round ascospore mutant, R. In heterozygous crosses of this mutant, all eight ascospores are usually "round" (spherical, of course) instead of the normal spindle-shape, and in the rare exceptional ascus, all eight are spindle-shaped (Mitchell, 1966); in what appears to be the corresponding mutant in N. tetrasperma, all four spores in the exceptional asci are spindle-shaped (Novak & Srb, 1969). The outcome does not depend on which parent is male or female, nor on which parent is of the A or of the a mating type. This, and the other ascus-dominant mutations, could be explained as overproducers of some substance that causes the characteristic phenotype in all segregants of a particular meiotic event — i.e., they could be explained away as hypermorphs. However, this explanation runs into difficulties, at least in the case of the R mutation. Turner (1977) showed that in heterozygous partial diploids of the constitution R/r^+ , the wild type allele was dominant over the mutant, which is obviously not in accord with the hypermorph hypothesis. Furthermore, Jacobson (1992), studying strains which are most plausibly interpreted as new deletions in the region near the r^+ locus, found a high frequency of new round-spore mutations, with the typical pattern of ascus dominance. If these are small deletions, as they appear to be, and that deleting the r^+ locus results in the round spore phenotype, a less conventional model for ascus-dominant mutations is needed.

What mechanism can produce such results? One possibility is that R could be a mutant in which there is faulty communication between the nuclei in some stage of perithecial development, or faulty communication between chromosomes in the zygote. We can hypothesize that the default shape of ascospores is round and that this shape is modified as a result of a two-way exchange of information between the nuclei, or between homologs in the diploid zygote. One simple way a nucleus could "know" that it had received the signal from another nucleus is to receive a macromolecule processed in a way that could not have occurred in its own confines. Since all the nuclei in a perithecium are or can be isogenic except for at the mating type locus, non-identical processing of the r^+ - derived signal in the two could depend on differences at the mating type locus. What can be said about the time and place of the exchange of information? It seems unlikely to be any time after the first meiotic division; by that time, some meiotic events will have created every combination of mating types in coupling with r^+ and with R , and one could expect autonomous expression rather than ascus dominance. At least in the case of a highly similar-appearing mutation in the pseudohomothallic species, *N. tetrasperma*, having both mating type idiomorphs in the same spore with r^+ is not enough to produce spindle-shaped spores if the cross in question was heterozygous r^+/R. Novak & Srb (1969) reported that even in crosses that segregated four self-fertile $A + a$ ascospores, the ascospores were round; hence, having a full complement of information for both mating types in a spore plus wild type information at the r locus is not enough — the critical stage for determination of shape must already have been passed. A reasonable time and place for it seems to be in the pre-caryogamic nuclei of the ascogenous hyphae or croziers. It also seems possible for the critical stage to be in the zygote nucleus. Every necessary kind of genetic information is present in this nucleus even in a heterozygous cross. However, allele pairing between homologs may constitute a necessary signal. It will be a challenge to design experiments to test these alternatives.

We cannot yet reject the possibility that ascus dominant mutations like *Ban*, *Pfr* , and *Iasc* are simply hypermorphs, as we have with R , because there are no mutants known that are convincingly null. However, R does not stand as the sole example. Deletion mutants called *Asm-1* (ascus maturation) give a vegetative phenotype of slow, mat-like growth with delayed conidiation, not qualitatively unlike the strains described above (Aramayo, R., & Metzenberg, R. L., unpublished results). In heterozygous crosses to $asm-1^+$, ascospores are produced copiously, but nearly 100% of them are white and inviable, though of normal shape. The rare, viable black spores that have been tested have always proved to be $asm-1^+$ except for a single heterocaryotic segregant which must have come through meiosis as a heterozygous disomic. The vegetative phenotype of heterocaryons between $asm-1^+$ and *Asm* is that of wild type, as befits the complementation of a deletion mutant.

DISCHARGE OF ASCOSPORES

General aspects of ascospore discharge appear to be very similar in the genera *Neurospora*, *Gelasinospora*, and *Sordaria*, but have been most extensively studied in *Sordaria* (Ingold, 1971). A ripe ascus approaching its moment of discharge will become positioned with its apex projecting through the neck canal and into the ostiole of the perithecium. The ascus becomes very much elongated, with the eight spores crowded toward the tip as if they were under pressure from fluid or gas that has accumulated in the base of the ascus. Finally, the spores are liberated explosively, being shot several centimeters upward from the perithecium. In nature, they are probably lofted by breezes and thus carried to new sites. The spores are usually discharged in the direction of the light, at least partly because the neck of the perithecium is phototropic. The empty envelope from which the spores had been shot is retracted back into the perithecium and soon degenerates, while the next ascus in line for discharge is prepared for its turn.

The motive engine behind ascospore discharge is unclear. Ingold (1971) emphasized that "in sordariaceous fungi packing is so tight that there is virtually no free space" and that there is no gas phase present. Presumably fluid pressure from the paraphyses, especially osmotic pressure, plays a role. However, a simple aqueous phase, which should be almost incompressible, seems an implausible way of imparting the observed muzzle velocity to an octad of ascospores. Their situation may be analogous to that of a cork departing from a bottle of champagne. Here the driving force is obviously the expansion of a gas, carbon dioxide; yet an observer describing an unopened bottle of champagne might say there was no gas phase at all, or a neglegible one compared to the liquid phase. Mutations that prevent individual spores from being cut out during the formation of asci, e.g., *Banana* (Raju & Newmeyer, 1977) or *Perforated* (Raju, 1987) do not prevent the giant spores/asci from being forcibly discharged. Abnormal morphology of the ascus in homozygous crosses of the *pk-1* mutant interferes with ascospore discharge, but probably only mechanically because no apical pore is present (Srb et al. 1973). Ascospores are neither shot nor exuded. In contrast, Maling (1960) reported that crosses homozygous for the *crisp-1* mutation (*cr-1* x *cr-1*) exude intact asci from the perithecial ostioles rather than discharging them forcibly. Esser & Straub (1958) have also described a mutant (*n*, for *non eiaculans*) of *Sordaria macrospora* that does not discharge its ascospores; it is not clear whether it has the same genetic basis as *cr-1*. In the case of *cr-1*, there is useful molecular information on the nature of the lesion. The mutant is deficient in adenyl cyclase, and the morphological and nutritional phenotype of the mutant can be abolished or strongly modified by exogenous cAMP (Terenzi et al., 1979) and by a plasmid carrying the gene coding for this enzyme (Kore-eda, Murayama & Uno, 1991a, b). The dependency of spore discharge on cAMP suggests that it could be process controlled by a protein kinase A-mediated signalling pathway.

SUMMARY

The developing perithecium, asci, and ascospores offer an attractive system for studying mechanisms of differentiation in a well-characterized filamentous fungus. The perithecium appears to be neither fully closed nor fully open to exogenous nutrients. The classical concepts of dominance, recessiveness, complementation, and cell autonomy offer useful models for interpreting some of the known facts, but not others. New insights are likely to be necessary for explaining phenomena such as "ascus dominance".

ACKNOWLEDGEMENTS

I am grateful to Rodolfo Aramayo, especially for stimulating discussions about protoperithecia, and for a reading of this manuscript. This work was supported by U.S.P.H.S. grant GM08995 to the author.

REFERENCES

Ahmad, M, Mozmadar, M., Baset, A., Fayaz, M., Rahman, M. A. and Saha, B., 1979. Studies on the organization of genes controlling lysine biosynthesis in Neurospora crassa. III. Studies on the organization of loci *lys-3* and *lys-4*. *Pak. J. Bot.* 11:179-184.

Bistis, G. N., 1983, Evidence for diffusible, mating type-specific trichogyne attractants in *Neurospora crassa*. *Exptl. Mycol.* 7:292-295.

Davis, R. H., 1979, Genetics of arginine biosynthesis in *Neurospora crassa*. *Genetics* 93:557-575.

Davis, R. H., and Thwaites, W. M. 1963, Structural gene for ornithine transcarbamylase in Neurospora. *Genetics* 48:1551-1558.

Davis, R. H., and DeSerres, F. J., 1970, Genetic and microbiological research techniques for *Neurospora crassa*. *Methods in Enzymol.* 27A:79-143.

Degli Innocenti, F., Pohl, U., and Russo, V. E. A., 1983, Photoinduction of protoperithecia in *Neurospora crassa* by blue light, *Photochem. Photobiol.* 37:49-51.

DeLange, A. M., and Griffiths, A. J. F., 1980. Meiosis in *Neurospora crassa*. 1. The isolation of recessive mutants defective in the production of viable ascospores. *Genetics* 96:367-378.

DeSerres, F J , 1956, Studies with purple adenine mutants in *Neurospora crassa* I Structural and functional complexity in the *ad-3* region *Genetics* 41 668-676

Esser, K , and Straub, J , 1958, Genetische Untersuchungen an *Sordaria macrospora* Auersw , Kompensation und Induktion bei genbedingten Entwicklungsdefekten, *Z Vererbungslehre* 80 729-746

Fairfield, A , and Turner, B C , 1993, Substitution of paper for sucrose can reverse apparent male sterility in Neurospora *Fungal Genet Newsl* 40 31-32

Frost, L C and Greenhill, A D , 1963, Multispored asci in *Neurospora crassa* *Neurospora Newsl* 4 6-7

Glass, N L , and Lee, L , 1992, Isolation of *Neurospora crassa A* mating type mutants by repeat induced point mutation *Genetics* 132 125-133

Glass, N L , Vollmer, S J , Staben, C , Grotelueschen, J , Metzenberg, R L , and Yanofsky, C , 1988, DNAs of the two mating-type alleles of *Neurospora crassa* are highly dissimilar *Science* 241 570-573

Griffiths, A J F , 1970, Topography of the *ad-3* region of *Neurospora crassa* *Can J Genet Cytol* 12 420-424

Griffiths, A J F and DeLange, A M , 1978, Mutations of the *a* mating time in *Neurospora crassa* *Genetics* 88 239-254

Hall, J G , 1990, Genomic imprinting review and relevance to human disease *Am J Hum Genet* 46 857-875

Howe, H B , Jr and Johnson, T E , 1976, Phenotypic diversity among alleles at the *per-1* locus of *Neurospora crassa* *Genetics* 82 595-603

Ingold, C T , 1971, Fungal spores Their liberation and dispersal Clarendon Press, Oxford

Ishikawa, T , 1962, Genetic studies of *ad-8* mutants in *Neurospora crassa* I Genetic structure of the *ad-8* locus *Genetics* 47 1147-1161

Jacobson, D J , 1992, New round spore mutations in *Neurospora crassa* accompanying changes in a duplication closely linked to the *R* locus *Fungal Genet Newsl* 39 24-27

Johnson, T E , 1976, Analysis of pattern formation in Neurospora perithecial development using genetic mosaics, *Developmental Biol* 54 23-36

Johnson, T E , 1978, Isolation and characterization of perithecial development mutants in Neurospora *Genetics* 88 27-47

Johnson, T E , 1979, A Neurospora mutation that arrests perithecial development as either male or female parent, *Genetics* 92 1107-1120

Kore-eda, S , Murayama, T , and Uno, I , 1991a, Suppression of the *cr-1* mutation in *Neurospora crassa, Japan J Genet* 66 77-83

Kore-eda, S , Murayama, T , and Uno, I , 1991b, Isolation and characterization of the adenylate cyclase structural gene of *Neurospora crassa, Japan J Genet* 66 317-334

Maling, B D , 1960, Replica plating and rapid ascus collection of Neurospora, *J Gen Microbiol* 23 257-260

Malone, R E , 1991, Dual regulation of meiosis in yeast *Cell* 61 375-378

McNelly-Ingle, C A , and Frost, L C , 1965, The effect of temperature on the production of perithecia by *Neurospora crassa J Gen Microbiol* 39 33-42

Metzenberg, R L and Glass, N L , 1990, Mating type and mating strategies in *Neurospora BioEssays* 12 53-59

Mitchell, M B , 1966 A round-spore character in *N crassa Neurospora Newsl* 10 6

Mitchell, M B and Mitchell, H K , 1952, A case of maternal inheritance in Neurospora crassa Proc Nat Acad Sci U S A 38 442-449

Nelson, M A , and Metzenberg, R L , 1992, Sexual development genes of *Neurospora crassa Genetics* 132 149-162

Novak, D R , and Srb, A M , 1969,Spore and ascus mutants in *N tetrasperma, Neurospora Newsl* 15 23

Perkins, D D , 1984, Advantages of using the inactive-mating-type *a^{m1}* strain as a helper component in heterokaryons, *Neurospora Newsl* 31 41-42

Perkins, D D , and Barry, E G , 1977, The cytogenetics of Neurospora, *Adv Genet* 19 1133-285

Perkins, D D , Radford, A , Newmeyer, D , and Bjorkman, M , 1982, Chromosomal loci of *Neurospora crassa Microbiol Rev* 46 426-570

Raju, N B , 1979, Cytogenetic behavior of spore killer genes in Neurospora, *Genetics* 93 607-623

Raju, N B , 1987, A *Neurospora* mutant with abnormal croziers, giant ascospores, and asci having multiple apical pores *Mycologia* 79 696-706

Raju, N B , 1992, Genetic control of the sexual cycle in Neurospora *Mycol Res* 96 241-262

Raju, N B , and Newmeyer, D , 1977, Giant ascospores and abnormal croziers in a mutant of *Neurospora crassa, Exp Mycol* 1 152-165

Reich, E , and Luck, D J L 1966, Replication and inheritance of mitochondrial DNA *Proc Nat Acad Sci U S A* 55 1600-1608

Ricci, M., Krappmann, D., and Russo, V.E.A., 1991, Nitrogen and carbon starvation regulate conidia and protoperithecia formation of *Neurospora crassa* grown on solid media. *Fungal Genet. Newsl.* 38:87-88.

Rothschild, H., and Suskind, S. R., 1966,Protoperithecia in Neurospora crassa: technique for studying their development. *Science* 154:1356-1357.

Selker, E.U., Cambareri, E., Garrett, P., Jensen, B., Haack, K. , Foss, E., Turpen, C., Singer, M. and Kinsey, J., 1989, Use of RIP to inactivate genes in *Neurospora crassa. Fungal Genet. Newsl.* 36:76-77.

Selker, E.U., Cambareri, E. B., Jensen, B. C. and Haack, K. R., 1987, Rearrangement of duplicated DNA in specialized cells of Neurospora. *Cell* 51:741-752.

Sommer, Th., Chambers, J. A. a., Eberle, J., Lauter, F. R., and Russo, V. E. A. , 1989, Fast light-regulated genes of *Neurospora crassa, Nucl. Acids Res.* 17:5713-5723.

Srb, A. M., Basl, M., Bobst, M., and Leary, J. V., 1973, Mutations in *Neurospora crassa* affecting ascus and ascospore development, *J. Hered.* 64:242-246.

Terenzi, H. F., Jorge, J. A., Roselino, J. E., and Migliorini, R. H., 1979, Adenyl cyclase deficient *cr-1* (crisp) mutant of *Neurospora crassa*: cyclic AMP-dependent nutritional deficiencies. Arch. Microbiol. 123:251-258.

Turner, B. C., 1977, Euploid derivatives of duplications from a translocation in Neurospora. *Genetics* 85:439-460.

Turner, B. C., and Perkins, D. D., 1991, Meiotic drive in *Neurospora* and other fungi, *Am. Naturalist* 137:416-429.

Vigfusson, N. V. and Weijer, J., 1972, Sexuality in *Neurospora crassa*. II. Genes affecting the sexual development cycle. *Genet. Res.* 19:205-211.

Vollmer, S. J. and Yanofsky, C., 1986, Efficient cloning of genes of *Neurospora crassa. Proc. Nat. Acad. Sci. U.S.A.* 83:4869-4873.

Westergaard, M., and Mitchell, H. K., 1947, Neurospora V. A synthetic medium favoring sexual reproduction, *Am. J. Botany* 43:573-577.

Yang, X., and Griffiths, A. J. F., 1993, Male transmission of linear plasmids and mitochondrial DNA in the fungus *Neurospora, Genetics* 134: 1055-1062.

9

GENETICS OF ECTOMYCORRHIZAL FUNGI AND THEIR TRANSFORMATION

R. Marmeisse, G. Gay, and J. C. Debaud

Université Claude Bernard Lyon I
Laboratoire d'Ecologie Microbienne du Sol (URA CNRS 1977)
Bâtiment 405, 43 Boulevard du 11 Novembre 1918
F-69622 Villeurbanne
France

INTRODUCTION

The ectomycorrhizal symbiosis between fungal species and the roots of woody plants is a widespread association especially in temperate and boreal ecosystems, but also in arid and tropical ones. It has long been recognized that such associations increase the fitness of the host plants (see Harley and Smith, 1983) and the most studied property is the improvement of the host-plant mineral nutrition. One of the most simple models used to explain this improvement takes into consideration the ability of the fungal hyphae to explore a larger volume of soil than the host plant root system as well as the ability of the fungus to solubilize and assimilate nutrient sources normally not accessible to the plant (e.g. proteins or some insoluble forms of phosphate). The nutrients assimilated by the fungus are translocated through the extramatrical hyphal network and transferred to the plant which, in exchange, supplies the fungus with part of its photosynthates. Other beneficial effects of the symbiosis on the host plant include protection against root pathogens, heavy metals and/or drought (see Harley and Smith, 1983). Altogether, these fungal properties increase the survival rate of plants in extreme environments. Another reason for economic interest in the ectomycorrhizal symbiosis concerns the fungus itself: several of the most valuable edible mushrooms (truffles, *Boletus spp.*, chanterelles) are mycorrhizal fungal species whose sporocarps have to date only been only collected in the field under host plants which appear necessary for the formation of the fruit bodies.

Although some aspects of the physiology and biochemistry of mycorrhizal fungi have been extensively studied, genetic study and genetic manipulation of the fungal partner could greatly improve our understanding of the mycorrhizal symbiosis. Selection of performant fungal strains can be done through the realization of breeding programmes. Most of the fungal properties which have to be improved are complex characters under polygenic control. Quantitative genetic analyses need to be carried out in order to understand how these characters are inherited and to choose the most appropriate selection strategy.

Biotechnology of Ectomycorrhizae, Edited by Vilberto Stocchi et al.
Plenum Press, New York, 1995

In order to precisely evaluate the relative importance of, for instance, a fungal metabolic pathway implicated in nutritional interactions between the two partners of the symbiosis, a genetic dissection of the different steps of this pathway can be carried out. This includes the production and the characterization of modified strains altered in either the structural or regulatory genes of this metabolic pathway. Modified strains include not only mutants but also transformants resulting from the introduction into their genomes of one or several copies of a cloned gene. The phenotype of the mycorrhizae formed by a modified strain and the corresponding wild-type can then be compared, and the function of the gene whose activity has been modified can be evaluated.

Genetic studies have, until recently, seldom been carried out on mycorrhizal fungi. This can be explained by the lack of suitable fungal model species which could easily be manipulated. Among the limitations for genetic research, the most frequently encountered with ectomycorrhizal fungi are their slow growth rates, their unknown mating systems and the impossibility to obtain their sexual stage under laboratory conditions.

MODEL SPECIES FOR GENETICAL AND MOLECULAR ANALYSES

General Considerations

There is no strict definition of a good model mycorrhizal species, but it should fulfill at least two basic requirements. It should firstly be easily grown in pure culture and the basic features of its life cycle must be known. Secondly, the model organism must share with other mycorrhizal species some common characteristics so that the conclusions drawn from the investigations carried out on the model organism also apply to other symbionts or at least can be used to design new experiments to be carried out on comparable organisms. Ectomycorrhizal and ericoid fungi are the main types of fungal root symbionts which fulfill the first requirement.

Although great differences do exist between mycorrhizal fungi, a large majority of them share some basic features: they all associate with plant roots, with some exceptions they have a broad host range and they improve, to varying levels, their host plant mineral nutrition. Peterson and Farquhar (1994) were able to divide the process which leads to the differentiation of the mycorrhiza into seven steps (starting at the contact between the hyphae and the root surface and finishing with the establishment of the nutrient exchange interface), observed in any mycorrhizal type (arbuscular, ecto-, ectendo-, ericoid, orchid, arbutoid or monotropoid). Although the underlying cellular and molecular mechanisms which intervene in the realization of each of these events certainly differ between the different mycorrhizal types, they might be similar for species belonging to the same type. For example, all ectomycorrhizae are characterized by a fungal sheath surrounding the root and by a Hartig net developing around the host cortical cells. Some more subtle common features are also shared by taxonomically unrelated ectomycorrhizal fungi, such as the increased branching of the mycorrhizal roots and the increase in size of the root cortical cells in close contact with the fungal hyphae, a morphogenetic effect described for both angiosperm (Peterson and Farquhar, 1994) and gymnosperm species (Gay et al., unpublished results).

Numerous studies carried out on different ectomycorrhizal species also show that most of them present similar functional properties, particularly in terms of growth improvement of their host plant, the differences being mostly quantitative rather than qualitative (see Harley and Smith, 1983). Hence, it can be estimated that results obtained with a given model organism should generally apply to another species. Therefore there is no absolute necessity to choose, from a functional point of view, one given species as a model organism to carry out genetical and molecular studies. In contrast, significant differences exist between ectomycorrhizal species in the ease with which they can be manipulated in the laboratory and also the knowledge we have of their life cycles and sexual comportment.

The best known filamentous fungal species used in genetics, the ascomycetes (e.g. *Aspergillus nidulans*, *Neurospora crassa* or *Podospora anserina*) and basidiomycetes (*Coprinus cinereus* or *Schizophyllum commune*), share some of the following characteristics which have undoubtedly contributed to their recognition as model organisms (see Esser and Kuenen, 1967; Fincham and Day, 1963; Perkins, 1991):

- Fast mycelial growth rates on simple minimal media allowing the building up of a large biomass in a minimum of time
- Extended haploid mycelial phase with the production of uninucleate conidia (with the exception of *Podospora* sp. and *S. commune*) allowing the direct selection of recessive mutations
- Known mating system allowing crosses between different strains to be made and formation of the fruit bodies *in vitro*. The obtaining of viable meiotic spores allows genetic analyses to be performed and genetic maps to be constructed (see King, 1974).

These characteristics have also facilitated the use of these species as model organisms for molecular studies. A comparison of the characteristics listed above with the corresponding features of most ectomycorrhizal species shows that very few of them are good for genetic and molecular studies. There is however a handful of exceptions such as the basidiomycetes *Laccaria laccata*, *L. bicolor* and *Hebeloma cylindrosporum*.

Basidiomycete Life Cycles and Breeding Systems

The life cycles of many saprophytic and parasitic homobasidiomycetes have been described and several breeding systems have been characterized (see Raper, 1966 and Kühner, 1977). Homobasidiomycete species can either be homothallic or heterothallic. Studies on ectomycorrhizal species have shown that they can belong to either of these classes.

The life cycle of a typical heterothallic species is divided into two phases (fig. 1). The germination of haploid basidiospores (meiospores) gives rise to haploid homokaryotic mycelia (with all nuclei genetically identical). When the cells are uninucleate (except sometimes for the apical ones) these mycelia are called monokaryons as in the case of *H. cylindrosporum* (Debaud *et al.*, 1986). Mating between two compatible homokaryons gives a heterokaryon, the cells of which contain the two different haploid nuclei from the parental homokaryons. The cells of these mycelia are typically binucleate (dikaryons) and, for some species, a clamp connection, formed during the synchronous division of the two nuclei, can be observed on each septum (as in the case of *H. cylindrosporum* or *L. laccata*). The fruit bodies differentiate on the dikaryon, the nuclear fusion between the two parental haploid nuclei takes place in the basidia and is immediately followed by meiosis. The recombined haploid nuclei migrate into the basidiospores which form outside the basidia. Among mycorrhizal fungi, the reproducible formation of sporophores under axenic conditions has only been mastered in the case of *H. cylindrosporum* (Debaud and Gay, 1987). Sporulating fruiting bodies of this species only differentiate after the dikaryon has formed mycorrhizae with *Pinus pinaster* seedlings. Many different wild dikaryons of this species have been induced to fruit as well as dikaryons resulting from laboratory crosses between wild type, mutant or transformed monokaryons (Debaud and Gay, 1987; Durand *et al.*, 1992 and unpublished results from the authors' laboratory). For other ectomycorrhizal species, fruit

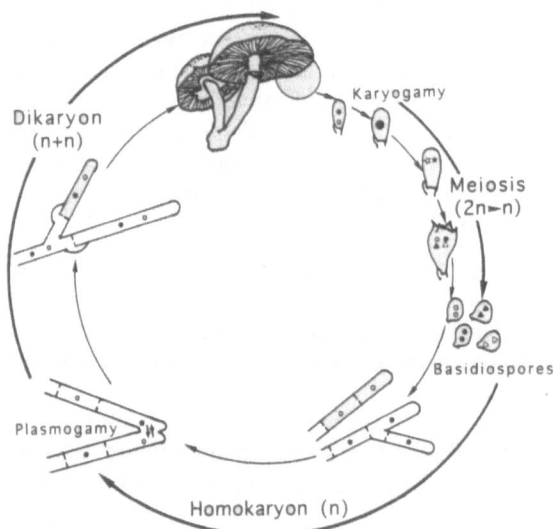

Figure 1. Life cycle of a heterothallic basidiomycete species.

body formation has either never been obtained or has only been obtained with some success under greenhouse conditions after non-sterile inoculation of the host plant as in the case of *L. bicolor* (Godbout and Fortin, 1990) or *L. laccata*.

Compatibility between homokaryotic mycelia is genetically controlled by mating-type genes. Two homokaryons are compatible when they have different alleles at their different mating-type loci. Some heterothallic homobasidiomycete species have a single mating-type locus (usually called *A*) and are called bipolar: only two mating-types segregate in the homokaryotic progeny of a dikaryon (Fig. 2 I). Other species have two mating-type factors (called *A* and *B*) and are called tetrapolar: four different mating-types segregate in the progeny of a fruiting dikaryon (Fig. 2 II). These breeding systems are also characterized by the existence of multiple alleles at the mating type loci. In the tetrapolar species *S. commune*, which has an almost worldwide distribution, Raper (1966) estimated that as many as 288 and 81 alleles at respectively the *A* and the *B* loci may exist. When two dikaryons of a tetrapolar species have different alleles at both their *A* and *B* loci, any cross between a

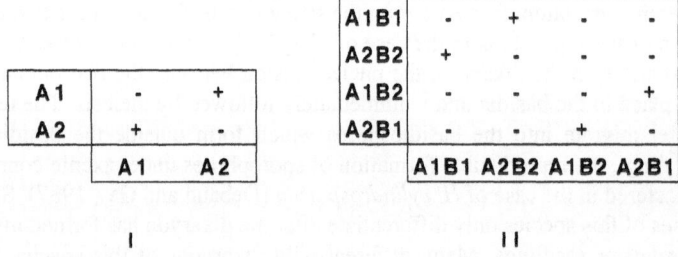

Figure 2. Mating-type segregation within the sexual progeny of (I) a bipolar species and (II) of a tetrapolar species. + and - indicate mating compatibility and incompatibility between the monokaryons belonging to the different mating types.

homokaryon belonging to the progeny of one of the dikaryons and a homokaryon belonging to the progeny of the other one will give rise to a dikaryon.

Among ectomycorrhizal basidiomycetes several species have been shown to have a typical tetrapolar breeding system with multiple alleles at the mating-type loci. This is the case for several *Laccaria* species (*L. laccata, L. amethystina, L. bicolor* and *L. proxima*) (Fries and Mueller, 1984), for *H. cylindrosporum* (Debaud *et al.*, 1986) or for *Pisolithus arhizus* (Kope and Fortin, 1990). Other species have typical bipolar mating systems as for example *Suillus luteus* and *S. granulatus* (Fries and Neumann, 1990).

Molecular analyses of the mating-type genes have been undertaken in the two tetrapolar model species *C. cinereus* and *S. commune* and the *A* mating-type locus from these two organisms has been cloned (Mutasa *et al.*, 1990; May *et al.*, 1991; Giasson *et al.*, 1989). Mating-type genes have not yet been characterized in ectomycorrhizal species but given the similarities found between the mating behaviours of many different homobasidiomycetes at both the morphological and genetic levels (Raper, 1966), we can imagine that the similarities already found at the molecular level between *C. cinereus* and *S. commune* (Casselton and Kües, 1994) could also be found in most of the symbiotic species having a bifactorial mating system.

QUANTITATIVE GENETIC VARIATION

Most quantitative traits vary continuously between individuals sampled in a natural population. Such characters are said to be under polygenic control, i.e. the expression of several genes controls the final phenotype which can be measured.

In the case of ectomycorrhizal fungi, most of their characters supposed to play an important role in the improvement of their host plant growth or fitness are likely to be under polygenic control. Thus, if these characters are to be improved, breeding programmes should be carried out. The following sections will review, firstly, the most appropriate experimental designs and mathematical models to study quantitative genetic variation in heterothallic fungal species and, secondly, give examples of such analyses carried out with ectomycorrhizal species.

Quantitative Genetic Analyses

The genetic control of a quantitative character is studied after performing a series of controlled crosses between known parental lines and measuring the quantitative values of the resulting hybrid lines. When working with basidiomycete species, a parental (breeding) line can either be a dikaryotic strain or even a single homokaryotic strain (the equivalent of a gamete in higher plants) which can be propagated indefinitely in pure culture. Hybrid lines are dikaryotic mycelia resulting from laboratory crosses between these 'parental' mycelia. For these organisms (table 1), n x n' crosses between a group of n lines of mating-type X and a group of n' lines of mating-type Y (X compatible with Y) are the preferred experimental design (see Caten, 1979).

The numerical values of the measurements made on the n x n' hybrid strains can be used in a two-way analysis of variance and estimates of the different components of the total (i.e. phenotypic) variation observed can be computed (table 2). The total phenotypic variation (σ^2_p) is the sum of the environmental (or residual: σ^2_e) and genotypic (σ^2_g) variations. The estimated genotypic variance is itself the sum of two components: the additive one (σ^2_a resulting from the additive effects of the n+n' parents) and σ^2_i, the interactive part of the variation resulting from the interactions in the hybrid lines between the genomes of the different homokaryons. These different components can be used to

Table 1. Factorial crosses between a group of n monokaryons of
mating-type X (monokaryons X1 to Xn) and a group of n'
monokaryons of mating-type Y (Y1 to Yn')

Mating-type Y	Mating-type X			
	X1	X2	...	Xn
Y1	X1 x Y1	X2 x Y1	... x Y1	Xn x Y1
Y2	X1 x Y2	X2 x Y2	... x Y2	Xn x Y2
...	X1 x ...	X2 x ...	...	Xn x ...
Yn'	X1 x Yn'	X2 x Yn'	... x Yn'	Xn x Yn'

calculate other parameters such as heritability coefficients: broad sense heritability $H^2 = \sigma^2_g/\sigma^2_p$ and narrow sense heritability $h^2 = \sigma^2_a/\sigma^2_p$.

From a practical point of view, the relative contributions of these different effects (additive, interactive and environmental) to the total phenotypic variation can be used to devise the most effective selection strategy in order to improve the studied character. When the additive component contributes to most of to the total variation (h^2 value high), the performance of a hybrid line can, to some extent, be deduced from the performances of its two parental ones. Therefore, in this case, it is advisable to study many potential breeding lines and to select the "best" ones as parental lines. On the contrary, a strong contribution of the interactive effects means that, in most cases, the value of a hybrid line cannot be predicted from the analysis of its parents and many independent crosses will have to be performed to select improved strains.

These interpretations should be taken with caution: at least two restrictions in particular should be considered (see Falconer, 1981). Firstly, the heritability coefficients can significantly vary according to the origins of the parental lines used to make the crosses. For instance, it is often the case that the more genetically related the parental lines are, the lower the heritability values, and vice versa. Secondly, as is often the case with higher plants (see Silvertown and Lovett Doust, 1993), the value of a character can dramatically vary from one environment to another and hybrid lines should best be tested under different environmental conditions.

Quantitative Genetic Variation in Ectomycorrhizal Fungi

Variations in Metabolic Activities between Strains of H. cylindrosporum. As previously mentioned, ectomycorrhizal fungi generally improve host plant mineral nutrition. This is for example the case for phosphate and nitrogen nutritions. It can be postulated that the ability of a fungal strain to assimilate a particular mineral form of phosphate or nitrogen

Table 2. Analysis of variance for n x n' dikaryons (as obtained in table 1)
z: number of replicates, $\sigma^2_a = \sigma^2_X + \sigma^2_Y$ (total additive variance, see text)

Source of variation	Degrees of freedom	Expected mean square
Between monokaryons X (additive)	n-1	$\sigma^2_e + n \times \sigma^2_i + n \times n' \times \sigma^2_X$
Between monokaryons Y (additive)	n'-1	$\sigma^2_e + n \times \sigma^2_i + n \times n' \times \sigma^2_Y$
X x Y interaction	(n-1) x (n'-1)	$\sigma^2_e + n \times \sigma^2_i$
Environmental	n x n' x (z-1)	σ^2_e

(ammonium or nitrate) depends *pro parte* on the activity levels of key enzymes implicated in the transformation of these compounds, as for example acid Phosphatases (Pases) for phosphates, of the NADP-Glutamate Dehydrogenase (GDH) for ammonium (see Martin and Botton, 1993) or of the Nitrate Reductase (NR) for nitrate. The efficiency of an ectomycorrhizal fungus also depends on its ability to form a large number of mycorrhizae on the root system. This mycorrhization ability is related, to some extent, to the production of auxin by the fungus (Gay *et al.*, 1994b, and paragraph, this chapter, on genetic analyses). For these reasons, the intraspecific variation of these different activities was studied in different populations of mono- and dikaryotic mycelia of *H. cylindrosporum*.

Indole-3-acetic acid production (Gay and Debaud, 1987), GDH activity (Wagner *et al.*, 1988), NR activity (Wagner *et al.*, 1989) and Pase activity (Meysselle *et al.*, 1991) were studied within several populations of strains whose mycelia were grown in pure culture. An estimate of the natural variability was made using a group of eleven wild dikaryons collected in different places in Les Landes forest, SW France. From one of these fruiting dikaryons (HC1), 20 sib-monokaryotic progenies were selected, five for each mating type: A1B2, A2B1, A1B1 and A2B2. The different activities were measured for each of the 20 monokaryons. These different monokaryons were crossed in all compatible combinations to give two groups of 25 (=5x5) synthesized dikaryons. The variations observed within the populations of synthesized dikaryons were used to estimate the different components of the phenotypic variation and to study the mode of inheritance of the four physiological activities measured.

For all of these activities, high levels of variation were recorded in the different groups of wild dikaryotic strains, sib-monokaryotic mycelia and synthesized dikaryons. For example, in the case of NR specific activity (Wagner *et al.*, 1989), more variability was observed within the group of 20 sib-monokaryons (the highest activity was ten times the lowest one with a mean value of 211 nmol NO_2^- synthesized $h^{-1}mg^{-1}$) than in the groups of synthesized dikaryons or of wild ones (ratios of highest to lowest activities of respectively 9.6 and 3.48 and mean values of 344 and 345 nmol NO_2^- synthesized $h^{-1}mg^{-1}$). These results show that the magnitude of physiological variability in *H. cylindrosporum* does not correlate with genetic relationships between the strains.

From the results obtained with the two groups of 25 synthesized dikaryons, the different components of the total phenotypic variation were estimated using the biometrical model described in the previous section. In most cases the environmental component of the variation represented *ca.* 10% of the total observed variation except in one case for auxin production where it accounted for almost 60% of it (Gay and Debaud, 1987). The genetic variation could be partitioned into additive (parental) and interactive components, the relative parts taken by each of them varying considerably according to the character. For GDH activity and IAA production about half of the genetic variation resulted from parental effects, the other half being interactive effects between the genomes of the 'parental' homokaryons. These estimates contrast with the results obtained with the NR activity where the parental effects accounted for only *ca.* 1-2% of the genetic variation. The low parental effects in the total phenotypic variation imply that the value of a dikaryon cannot in most cases be predicted from the activities of its two 'parental' homokaryons.

Heritability of Mycorrhizal Traits in Pisolithus tinctorius. The variability and the heritability of several fungal traits measured in the presence of the host plant were studied by Rosado *et al.* (1994) using a group of 16 half-sib dikaryons of *P. tinctorius* in combination with *Pinus elliottii* plants belonging to three different open-pollinated families. The 16 dikaryons used were synthesized by crossing two compatible groups of four monokaryons obtained from basidiospores collected on a single fruit body (as presented in table 1). The traits quantified were the number and percentage of mycorrhizae formed on the root system and the extramatrical mycelial growth.

The highest amount of variation was recorded for total mycelial strand growth, the highest value (144 cm) being almost twice the lowest one (77 cm). From these results, the different components of the genetic variation were estimated. Interactive effects accounted for most of the genetic ones and the additive effects were found to be significant only for one of the studied traits: the percentage of ectomycorrhizal colonization. Since the different mycelia were used to inoculate plants belonging to three open-pollinated families, the interaction between the fungal and the plant genomes in the expression of the fungal mycorrhizal ability was tested. Although there were significant differences between the three *Pinus* families for several of the fungal traits studied, no significant family x dikaryon interaction was detected.

Two conclusions were drawn from these results. Firstly, mycorrhizal traits of *P. tinctorius* can be improved through the realization of breeding programmes. Secondly, a fungal breeding programme can be conducted independently from the host-plant breeding programme.

However, the absence of interaction between plant and fungal genomes in the expression of fungal traits is not the rule. Rosado (1993), studying the ability of the same 16 dikaryons of *P. tinctorius* to form sclerotia in association with *P. elliottii*, showed that in this case significant levels of interaction existed. Some dikaryons produced large numbers of sclerotia when associated with trees belonging to one open-pollinated family but produced fewer sclerotia with plants of another family, whereas other dikaryons showed the opposite response to the tree genotypes. As a result, breeding programmes designed to improve sclerotia formation (which could be used as inoculum sources) have to take into account the genotypes of both plants and fungi.

From these results obtained with *H. cylindrosporum* and with *P. tinctorius* it can be concluded that the quantitative genetic approach to the variation of mycorrhizal traits has to be pursued, as it offers an opportunity to improve the performances of fungal strains for their use in forestry.

MUTANTS AND GENETIC ANALYSES

The comprehension of many fungal physiological functions has taken advantage of the use of mutant strains specifically impaired in one or several steps of the studied function. We believe that a genetic approach to mycorrhizal research should not be neglected. The transposition, with some modifications, of analytical methods developed with other fungi could greatly benefit from the study of some aspects of mycorrhiza differentiation and functioning, as demonstrated by the use of *H. cylindrosporum* mutants overproducing IAA.

Protoplasts and Mutagenesis

In the absence of vegetative spores, protoplasts or even hyphal macerates can be used as biological material for the induction of mutations in ectomycorrhizal fungi. Several protocols for the isolation of protoplasts from ectomycorrhizal fungi have been published. As expected, the different fungi used were species whose mycelia are easy to grow in pure culture and which do not produce thick cell walls and/or pigments. Such is the case for *L. bicolor* (Kropp and Fortin, 1986), *H. cylindrosporum*, *H. edurum*, *H. sinapizans*, *Suillus bellinii* (Hébraud and Fèvre, 1988), *L. laccata* and *H. circinans* (Barrett *et al.*, 1989). For these species, significant amounts of protoplasts (more than 10^6/ml of lytic solution) which could revert to hyphal growth on a suitable medium were produced. This is in contrast with

Cenococcum geophilum and *P. tinctorius*, two species for which protoplasts were produced but could not be regenerated (Barrett *et al.*, 1989).

As yet, mutants have only been selected with *H. cylindrosporum*. Using U.V. light as a mutagen, Hébraud and Fèvre (1988) selected monokaryotic mutant strains resistant to different fungicides. Although many of the protoplasts produced are bi- or plurinucleated, protoplasts from monokaryotic strains can be used to select recessive mutations such as auxotrophic ones. Auxotrophic mutants requiring an aminoacid for growth could be used to study the role of some of these molecules in the transfer of nitrogen from the fungus to the plant in the mycorrhiza. The screening of more than 3000 mycelia of *H. cylindrosporum* recovered after U.V. irradiation of protoplasts yielded auxotrophic mutants requiring molecules like adenine, lysine or methionine (Marmeisse, unpublished results). Unfortunately, none of these molecules is suspected to represent one of the main storage forms of nitrogen in the fungal cells and to be used in the transfer of nitrogen to the plant.

Auxin-Overproducer Mutants of *H. cylindrosporum*

The phytohormone Indole-3-Acetic Acid (IAA or auxin) is a compound which is produced by most saprophytic and symbiotic fungi grown in pure culture (Gay *et al.*, 1994a). Detectable amounts of IAA are usually found when tryptophan, which is the precursor of auxin, is added to the culture medium. Earlier studies (Slankis, 1973) have suggested that auxin produced by ectomycorrhizal fungi might play a role in ectomycorrhiza differentiation and functioning. To clarify this role, mutants affected in the auxin biosynthetic pathway were selected in *H. cylindrosporum* (Durand *et al.*, 1992). The mutants produced by Durand *et al.* (1992) were over-producers and the two-step protocol used for their selection was analogous to that used to select similar mutants in higher plants (Widholm, 1977). Mutants were first selected for resistance to 5-fluoroindole which is a toxic analog of indole, a precursor of tryptophan. Among the FluoroIndole Resistant (FIR) mutants obtained, approximately 10% produced IAA in pure culture in the absence of tryptophan in the medium. IAA overproduction was correlated with accumulation of tryptophan in the cells; such an accumulation was never observed in the wild-type mycelia. The different FIR mutations were shown to be recessive: dikaryons heterozygous for the mutations were sensitive to fluoroindole. Furthermore, study of the monokaryotic progenies of these fruiting heterozygous dikaryons indicated that the FIR phenotypes resulted from single-gene mutations.

Further studies (Gay *et al.*, 1994b) demonstrated that IAA overproduction affected the symbiotic properties of the strains. IAA-overproducer mutants formed significantly more mycorrhizae on *P. pinaster* roots than did the corresponding wild-type mycelia. IAA over production also has a profound effect on the differentiation of the Hartig net. *H. cylindrosporum* wild-type strains form a uniseriate Hartig net characterized by a single layer of hyphae between the outermost layers of the root cortical cells. In contrast, mutants overproducing IAA tend to form a pluriseriate Hartig net (up to seven layers of hyphae) which can reach the endodermis (Gea *et al.*, 1994). Ultrastructural modifications and the possible changes in the plant and fungal cell wall metabolisms will have to be compared between mycorrhizae formed by wild type and mutant fungal strains.

These examples demonstrate the feasibility of genetic analyses in ectomycorrhizal fungi with the possibility of selecting mutants, of studying their biochemical properties, the dominance of the mutations and the number of genes affected by looking at segregation ratios in the monokaryotic progenies of fruiting dikaryons. Such analyses can only be performed on model species like *H. cylindrosporum* or *Laccaria* spp. for which the mating system has been characterized and fruiting obtained in controlled conditions.

GENETIC TRANSFORMATION

Through its different applications — from the initial complementation of auxotrophic mutants to the targeted inactivation of cloned genes — DNA-mediated transformation has had a tremendous impact on our understanding of different fungal processes such as metabolic regulation, developmental biology or plant-pathogen interactions. Such DNA manipulations could also greatly improve our knowledge of the symbiotic relationship.

To achieve transformation of a given fungal species two prerequisites have to be fullfilled:

- Viable competent fungal cells which upon a chemical or physical treatment will take up the transforming DNA
- A cloned gene whose expression in the fungal cell will allow the selection of the transformed mycelia among the untransformed ones.

Fungal transformations are usually performed on protoplasts and the DNA is introduced either by using polyethylene glycol and calcium ions or by electroporation. As described in the previous section, viable protoplasts which can be regenerated to give new thalli have been obtained for a limited number of ectomycorrhizal species (Barrett et al., 1989).

Having a cloned gene which can be expressed and therefore used as a selectable marker in a species poorly characterized at the molecular level (as all mycorrhizal species) is certainly the most critical problem. In the following section, the development of transformation systems in basidiomycete fungi and their potential applications to ectomycorrhizal ones will be reviewed.

Transformation of Homobasidiomycete Fungi

Selection Systems. To date, successful transformation has been reported for less than ten different homobasidiomycete species (Table 3). The first two species to be transformed were the saprophytic *S. commune* (Munoz-Rivas et al., 1986) and *C. cinereus* (Binninger et al., 1987). For these two species, known auxotrophic mutant strains requiring tryptophan for growth were complemented with the cloned genes coding for the corresponding functional enzymes of the tryptophan biosynthetic pathway. For both species more than one prototrophic transformant per 10^4 viable protoplast was recovered and Southern blot analyses of their genomic DNAs showed that the transforming plasmids were stably integrated in the nuclear genomes.

Table 3. Selection systems used to transform saprophytic and (*) ectomycorrhizal homobasidiomycete fungi

Species	Selection systems	References
Agrocybe aegerita	auxotroph complementation	Noël and Labarère, 1994
Coprinus bilanatus	auxotroph complementation	Burrows et al., 1990
Coprinus cinereus	auxotroph complementation	Binninger et al., 1987, 1991
*Hebeloma cylindrosporum	hygromycin resistance	Marmeisse et al., 1992
*Laccaria laccata	hygromycin resistance	Barrett et al., 1990
Phanerochaete chrysosporium	auxotroph complementation	Alic et al., 1989, 1990
	phleomycin resistance	Gessner and Raeder, 1994
Pleurotus ostreatus	hygromycin resistance	Peng et al., 1992
Schizophyllum commune	auxotroph complementation	Munos-Rivas et al., 1986
	phleomycin resistance	Schuren and Wessels, 1994

The strategy of transforming auxotrophic mutant strains to prototrophy using cloned wild-type genes has been used successfully for other basidiomycete species. This is the case for adenine mutants of the lignin-degrading fungus *Phanerochaete chrysosporium* complemented with adenine genes cloned from either *S. commune* (Alic *et al.* 1989; 1990) or from *P. chrysosporium* (Alic *et al.*, 1991). This is also the case for a tryptophan mutant of *Coprinus bilanatus* (Burrows *et al.*, 1990) and for a uracil mutant of the edible species *Agrocybe aegerita* (Noël and Labarère, 1994).

To overcome the need to select auxotrophic mutants, plasmids which allow the transformation of a wide range of fungal species have been constructed. Ideally these plasmids should contain a gene conferring resistance to a compound which is toxic for most fungal species. This gene should be under the control of promoter sequences which allow its expression in different genomic backgrounds. Examples of such plasmids are the constructs made by Drocourt *et al.* (1990) and Punt *et al.* (1987) to transform ascomycete species. These plasmids contain respectively the prokaryotic genes *ble*, conferring resistance to phleomycin, and *hph*, conferring resistance to hygromycin B, fused to promoter sequences of ascomycete genes. The pAN7.1 plasmid made by Punt *et al.* (1987) contains the promoter of the glyceraldehyde-phosphate-dehydrogenese gene from *Aspergillus nidulans*. This plasmid has successfully been used to transform two ectomycorrhizal basidiomycetes: *L. laccata* (Barrett *et al.*, 1990) and *H. cylindrosporum* (Marmeisse *et al.*, 1992). Transformation of these two species, although successful, was poorly efficient: between one to five transformants were obtained per μg of transforming DNA for *H. cylindrosporum* and up to 50 for *L. laccata*. One possible explanation for the low efficiency of these transformations is that the antibiotic resistance gene fused to ascomycete regulatory sequences is poorly transcribed in basidiomycete cells.

To ensure a better expression, the resistance genes should be fused to promoters of cloned, strongly expressed, basidiomycete genes. Two such plasmid constructs have been made using the *ble* gene fused to either a histone H4 promoter of *P. chrysosporium* (Gessner and Raeder, 1994) or to the glyceraldehyde-3-phosphate-dehydrogenase promoter of *S. commune* (Schuren and Wessels, 1994). The first plasmid transformed *P. chrysosporium* to phleomycin resistance and the second one was shown to transform *S. commune* very efficiently (up to 10^4 transformants per μg DNA and per 10^7 protoplasts).

A powerful method called co-transformation allows the introduction, in the fungal genome, of plasmids devoided of any selection marker. In this case, such plasmids are mixed, prior to the transformation, with another one which does contain a selectable marker. The transformants are initially selected for the expression of the selectable marker and then screened for the presence of the second plasmid integrated into their genomes. With many species, as for example *H. cylindrosporum* (Marmeisse *et al.*, 1992), a large proportion of the transformants integrated the second plasmid in their genomes, more than would be expected if the integrations of the two plasmids were independent events.

Fate of the Transforming DNA, Gene Targeting and Gene Replacement. In eukaryotic genomes a transforming plasmid can either integrate in the nuclear DNA via a recombination event or, in exceptional cases, replicate autonomously. Transformants having replicating plasmids are usually unstable once the selection pressure has been removed and are of limited interest for the genetic manipulation of mycorrhizal species.

Fungal transformants can have either a single copy of the transforming plasmid integrated in their genomes or several copies. Furthermore, when several copies are present they can be integrated at either a single genomic site (in tandem) or at several sites dispersed throughout the genome. A knowledge of the fate of the transforming plasmid is essential for some applications of the genetic transformation. For example, a high number of plasmids

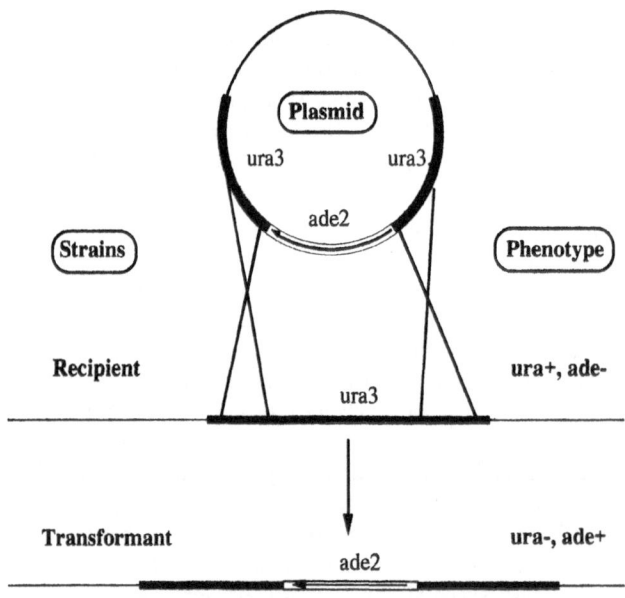

Figure 3. Events leading to single-step gene disruption.

integrated into a genome is most likely to lead to a strong increase in production of an enzyme encoded by a plasmid-born gene.

The different integration events described above have been found in basidiomycete fungi. This is the case for *TRP 1* transformants of *C. cinereus* where single-copy, multiple-dispersed and multiple copies arranged in tandem have been found (Binninger *et al.*, 1987; 1991). In contrast, only transformants with multiple copies of the transforming plasmid have been described in pAN7.1-transformed *H. cylindrosporum* (Marmeisse *et al.*, 1992) and *L. laccata* strains (Barrett *et al.*, 1990).

When the plasmid used in the transformation experiments has sequences identical to genomic sequences of the transformed organism, the frequency at which the DNA molecule integrates at its homologous site on the chromosome can be studied. In homobasidiomycetes, homologous integration events were recorded in three of the species for which plasmid containing homologous sequences have been used: *C. cinereus* (Binninger *et al.*, 1987; 1991), *S. commune* (Giasson *et al.*, 1989) and *P. chrysosporium* (Alic *et al.*, 1993). In *C. cinereus*, with a plasmid containing the homologous *TRP 1* gene, five percent of the transformants resulted from a homologous integration event at the resident *TRP 1* chromosomal locus (Binninger *et al.*, 1991). A significantly high frequency of homologous integrations can be used to perform gene disruption experiments (Fig. 3) which lead to the induction of mutations in a cloned gene whose functions are not necessarily known (a process called reverse genetics). This can be illustrated by the disruption of the *P. chrysosporium ura3* gene (Alic *et al.*, 1993). In this experiment, a cloned copy of the *ura3* gene was inactivated by inserting the *S. commune ade2* gene. The resulting plasmid containing the adenine gene flanked on each side by non-functional *ura3* sequences was used to complement a *P. chrysosporium* strain auxotrophic for adenine but prototrophic for uridine. Among the adenine prototrophic transformants, some were auxotrophic for uridine. Molecular analyses demonstrated that these transformants had their resident *ura3* gene replaced by the modified *ura3* gene present on the transforming plasmid. This resulted from a double crossing-over between plasmid and chromosomal *ura3* sequences, one on each side of the adenine gene (Fig. 3).

Applications to Ectomycorrhizal Species

Transformation of ectomycorrhizal fungi has only been achieved with two species: *L. laccata* and *H. cylindrosporum*. For these two species, the same selection system using the pAN7.1 plasmid has been used. Transformation of other species might be achieved in the future once reliable protocols for protoplast production will have been described and once more versatile plasmids allowing the transformation of a large number of homo-basidiomycete species will have been developed.

Other plasmid constructs will also have to be used with *H. cylindrosporum* and *L. laccata* in order to increase transformation frequencies and to obtain transformants with a single copy of the transforming plasmid integrated in their genomes. Of particular interest would be the construction of plasmids with resistance genes fused to promoters of genes isolated from these species. Moreover, the study of homologous recombination is needed in order to perform gene disruption experiments.

Gene disruption and reverse genetic analyses, if feasible, will certainly constitute a breakthrough in our understanding of the molecular mechanisms leading to mycorrhiza differentiation and functioning. Recently, Tagu *et al.* (1993) reported the isolation of cDNA clones differentially expressed in *Eucalyptus/Cenoccocum* mycorrhizas. In order to evaluate the functions of fungal genes whose expression is either specifically induced or repressed during mycorrhiza formation, these genes will have to be inactivated and the phenotype of the resulting mutant strains studied. Some of the data from the plant-pathogen interactions can be used to illustrate the experimental approach which could be applied to mycorrhizal species. Fungal genes specifically expressed during the infection cycle of various fungal pathogens have been isolated. This is the case of the *ecp* genes from the tomato pathogen *Cladosporium fulvum* coding for low molecular-weight proteins which accumulate in the intercellular spaces of the infected leaves (Van den Ackerveken *et al.*, 1993). This is also the case for the *MPG1* gene from the rice pathogen *Magnaporthe grisea*; this gene encoding a hydrophobin-like protein is expressed early in the infection cycle (Talbot *et al.*, 1993). Disruption of the *ecp2* gene of *C. fulvum* (Marmeisse *et al.*, 1994) gave mutant strains still fully pathogenic on tomato seedlings whereas disruption of the *MPG1* gene of *M. grisea* (Talbot *et al.*, 1993) gave strains no longer able to infect rice leaves, demonstrating that the MPG1 protein acts as a genuine pathogenicity factor.

Despite the different possible improvements in the transformation systems of *L. laccata* and *H. cylindrosporum*, the protocols described can already be used to introduce in the genome of these two species other genes of interest whose expression could modify the symbiotic properties of the strains.

Several additional genes have been introduced in the genomes of these two organisms by co-transformation. Two different plasmids were used simultaneously: one contained the selection marker (pAN7.1 in this case) and the other one contained the gene of interest. The transformants were initially selected on hygromycin plates and their genomic DNAs analyzed by Southern blot hybridization for the presence of the second plasmid. It was shown with *H. cylindrosporum* (Marmeisse *et al.*, 1992) that a large number (up to 70%) of the hygromycin resistant transformants had also integrated in their genomes several copies of the co-transforming plasmids.

The additional genes introduced into the genome of *H. cylindrosporum* were the NADP-glutamate dehydrogenase (GDH) gene and two genes of the tryptophan pathway of the saprophytic *C. cinereus* (Marmeisse *et al.*, 1992). Lemke (1994) reported the introduction of the functional *E. coli* *pro*B74 gene in *L. laccata*. These genes were chosen because they code for proteins which may play a role in the symbiosis. The NADP-GDH, which contributes to ammonium assimilation by the fungus (see Martin and Botton, 1993), could play a role in the improvement of the nitrogen nutrition of the host plant. As described in the previous section, tryptophan is the precursor of auxin which participates in the differentiation of the mycorrhizae. Expression of the *pro*B74 gene in *L. laccata* increases its salt tolerance (Lemke,1994) and could therefore constitute a model to study the salt tolerance of mycorrhizal plants.

It is difficult to give a comprehensive list of the different genes of interest which could be introduced into the genomes of different mycorrhizal fungi in order to modify their symbiotic properties. However, with the ever-increasing number of genes available, especially of genes cloned from basidiomycete species, almost any function related to the symbiosis could be altered in the near future by the expression of an appropriate foreign gene in the fungal partner.

CONCLUSIONS

Although still in their infancy, the genetic and molecular manipulation of ectomycorrhizal fungi may in the future significantly contribute to our understanding of the complex interactions between plants and their symbiotic fungal partners. Recent work has demonstrated that a limited number of ectomycorrhizal fungal species, which had been used as model species to study mycorrhiza differentiation and functioning, can be manipulated *in vitro* to modify their symbiotic properties. This is the case of *H. cylindrosporum* and of different closely related species of *Laccaria*. We can expect that, in the future, other species will be added to this list; not only other homobasidiomycetes but also ascomycetes, as well as species forming other types of mycorrhizae like ericoid fungi which can be grown in pure culture. The plant-dependence for growth of arbuscular endomycorrhizal fungi constitutes an obstacle to the genetic manipulation of these fungal species which can associate with most crop species.

The range of feasible manipulations on the existing potential model species also has to be extended to meet the demands of new applications. This is particularly the case for DNA-mediated transformation. New plamid vectors have to be developed to increase transformation frequency and to achieve gene replacement, and the expression of reporter genes (such as the GUS or the *lux* genes) should be tested for the study of functional promoter elements.

In conjunction with the analysis of fungal and plant gene expression during mycorrhiza differentiation (see corresponding chapters, this book), genetic manipulation of mycorrhizal fungi appears to be one of the most promising developments in the field of mycorrhizal research.

REFERENCES

Alic, M., Akileswaran, L. and Gold, M. H., 1993, Gene replacement in the lignin-degrading basidiomycete *Phanerochaete chrysosporium*, *Gene* 136: 307-311.

Alic, M., Clark, E. K., Kornegay, J. R. and Gold, M. H., 1990, Transformation of *Phanerochaete chrysosporium* and *Neurospora crassa* with adenine biosynthetic genes from *Schizophyllum commune*, *Curr. Genet.* 17: 305-311.

Alic, M., Kornegay, J. R., Pribnow, D. and Gold, M. H., 1989, Transformation by complementation of an adenine auxotroph of the lignin-degrading basidiomycete *Phanerochaete chrysosporium*, *Appl. Environ. Microbiol.* 55: 406-411.

Alic, M , Mayfield, M B Akıleswaran, L and Gold, M H , 1991, Homologous transformation of the lignin-degrading basidiomycete *Phanerochaete chrysosporium*, *Curr Genet* 19 491-494

Barrett, V, Dıxon, R K and Lemke, P A , 1990, Genetic transformation of a mycorrhizal fungus, *Appl Microbiol Biotechnol* 33 313-316

Barrett, V, Lemke, P A and Dıxon, R K , 1989, Protoplast formation from selected species of ectomycorrhizal fungi, *Appl Microbiol Biotechnol* 30 381-387

Binninger, D M , Le Chevanton, L , Skrzynıa, C , Shubkın, C D and Pukkıla, P J , 1991, Targeted transformation in *Coprinus cinereus*, *Mol Gen Genet* 227 245-251

Binninger, D M , Skrzynıa, C , Pukkıla, P J and Casselton, L A , 1987, DNA-mediated transformation of the basidiomycete *Coprinus cinereus*, *EMBO J* 6 835-840

Burrows, D M , Ellıott, T J and Casselton, L A , 1990, DNA-mediated transformation of the secondarily homothallic basidiomycete *Coprinus bilanatus*, *Curr Genet* 17 175-177

Casselton, L A and Kues, U , 1994, Mating-type genes in homobasidiomycetes, *In Wessels, J G H and Meinhardt, F, (Eds), The Mycota I Growth, differentiation and sexuality, Springer-Verlag, Berlin, Heidelberg* 307-321

Caten, C E , 1979, Quantitative genetic variation in fungi, *In Thompson, J N and Thoday, J M , (Eds), Quantitative genetic variation, Academic Press* 35-59

Debaud, J C and Gay, G , 1987, *In vitro* fruiting under controlled conditions of the ectomycorrhizal fungus *Hebeloma cylindrosporum* associated with *Pinus pinaster*, *New Phytol* 105 429-435

Debaud, J C , Gay, G and Bruchet, G , 1986, Intraspecific variability in an ectomycorrhizal fungus *Hebeloma cylindrosporum* 1-Preliminary studies on *in vitro* fuiting, spore germination and sexual comportment, *In Gianinazzi-Pearson, V and Gianinazzi, S (Eds), Physiological and genetic aspects of mycorrhizae, INRA Publications, Paris* 107-110

Drocourt, D , Calmels, T P G , Reynes, J P , Baron, M and Tıraby, G , 1990, Cassettes of the *Streptoalloteichus hindustanus ble* gene for transformation of lower and higher eukaryotes to phleomycin resistance, *Nucleic Acids Res* 18 4009

Durand, N , Debaud, J C , Casselton, L A and Gay, G , 1992, Isolation and preliminary characterization of 5-fluoroindole resistant and IAA overproducer mutants of the ectomycorrhizal fungus *Hebeloma cylindrosporum*, *New Phytol* 12 545-553

Esser, K and Kuenen, R , 1967, Genetics of fungi, *Springer-Verlag, Berlin, Heidelberg, New York*

Falconer, D S , 1981, Introduction to quantitative genetics, 2nd ed, *Longman, London, New York*

Fincham, J R S and Day, P R , 1963, Fungal Genetics, *Blackwell Scientific Publications*

Fries, N and Mueller, G M , 1984, Incompatibility systems, cultural features and species circumscriptions in the ectomycorrhizal genus *Laccaria* (Agaricales), *Mycologia* 76 633-642

Fries, N and Neumann, W , 1990, Sexual incompatibility in *Suillus luteus* and *S granulatus*, *Mycol Res* 94 64-70

Gay, G , Bernıllon, J and Debaud, J C , 1994a, Comparative analysis of IAA production in ectomycorrhizal, ericoid and saprophytic fungi in pure culture, *In Read, D J , Lewis D H , Fitter, A H and Alexander, I J , (Eds), Mycorrhizas in ecosystems, C A B International U K* 356-366

Gay, G and Debaud, J C , 1987, Genetic study on indole-3-acetic acid production by ectomycorrhizal *Hebeloma* species inter- and intraspecific variability in homo- and dikaryotic mycelia, *Appl Microbiol Biotechnol* 26 141-146

Gay, G , Normand, L , Marmeisse, R , Sotta, B and Debaud, J C , 1994b, Auxin overproducer mutants of *Hebeloma cylindrosporum* Romagnesi have increased mycorrhizal activity, *New Phytol* 128 645-657

Gea, L , Normand, L , Vian, B and Gay, G , 1994, Structural aspects of ectomycorrhizae of *Pinus pinaster* (Ait) Sol formed by an IAA-overproducer mutant of the fungus *Hebeloma cylindrosporum*, *New Phytol* 128 659-670

Gessner, M and Raeder, U , 1994, A histone *H4* promoter for expression of a phleomycin-resistance gene in *Phanerochaete chrysosporium*, *Gene* 142 237-241

Gıasson, L , Specht, C A , Mılgrım, C , Novotny, C P and Ullrich, R C , 1989, Cloning and comparison of *A*a mating-type alleles of the basidiomycete *Schizophyllum commune*, *Mol Gen Genet* 218 72-77

Godbout, C and Fortın, J A , 1990, Cultural control of basidiome formation in *Laccaria bicolor*, *Mycol Res* 94 1051-1058

Harley, J L and Smith, S E , 1983, Mycorrhizal symbiosis, *Academic Press, London*

Hebraud, M and Fevre, M , 1988, Protoplast production and regeneration from mycorrhizal fungi and their use for isolation of mutants, *Can J Microbiol* 34 157-161

Kıng, R C , 1974, Handbook of Genetics Volume 1 Bacteria, bacteriophages and fungi, *Plenum Press, New York*

114   R. Marmeisse et al.

Kope, H H and Fortin, J A , 1990, Germination and comparative morphology of basidiospores of *Pisolithus arhizus*, *Mycology* 82 350-357

Kropp, B R and Fortin, J A , 1986, Formation and regeneration of protoplasts from the ectomycorrhizal basidiomycete *Laccaria bicolor*, *Can J Bot* 68 1224-1226

Kuhner, R , 1977, Variation of nuclear behaviour in the homobasidiomycetes, *Trans Br Mycol Soc* 68 1-16

Lemke, P, 1994, Basidiomycetes and the new genetics, *Mycologia* 86 173-180

Marmeisse, R , Gay, G , Debaud, J C , and Casselton, L A , 1992, Genetic transformation of the symbiotic basidiomycete fungus *Hebeloma cylindrosporum*, *Curr Genet* 22 41-45

Marmeisse, R , Van den Ackerveken, G F J M , Goosen, T , De Wit, P J G M and Van den Broek, H W J , 1994, The *in planta* induced *ecp2* gene of the tomato pathogen *Cladosporium fulvum* is not essential for pathogenicity, *Curr Genet* 26 245-250

Martin, F and Botton, B , 1993, Nitrogen metabolism of ectomycorrhizal fungi and ectomycorrhiza, *Adv Plant Pathol* 9 83-102

May, G , Le Chevanton, L and Pukkila, P J , 1991, Molecular analysis of the *Coprinus cinereus* mating type *A* factor demonstrated an expectedly complex structure, *Genetics* 128 529-538

Meyselle, J P , Gay, G and Debaud, J C , 1991, Intraspecific genetic variation of acid phosphatase activity in monokaryotic and dikaryotic populations of the ectomycorrhizal fungus *Hebeloma cylindrosporum* , *Can J Bot* 69 808-813

Munoz-Rivas, A , Specht, C A , Drummond, B J , Froeliger, E , Novotny, C P and Ullrich, R C , 1986, Transformation of the basidiomycete, *Schizophyllum commune*, *Mol Gen Genet* 205 103-106

Mutasa, E S , Tymon, A M , Gottgens, B , Mellon, F M , Little, P F R and Casselton L A , 1990, Molecular organization of an *A* mating type factor of the basidiomycete fungus *Coprinus cinereus*, *Curr Genet* 18 223-229

Noel, T and Labarere, J , 1994, Homologous transformation of the edible basidiomycete *Agrocybe aegerita* with the *URA1* gene characterization of integrative events and of rearranged free plasmids in transformants, *Curr Genet* 25 432-437

Perkins, D D , 1991, In praise of diversity, *In Bennet, J W and Lasure, L L (Eds) More gene manipulations in fungi, Academic Press , London, U K* 3-26

Peterson, R L and Farquhar, M L , 1994, Mycorrhizas-Integrated development between roots and fungi, *Mycologia* 86 311-326

Punt, P J , Oliver, R P , Dingemanse, M A Pouwels, P H and Van den Hondel, C A M J J, 1987, Transformation of *Aspergillus* based on the hygromycin B resistance marker from *Escherichia coli*, *Gene* 56 117-124

Raper, J R , 1966, Genetics and sexuality in higher fungi, *Ronald Press, New York*

Rosado, S C S , 1993, Variations genetiques chez la symbiose ectomycorhizienne *Pinus elliottii-Pisolithus tinctorius*, *PhD thesis Universite Laval Quebec, Canada*

Rosado, S C S , Kropp, B R and Piche, Y , 1994, Genetics of ectomycorrhizal symbiosis II Fungal variability and heritability of ectomycorrhizal traits, *New Phytol* 126 111-117

Schuren, F H J and Wessels, J G H , 1994, Highly-efficient transformation of the homobasidiomycete *Schizophyllum commune* to phleomycin resistance, *Curr Genet* 26 179-183

Silvertown, J W and Lovett Doust, J , 1993, Introduction to plant population biology, *Blackwell Scientific Publications, London*

Slankis, V , 1973, Hormonal relationship in mycorrhizal development, *In Marks, G C and Koslowski, T T (Eds), Ectomycorrhizae, Academic Press London, New York* 231-298

Tagu, D , Python, M , Cretin, C and Martin, F , 1993, Cloning symbiosis-related cDNAs from eucalypt ectomycorrhiza by PCR-assisted differential screening, *New Phytol* 125 339-343

Talbot, N J , Ebbole, D J and Hamer, J E , 1993, Identification and characterization of *MPG1*, a gene involved in pathogenicity from the rice blast fungus *Magnaporthe grisea*, *The Plant Cell* 5 1575-1590

Van den Ackerveken, G F J M , Van Kan, J A L , Joosten, M H A J , Muisers J M , Verbakel, H M and De Wit, P J G M , 1993, Characterization of two putative pathogenicity genes of the fungal tomato pathogen *Cladosporium fulvum*, *Mol Plant-Microb Interact* 6 210-215

Wagner, F , Gay, G and Debaud, J C , 1988, Genetical variability of glutamate dehydrogenase activity in monokaryotic and dikaryotic mycelia of the ectomycorrhizal fungus *Hebeloma cylindrosporum*, *Appl Microbiol Biotechnol* 66 588-594

Wagner, F , Gay, G and Debaud, J C , 1989, Genetic variation of nitrate reductase activity in mono- and dikaryotic populations of the ectomycorrhizal fungus, *Hebeloma cylindrosporum* Romagnesi, *New Phytol* 113 259-264

Widholm, J M , 1977, Relation between auxin auxotrophy and tryptophan accumulation in cultured plant cells, *Planta* 134 103-108

10

INACTIVATION OF GENE EXPRESSION TRIGGERED BY SEQUENCE DUPLICATION

C. Cogoni, N. Romano, and G. Macino

Dipartimento Biopatologia Umana, Sezione di Biologia Cellulare
Università di Roma "La sapienza"
Policlinico Umberto I
I-00161, Rome
Italy

INTRODUCTION

In *Neurospora crassa* it has been shown that duplicated genes are irreversibly inactivated by a mechanism called R.I.P. (Repeat-Induced Point mutation). Duplicated sequences are extensively modified by point mutation, namely cytosine-thymine transitions. R.I.P. occurs at a precise moment during the premeiotic phase in the sexual cycle of *Neurospora*.

Recently we have shown that in *Neurospora crassa* inactivation of duplicated gene sequences also occurs in the vegetative cycle. We have termed this new phenomenon quelling. Like R.I.P., quelling seems to be a general phenomenon in *Neurospora crassa*. Several duplicated genes have been shown to be the target for gene silencing. Gene inactivation by quelling is not the result of mutagenesis as in R.I.P.; instead, it is the result of the suppression of gene expression, through a reduction of the steady-state mRNA level of the duplicated gene. Quelling is reversible and the function of the inactivated gene may be restored after prolonged culturing time.

Similar phenomena of reversible gene inactivation have been reported in plants. In several cases the introduction of a gene into a plant genome by transformation resulted in co-suppression of both the transgene and the endogenous gene.

In this paper we will describe the characteristics of quelling in *Neurospora crassa* and discuss the analogies and differences between quelling and other gene-inactivation phenomena associated with the presence of DNA duplications in fungi and plants.

Putative mechanisms proposed to explain the occurrence of gene inactivation will also be discussed.

REVERSIBLE SUPPRESSION OF GENE EXPRESSION IN *NEUROSPORA CRASSA* BY INTRODUCTION OF HOMOLOGOUS SEQUENCES

Wild type *Neurospora crassa* is bright orange in colour due to the presence of carotenoids. These pigments are synthesised by different enzymes encoded by three structural genes: albino-1 (*al-1*) encoding the phytoene dehydrogenase (Schmidhauser et al. 1990), albino-2 (*al-2*) encoding phytoene synthetase (Schmidhauser et al. 1994) and albino-3 (*al-3*) which codes for geranylgeranyl pyrophosphate synthase (Nelson M.A. et al. 1989). Mutation of the carotenoid biosynthetic genes results in an albino phenotype. Quelling was discovered (Romano and Macino, 1992) as a result of transforming wild type *Neurospora crassa* spheroplasts with a construct containing the *al-3* gene. Unexpectedly, several transformants showed an albino phenotype. Analysis by Southern blot hybridisation showed that no rearrangements had occurred in the endogenous *al-3* gene. Similar results have been obtained transforming a wild type strain with a construct containing different portions of *al-1* gene, indicating that the observed phenomenon was more general and not restricted to the *al-3* gene.

Quelling Results in Specific Inhibition of the Accumulation of mRNA of the Duplicated Gene

Northern blot analysis performed on RNA extracted from albino transformants revealed that quelling acts through a reduction of the steady-state level of the mRNA of the gene duplicated by transformation. The albino transformants recovered from transformation with the construct containing the albino-1 gene sequence showed a variety of phenotypes that ranged from white to yellow to dark yellow. This result suggests that more or less severe albino phenotypes are characterised by a stronger or a weaker suppression of gene expression. In fact, RNA analysis of transformants with an intermediate phenotype showed that there is a intermediate level of mRNA. Moreover, the reduction of the mRNA level is absolutely sequence specific; the expression of the duplicated gene only is affected.

Gene Silencing by Quelling Is Reversible

All the albino transformants produced by quelling have an unstable phenotype and were observed to revert progressively to wild type or an intermediate phenotype over a prolonged culturing time. The reversion took place in a fraction of growing cells; about 25% of microconidia (mononuclear cells) from a culture of a purified albino transformant showed reversed phenotype, whereas the remaining 75% had the same albino phenotype as the primary transformant. Reversion, however, does not occur in an all-or-none fashion, since revertants showed a variety of intermediate phenotypes. The reversion of quelling appears to be monodirectional since once it has occurred the gene silencing cannot take place again. No albino colonies were isolated from many thousands of reversed orange or intermediate transformants. Northern analysis indicated that revertants present an increased level of mRNA of the duplicated gene. Reversion seems to be associated with a reduction of copy number of exogenous sequences. The mitotic instability of ectopic sequences could be a phenomenon naturally occurring in *Neurospora crassa* independently from quelling. A case of instability of exogenous sequences has been observed in a *fluffy* strain transformed with a plasmid containing the benomyl resistance gene (Rossier et al. 1992). Loss of tandem arranged transforming sequences has been observed (Selker et al 1987) after prolonged passages corresponding to 50-100 nuclear doublings. Loss involving deletion of plasmid sequences, is probably the result of homologous recombination occurring in the vegetative phase.

R.I.P. AND M.I.P.: PREMEIOTIC GENE INACTIVATION IN FUNGI

R.I.P. and M.I.P. are two related processes of modification of duplicated sequences that have been found respectively in *Neurospora crassa* and in *Ascobolous immersus*.

R.I.P.: Repeat-Induced Point Mutation in *Neurospora crassa*

The capability of *Neurospora crassa* to modify duplicated sequences was first reported by Selker et al. (1987) from a study of transformants obtained by the introduction of a plasmid containing 6kb of a *Neurospora crassa* homologous sequence. The 6kb segment was present either as an unlinked duplication after non-homologous integration or in a tandem direct repeat after homologous integration. Restriction analysis of the duplicated sequences in the sexual progeny of such transformants showed a high frequency of rear-rangement of the 6kb sequence. When the duplication was present as a tandem repeat it was subject to rearrangement in 100% of the progeny. Unlinked copies of the 6kb sequence were also altered, but rearrangement occurred at a lower frequency and generally appeared less radical.

Tetrad analysis indicated that the products of sister chromatids displayed common alteration of duplicated sequences. This strongly suggests that rearrangements occurred before premeiotic DNA replication (Selker et al., 1987). The analysis of ascospores derived from the same perithecium showed that they could display different patterns of modification of duplicated DNA sequences. Since each perithecium usually results from a single fertilisation event , R.I.P. must occur after fertilisation. Thus R.I.P. seems to occur at a precise moment of the sexual phase of *Neurospora* between fertilisation and karyogamy.

The nature of the alteration of duplicated sequences by R.I.P. was clarified through DNA sequencing. The R.I.P. phenomenon introduces only one type of mutation: C-G pairs are converted to T-A pairs. The frequency of transition induced by R.I.P. is very high; about 10% of G-C pairs were mutated in a 900 bp stretch of an unlinked duplication, whereas up to 50% of C-G pairs were changed to T-A in a 6kb segment of a linked duplication (Cambareri et al. 1989). Furthermore, mutagenesis is strictly confined to duplicated sequences.

M.I.P.: Methylation Induced Premeiotically in *Ascobolous immersus*

M.I.P. was first observed in *Ascobolous* from a study of transformants harbouring tandem duplications of the *met-2* gene (Goyon et al., 1989). Unexpectedly, about 90% of the progeny of a sexual cross with a wild type strain lost the Met⁺ phenotype. Genetic analysis of Met⁻ progeny showed that they could revert to prototrophy after culturing in selective conditions. Southern analysis showed that artificially-repeated *met-2* genes were heavily methylated at cytosine residues. In a subsequent study using transformants in which *met-2* fragments of different sizes were inserted in positions unlinked to the resident gene, Barry et al. (1993) demonstrated by Southern analysis that cytosine methylation is limited to duplicated regions. As in R.I.P. in *Neurospora*, the unlinked duplicated sequences were modified at a lower frequency than tandem repeats.

Both M.I.P. and R.I.P. occur with very similar frequencies and both take place in dikaryotic ascogenous tissue. The simultaneous alteration of the two duplicated elements and the fact that modification is exclusively confined to the duplicated area, strongly suggests a role of direct DNA-DNA interaction in these two phenomena.

QUELLING AND R.I.P.: TWO DIFFERENT PHENOMENA OF INACTIVATION OF DUPLICATED SEQUENCES IN *NEUROSPORA CRASSA*

R.I.P. and quelling act in *Neurospora crassa* when single copy sequences are duplicated. Both phenomena seem to be general: several genes have been shown to be the target for gene inactivation by quelling when they are duplicated by transformation and similarly, sequence duplication is sufficient for the occurrence of R.I.P. independently from the nature of the duplicated sequences. Therefore, in both cases inactivation is strictly confined to the duplicated sequences. For example, when the albino-1 gene was duplicated by transformation, silencing by quelling was observed (Romano and Macino, 1992), affecting *al-1* expression only, whereas the expression of correlated carotenoid biosynthetic genes was unmodified. In R.I.P. there is a coextension between the length of homology and the length of sequences affected by point mutation.

Despite the fact that R.I.P. and quelling occur in the same organism, both are general and specifically modify duplicated sequences, there are substantial differences that could suggest that these two phenomena may be unrelated and have different mechanisms of action.

R.I.P. takes place at a precise moment of the life cycle of *Neurospora*, between fertilisation and karyogamy, whereas quelling acts in coenocytic cells when exogenous DNA is introduced by transformation. The duplicated sequences are irreversibly inactivated by R.I.P., and this process leads to 50% of C-G pairs being converted to T-A. Cytosines present in a duplicated DNA fragment are probably first methylated and then converted to thymines by a second enzymatic step of deamination (Selker, 1990). Gene inactivation in quelling however, is not the result of mutagenesis as in R.I.P.; in fact, quelling is reversible and the function of the inactivated gene may be restored after prolonged culturing time. Gene inactivation is a result of a reduction of steady-state mRNA levels and duplicated DNA sequences appear to be unmodified.

The frequency of occurrence of R.I.P. and quelling are different. DNA duplication is necessary and sufficient to trigger R.I.P., whereas in the case of quelling, gene duplication is necessary, but not sufficient. It is therefore likely that the locus of integration and/or the arrangement of transforming DNA play a crucial role in gene silencing.

SILENCING OF PLANT GENES BY HOMOLOGOUS TRANSGENES

A number of cases of suppression of gene expression in plants as a consequence of the introduction of a gene into the genome by transformation, have recently been reported. The first example is the result of the introduction by transformation of the chalcone synthase (*chs*) gene into petunia plants (Napoli et al., 1990). The *chs* gene is involved in anthocyanine biosynthesis and in petal pigmentation. Unexpectedly, many transgenic plants, instead of over-accumulating the purple colour, produced flowers that were variegated purple and white, or even completely white. In white sectors, RNase protection analysis showed that both the transgene and the endogenous gene were coordinately suppressed, whereas both genes were expressed in purple coloured regions. This phenomenon was termed co-suppression. After this first observation, several examples of plant gene silencing have been reported, and co-suppression appears to be a general phenomenon. Co-suppression has also been observed in petunia plants transformed with a chimeric petunia dihydroflavonol-4 reductase (*dfr*) gene (van der Krol et al., 1990). A case of "regulated" suppression has also been reported (Hart et al., 1992): the introduction of chimeric tobacco chitinase gene silences the endogenous gene, but the incidence of silencing depends on growth conditions. Silencing occurs at a high frequency in tobacco plants raised as seedlings in closed culture vessels, but not in plants raised from seed in a greenhouse.

Silencing of endogenous genes has been similarly achieved by transformation with closely related, but not identical sequences: tobacco plants transformed with a bean phenylalanine ammonium- lyase (*pal*) gene showed inhibition of the native *pal* gene (Elkind et al., 1990). In *Arabidopsis* a mutant that displays co-suppression of both the transgenic *cab140* promoter *tms-2* gene transcriptional fusion construct and the native *cab140* gene has been isolated (Brussian et al. 1993). In this case the only duplicated sequence is 1300 bp of the *cab140* promoter.

Other examples of gene silencing are: the silencing of a complete nopaline synthase (*Nos*) transgene by the introduction of a truncated Nos gene into tobacco (Goring et al., 1991); in tomato plants, the introduction of a truncated polygalactouronase gene has been shown to be sufficient to silence the endogenous gene (Smith et al.,1991); the inactivation of kanamicine resistance and *Nos* transgenes by successive introduction in tobacco plants of a construct containing hygromicin phosphotransferase and octopine synthase genes (Matzke et al., 1989); the reversible inactivation of multiple copies of a hygromicin-resistance transgene introduced into *Arabidopsis* (Mittelsten Scheid et al., 1991).

QUELLING AND COSUPPRESSION: REVERSIBLE INHIBITION OF GENE EXPRESSION

Quelling acts through a heavy reduction of the steady-state level of the mRNA of the genes duplicated by transformation. A specific decrease of steady-state mRNA levels of silenced genes has also been observed in all cases of gene suppression in plants. Whether co-suppression works at the level of transcription or post-transcriptionally is still under debate since there are examples that involve both situations. In the co-suppression of the *chs* gene in petunia (Napoli et al., 1990)and of the β-1,3 glucanase gene in tobacco (de Carvalho et al., 1992), run on experiments indicate that, despite the absence of mRNA transcripts, the endogenous gene is normally transcribed, suggesting that gene suppression acts at a post-transcriptional level. However, in *Arabidopsis* , *in vitro* transcript elongation of the *cab140* gene has shown that the rate of transcription is greatly reduced (Brussian et al., 1993).

Both in *Neurospora crassa* and in every case of co-suppression reported in plants, gene silencing occurred after transformation and is not the result of mutation. Moreover, in a large number of observations the gene silencing in plants is reversible. The silencing of a transgene complex in stably transformed tobacco plants by introduction of a second transgene complex was reversible after segregation of the second transgene from the first (Matzke et al., 1989). In this example the persistence of inactivation of the first transgene is sometimes exhibited after removal of the second transgene locus by genetic separation, indicating that an "altered state" of the silenced gene persists after segregation. The inactivation of the *chs* and *dhf* genes in petunia , the *Nos* gene in tobacco and the *cab140* gene in *Arabidopsis* is reversible upon genetic segregation of the transgene from the endogenous gene. In these cases reversion of gene silencing is dependent on the presence of transforming DNA sequences in the same way reversion of quelling is correlated with loss of exogenous sequences. Another similarity between quelling and several examples of gene silencing in plants is the fact that duplication of DNA sequences is necessary, but not sufficient, for suppression of gene expression, indicating that other events are involved in gene silencing.

PUTATIVE MECHANISM INVOLVED IN THE PHENOMENA OF GENE SILENCING

Several mechanisms may be proposed to explain the occurrence of quelling in *Neurospora* and co-suppression in plants.

Biochemical Switch

This mechanism implies the existence of a regulatory circuit extremely sensitive to the concentration levels of a gene product and/or mRNA level. The over-expression of a gene after reaching a threshold, may induce a feedback response that inhibits gene transcription and/or increases mRNA turnover. The biochemical switch hypothesis is consistent with the cases of co-suppression observed in transgenic plants (Napoli et al., 1990; de Carvalho et al., 1992) where RNA transcribed from transgenes in addition to that transcribed from endogenous gene, could accumulate at abnormally high levels, triggering the expression of the gene to be switched off. This mechanism could also act in the case of virus-resistant tobacco plants containing a nuclear transgene encoding each virus coat protein (Lindbo et al., 1994). In this case, infection and accumulation of virus RNA in the cytoplasm induces the reduction of transgene expression and the inhibition of virus replication.

The hypothesis of a biochemical switch in the quelling mechanism in *Neurospora* is unlikely, as gene suppression was achieved by transformation with constructs without promoter regions.

Unintended Antisense Transcript Production

The suggestion that the unexpected production of antisense RNA in transgenic plants may play a role in the silencing of genes was based on the observation that there are strong similarities between co-suppression and silencing of gene expression by antisense constructs. In both cases there is a specific reduction of accumulation of mRNA from the endogenous gene. The determination that co-suppression, in several cases, works at a post-transcriptional level, is also consistent with the antisense hypothesis. Co-suppression was demonstrated to occur at a post-transcriptional level in the inactivation of β-1,3 glucanase tobacco gene (de Carvallo et al., 1992) and in the silencing of the polygalactouronase tomato gene (Smith et al., 1990). A possible role for antisense RNA in co-suppression of the *chs* gene in petunia was also hypothesized (Mol et al., 1991); in cosuppressed petunia lines a *chs* gene antisense transcript has been observed, but its role in gene suppression remains to be investigated.

A complication for the antisense hypothesis comes from the work of Goring et al. (1991). In this paper the authors describe a transformation of the same strain of tobacco with constructs containing partial nopaline synthase gene in the sense and antisense orientation with respect to the CaMV 35S promoter. In this experiment, the antisense transgene construct was much less effective in the suppression than the sense transgene construct. It is therefore difficult to understand in terms of an antisense hypothesis how a construct designed to produce a large amount of antisense RNA could be less efficient for gene silencing than a construct designed to produce sense RNA.

The production of antisense RNA by read-through transcription from a cryptic promoter in the vector or adjacent to the chromosome integration site can also explain the occurrence of quelling. However, the antisense RNA products in albino transformants obtained by duplication of *al-1* sequences have not been observed on RNA gel blots using double stranded probes (Romano and Macino, 1992). On the other hand, we cannot rule out the possibility that production of a very unstable antisense RNA, not detectable by Northern analysis, is sufficient to inactivate gene expression.

Methylation of Duplicated Sequences

Methylation is one of the most general mechanisms known in controlling regulation of eukaryotic gene expression (Bird, 1986; Cedar H., 1988; Doerfler, 1983; Razin & Riggs, 1980). Methylation of promoter sequences of duplicated genes could repress transcription initiation by preventing essential transcription factors from interacting with the promoter. Proteins which specifically bind to methylated sequences could also cause transcriptional inhibition by excluding some component of transcriptional machinery.

The erratic and reversible nature of quelling could suggest a role for DNA methylation. The state of DNA methylation was studied through Southern blot analysis (Romano and Macino, 1992). Genomic DNA extracted from albino transformants obtained by duplication of *al-3* or *al-1* sequences was digested with isoschizomeric pairs of methylation-sensitive restriction enzymes. This analysis showed that the ectopic sequences were heavily methylated at cytosine residues while no cytosine methylation could be detected in the silenced resident genes, indicating that methylation is not correlated with suppression of gene expression.

In tobacco plants the inactivation of the *Nos* transgene (Matzke et al., 1989) was correlated with the methylation of the promoter of the affected gene. Both methylation and inactivation depend on the presence of a second transgene in the same genome. However, it is important to keep in mind that this does not imply that methylation plays a causal role. Methylation may be a result of changes in chromatin structure which accompany changes in gene expression. A role of DNA methylation in gene inactivation has been reported in the filamentous fungus *Ascobolus immersus* (Barry et al., 1993). The duplication of the *met-2* gene in *A.immersus*, obtained by integrative transformation, produced methylation and inactivation of artificially repeated *met-2* genes. Methylation occurs during the sexual phase of the life cycle. When a portion of *met-2* gene is introduced by transformation, only the duplicated region of *met-2* gene is heavily methylated. DNA methylation of distinct *met-2* segments has different effects upon *met-2* transcription. RT-PCR analysis indicated that methylation might inhibit elongation of RNA transcripts.

CONCLUSION

The silencing of duplicated gene sequences seems to be a widely diffused phenomenon in plants and in fungi. The evolution of this complex mechanism suggests that it may play an important role in genome organisation and in the control of gene expression. A possible role for co-suppression may be the protection of plant genomes from transposable elements, viruses and other repeated elements that can accumulate in the genome. Similarly, the *Neurospora crassa* genome could be protected from the expansion of transposons and viruses by the evolution of two mechanisms of inactivation of duplicated sequences. Quelling could act in the vegetative phase through the silencing of transposons and virus genes, inhibiting their expansion and replication, while R.I.P. acts in the sexual phase through irreversible mutagenesis of duplicated sequences preventing their transmission to progeny.

The mechanisms of action of both co-suppression in plants and quelling in *Neurospora* are still unknown. Their understanding can increase our comprehension of the regulation of gene expression and genome organisation.

ACKNOWLEDGMENTS

We thank A. Pickford for critically reading the manuscript. This work was supported by grants from the Ministero Agricoltura e Foreste, Piano Nazionale Tecnologie Applicate alle Piante, the Progetti Finalizzati, Ingegneria Genetica del Consiglio Nazionale delle Ricerche and the Istituto Pasteur Fondazione Cenci Bolognetti.

REFERENCES

Barry C., Faugeron G. & Rossignol J. (1993) Methylation induced premeiotically in *Ascobolus* : Coextension with DNA repeat lengths and effect on transcrip.t elongation. Proc. Natl. Acad. Sci. USA 90: 4557-4561

Bird A.P. (1986) CpG rich islands and the function of DNA methylation. Nature 321:209-213

Brussian J.A., Karlin-Neumann G.A., Lu Huang & Tobin E.M. (1993) An Arabidopsis mutant with a reduced level of cab140 RNA is a result of cosuppression. Plant cell 5: 667-677

Bull J.H. & Wootton J.C. (1984) Heavily methylated amplified DNA in transformants of *Neurospora crassa*.. Nature 310: 701-704

Carattoli A., Romano N., Ballario P., Morelli G. & Macino G. (1991) The *Neurospora crassa* carotenoid biosynthetic gene (albino-3) reveals highly conserved regions among prenyltransferases. J. Biol. Chem. 266: 5854-5859

Cedar H. (1988) DNA methylation and gene activity. Cell 53: 3-4

de Carvalho F., Gheyesen G., Kushnir S.,van Montagu M., Inze D., & Castresana C. (1992) Suppression of β-1-3 glucanase transgene expression in homozygous plants. EMBO J. 11: 2595-2602

Doerfler W. (1983) DNA methylation and gene activity Ann Rev. Biochem. 52: 93-124

Ebbole D. & SachsM.S. (1990) A rapid and simple method for isolation of *Neurospora crassa* homokaryons using microconidia. Neurospora Newslett. 37:17-18

Elkind Y., Edwards R., Mavandad M., Hedrick S. A., Ribak O., Dixon R. A. & Lamb C. J. (1990) Abnormal plant development and down regulation of phenylpropanoid biosynthesis in trnsgenic tobacco containing a heterologous phenylalanine ammonia lyase gene. Proc. Natl. Acad. Sci. USA 87: 9057-9061

Faugeron G., Rhounim L. & Rossignol J.L. (1990) How does the cell count the number of ectopic copies of a gene in the premeiotic inactivation process acting in *Ascobolus immersus*? Genetics 124: 585-591

Goring D., Thomson L., & Rothstein S.J. (1991) Transformation of a partial nopaline synthase gene into tobacco suppresses the expression of a resident wild-type gene. Proc. Natl. Acad. Sci. USA 88: 1770-1774.

Goyon C. & Faugeron G. (1989)Targeted transformation of *Ascobolus immersus* and de novo methylation of the resulted duplicated sequences. Mol. Cell. Biol. 9 : 2818-2827

Hart C.M., Fischer B., Neuhaus J.M., & Meins F. (1992) Regulated inactivation of homologous gene expression in transgenic Nicotiana sylvestris plants containing a defense-releted tobacco chitinase gene. Mol Gen. Genet. 235:179-188

Jones P.A. (1985) Altering gene expression with 5-azacytidine. Cell 40:485-486

Lindbo J.A.,Silva-Rosales L., Proebsting W.M. & Dougherty W.J. (1993) Induction of a highly antiviral state in trangenic plants: implication for regulation of gene expression and virus resistance. Plant cell 5: 1749-1759

Matzke M.A., M. Primig, J. Trnovsky & A.J.M. Matzke (1989) Reversible methylation and inactivation of marker genes in sequentially transformed tobacco plants EMBO J. 8 : 643-648

Mittelsten Scheid O., Paszkowski J. & Potrykus I. (1991) Reversible inactivation of a transgene in *Arabidopsis thaliana* . Mol. Gen. Genet. 228: 104-112

Mol J., van Blockland R.V., & J. Kooter (1991) More about cosuppression . Trends Biotech 9: 182-183

Napoli C.,Lemieux C. & R. Jorgensen (1990) Introduction of a chimeric chalcone synthase gene into Petunia results in reversible co-suppression of homologous genes in trans. The Plant Cell 2 : 279-289

Nelson M.A., Morelli G., Carattoli A., Romano N. & Macino G. (1989) Molecular cloning of a *Neurospora crassa* carotenoid biosynthtic gene (albino-3) regulated by blue light and the products of white collar genes. Mol. Cell. Biol. 9:1271-1276

Nelson M. & McClelland (1989) Effect of site specific methylation or DNA modification methyltransferases and restriction endonucleases. Nucleic Acid Research vol.17 supplement r389-415.

Pandit N.N. & Russo V.E.A. (1992) Reversible inactivation of a foreign gene, hph, during the asexual cycle in *Neurospora crassa* transformants Mol. Gen. Genet. 234: 412-422

Razin A. & Riggs A.D. (1980) DNA methylation and gene function. Science 210:604-610

Romano N. & Macino G. (1992) Quelling: transient inactivation of gene expression in *Neurospora crassa* by transformation with homologous sequences. Molec. Microbiol. 6: 3343-3353

Rossier C., Brazil G. &Utz-Pugin A. (1992) Mitotic instability in benomyl-resistant tranformants of a fluffy strain of Neurospora crassa. Fungal Gen. Newsletter 39:72-73

Schmidhauser T.J., Lauter F.R., Russo V.E.A. & Yanofsky C. (1990) Cloning, sequence and photoregulation of al-1, a carotenoid biosynthetic gene of *Neurospora crassa*. Mol. Cell. Biol. 10: 5064-5070

Schmidhauser T.J., Lauter F.R., Schumaker, M., Zhou, W., Russo, V. E. A., Yanofsky, C. (1994) Characterization of *al-2*, the Phytoene Synthase Gene of *Neurospora crassa*. Jour. Biol. Chem., 269: 12060-12066.

Selker E.V.(1990) Premeiotic instability of repeated sequences in *Neurospora crassa*, Ann. Rev. Genet. 24 : 579-613

Selker E. V. , E. B. Cambareri, B. C. Jensen & K. K. Haach (1987) Rearrangement of duplicated DNA in specialized cells of Neurospora. Cell 51: 741-752

Smith C.J.S., Watson C.F., Bird C.R., Ray J., Schuch W., & Grierson D. (1990) Expression of a truncated tomato polygalacturonase gene inhibits expression of the endogenous gene in transgenic plants. Mol. Gen. Genet. 224: 477-481

van der Krol A.R., Mur L.A., Beld M., Mol J.N.M. & Stuitje A.R. (1990) Flavonoid genes in petunia: Addition of a limited number of gene copies may lead to a suppression of gene expression. Plant Cell 2: 291-299

11

GENE EXPRESSION IN ROOTS DURING ECTOMYCORRHIZA DEVELOPMENT

U. Nehls and F. Martin

Equipe de Microbiologie Forestière
Centre I.N.R.A de Nancy
54280 Champenoux
France

ABSTRACT[*]

The identification of plant genes, differentially expressed during mycorrhiza formation, was carried out using two approaches. The first approach involved the random selection of clones from a mycorrhizal cDNA library, followed by differential screening to identify the source of the isolated clones and their expression in three different tissues: uninfected roots, mycorrhiza and the free living fungus. Fifty-three out of 200 investigated clones were identified as plant clones. Ten of these plant clones were differentially expressed (up- or down-regulated) during mycorrhiza formation. To obtain some information on the biological function of the identified plant clones, 25 cDNAs were partially sequenced and the deduced protein sequences aligned to protein databases; 7 plant clones could be identified by their strong homology to known proteins. Identified genes encoded for proteins with various functions in the cellular metabolism, from protein folding and structural functions to carbon metabolism and components of signaling pathways. The second approach included direct screening of a mycorrhizal cDNA library, using a labeled cDNA probe enriched in plant cDNAs to identify highly expressed plant genes. Three hundred and eighty positive clones were isolated and investigated further by a differential screening procedure. Only 6 % of the isolated clones were identified as fungal cDNAs, while the majority (71 %) represented plant genes. The expression of 40 of the isolated plant clones was differentially regulated during mycorrhiza formation.

INTRODUCTION

Ectomycorrhiza formation is a complex process with many developmentally-regulated morphological and physiological changes in both the fungal and the plant partner (for

---

[*]Abbreviations used: IAA, 3-indol acetic acid; SDS, sodium dodecyl sulfate; SR, symbiosis-regulated; EDTA, ethylenediaminetetraacetic acid.

Biotechnology of Ectomycorrhizae, Edited by Vilberto Stocchi et al.
Plenum Press, New York, 1995

a review see Martin et al., this volume). While the morphology of this developmental process has been well investigated during the last decade (Dexheimer & Pargney, 1991), our knowledge at the molecular level of how mycorrhiza formation is controlled and develops is still very limited. Recently, considerable progress has been made by the identification of fungal genes differentially regulated during mycorrhiza formation (Symbiosis-Regulated genes) at both the protein (Hilbert et al., 1991, Burgess et al., 1995) and the mRNA level (Tagu et al., 1993, Martin et al., this book). To obtain similar progress for the plant partner, we investigated plant gene expression during mycorrhiza formation to identify differentially-regulated genes. The model system used for our investigation was *Eucalyptus globulus* associated with the basidiomycete *Pisolithus tinctorius* in a petri dish system (Malajczuk et al., 1990).

MATERIAL AND METHODS

Strains and Vectors

A mycorrhizal cDNA library (Tagu et al., 1993) in the bacteriophage Lambda Zap (Stratagene) was used. The bacterial strain Xl-1 blue (Stratagene) was used for plating the library and for cloning cDNA inserts. For excision of the KS-plasmid (Stratagene) containing the cDNA inserts out of Lambda Zap, the f1 helper-phage (Stratagene) was used. Strain 441 of *Pisolithus tinctorius* Coker and Couch and *Eucalyptus globulus* ssp. bicostata, (Kylisa Seeds, Australia) were used.

Mycorrhiza Formation

P. tinctorius was precultivated on Pachlewski medium (Martin et al., 1990) at 24 °C for 2 weeks. *E. globulus* seedlings were germinated on Pachlewski medium at 24 °C for 2 days in the dark, and precultivated for 4 days at 24 °C in a growth chamber (16 h day period with 200 μM photons x m^{-2} x s^{-1}). For mycorrhiza formation *E. globulus* seedlings were placed on the surface of *P. tinctorius* colonies (approx 6 plants per colony) and cultivated in the growth chamber at 24 °C for further 4 days (Malajczuk et al., 1990; Hilbert et al., 1991). Mycorrhiza were collected by cutting off the surrounding fungal mycelium and the plant shoot, and were then frozen in liquid nitrogen and stored at -80 °C. As a control, uninfected roots and the outer 4 mm of fungal colonies were isolated from material grown at the same time and under the same conditions as the mycorrhizal material.

Differential Screening

The cDNA library was plated on agar plates (ϕ 15 cm, approx. 500 phages per plate) and incubated for 14 h at 37 °C. Single clones were isolated from phage plaques and stored in 130 μl SM (Maniatis et al., 1982). The cDNA inserts were amplified by PCR (GeneAmp 9600, Perkin Elmer) using M13 sequencing and reverse primer (Eurogentec). Five μg of the PCR-amplified cDNA inserts were applied to nylon membranes (Zetaprobe, BioRad) using a slot-blot apparatus (Millipore) and hybridized with [^{32}P]dCTP-labeled first strand cDNA (2×10^6 cpm/ml) of uninfected roots, fungus or mycorrhiza for 18 h in 10 % SDS, 200 mM sodium phosphate (pH 7.0). The membranes were then exposed to Hyperfilm ß-max (Amersham) for 1-4 weeks at -70° C.

Total RNA from uninfected roots, mycorrhiza and fungus was isolated according to Logeman et al. (1993). First strand cDNA synthesis was carried out using 4 to 8 μg total RNA (approx. 0.2-0.4 μg mRNA), 2 μl oligodT18 primer (Pharmacia) and 10 U Moloney

Murine Leukemia Virus reverse transcriptase (Gibco BRL) at 37° C for 2 h. Forty μCi [^{32}P]dCTP (Amersham) were used for labeling the first strand cDNA.

For Northern blot analysis, 20 μg total RNA of either root, mycorrhiza and fungus were separated on formaldehyde gels (Maniatis et al., 1982) for 2 -3 h and transferred to a nylon membrane for 16 h with 3.6 M NaCl, 2 M sodium phosphate, 20 mM EDTA, pH 7.4 (20 x SSPE). The hybridization was carried out in 50 % formamide, 5 x SSPE, 0.2 % bovine serum albumin (fraction V, Sigma), 0.5 % SDS, 0.1 mg/ml Hering sperm DNA (Sigma) at 42 °C for 20 h using 400 ng [^{32}P]dCTP-labeled PCR fragment 2 x 10^6 cpm/ml). The membranes were exposed to X-OMAT XAR-5 (Kodak) for 3-20 days at -70° C. Quantification of the Northern blot signals was carried out using the densitometer unit of a DU 70 Spectrophotometer (Beckman).

Substraction-Hybridization

For the preparation of genomic DNA, 1 g fungal tissue (only the outer 1 cm of growing colonies) was pulverized in a mortar under liquid nitrogen. The powder was resuspended in 15 ml 50 mM EDTA pH 8.0, 0.2 % SDS, 5 mM ß-mercaptoethanol to which 15 μl diethyl pyrocarbonate (Sigma) was added. The lysate obtained was incubated at 68 °C for 15 min, cooled to room temperature and centrifuged at 10,000 x g for 15 min. The supernatant was mixed with 1 ml 5 M potassium acetate (pH 5.5), incubated at 4 °C for 1 h and centrifuged at 10,000 x g at 4 °C for 15 min. The supernatant was mixed with 0.7 vol. isopropanol and centrifuged at 10,000 x g for 15 min. The DNA-pellet was washed once with 70 % ethanol, dried, dissolved in 0.4 ml 10 mM Tris (pH 7.0), 1 mM EDTA and stored at 4 °C.

[^{32}P]dCTP-labeled first strand cDNA (4x10^7 cpm) obtained from approx. 0.4 μg mycorrhizal mRNA (see above) was hybridized with 80 μg fungal genomic DNA in 10 % SDS, 200 mM sodium phosphate (pH 7.0) for 14 h.

The cDNA library was plated on agar plates (φ 15 cm, approx. 1000 phage per plate) and incubated for 14 h at 37 °C. Nylon membrane replicates (Zetaprobe, BioRad) were made from the plates and hybridized with the subtracted mycorrhizal cDNA probe (see above) in 10 % SDS, 200 mM sodium phosphate (pH 7.0) For 18 h. Single phage clones, giving hybridization signals, were isolated and stored in 130 μl SM.

Phage Dot blotting was carried out by pipetting 1 μl phage solution per dot on a growing bacteria lawn (80 Dots per agar plate, φ 15 cm) and incubation for 14 h at 37 °C. Three identical nylon membrane replicates (Zetaprobe, BioRad) were made from the plates. Each set of membranes was hybridized with [^{32}P]dCTP-labeled first strand cDNA (2x10^6 cpm/ml) of either: uninfected roots, fungus or mycorrhiza in 10 % SDS, 200 mM sodium phosphate (pH 7.0) for 18 h. The membranes were exposed to X-OMAT LS (Kodak) for 1-4 weeks at -70 °C.

Miscellaneous

The ergosterol content in mycorrhiza was determined according to Martin et al. (1990). Sequencing (Sanger et al, 1977) was performed using either T7 DNA-polymerase (Pharmacia) or *Tth* DNA-polymerase (Gibco BRL).

The sequence data were compared to databases (EMBL and Genbank) using Blastp or Blastx (Altschul et al., 1990) and homologies of the deduced protein sequences with known proteins were further analyzed using ClustalV (Higgins et al., 1992).

RESULTS

Mycorrhiza formation encompasses a complex series of ontogenic steps and therefore the current investigation was restricted to the developmental step where the fungal mantle is formed and development of the Hartig net is initiated. Two approaches were used to identify differentially expressed plant genes:

1. Random selection of clones from a mycorrhiza cDNA library (Tagu et al., 1993), followed by a differential screening procedure to identify plant genes and their expression during mycorrhiza formation.

2. Screening of the mycorrhiza cDNA library with a labeled cDNA probe enriched in plant cDNAs to identify highly expressed plant genes, followed by a differential screening procedure to investigate their expression during mycorrhiza formation.

Cloning of SR-Genes by Differential Screening

A flow chart summarizing this procedure is shown in Fig 1. The cDNA inserts of 600 randomly isolated clones from the cDNA library were amplified by the polymerase chain reaction (see Material and methods). Two hundred clones with inserts longer than 500 bp were selected for further analysis. Equal amounts of amplified cDNA were applied to 3 identical sets of membranes using a slot blot apparatus. Each set of membranes was hybridized with radioactive-labelled first strand cDNA made from mRNA of either uninfected plant roots, mycorrhiza or fungus. This enabled us to obtain 3 pieces of information:

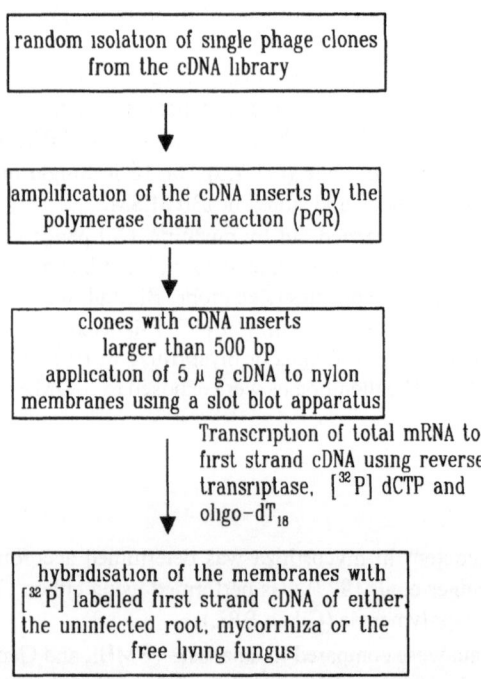

Figure 1. Flow chart of the strategy for cloning of SR-genes by differential screening

1. By hybridization of a given clone to the first strand cDNA probe corresponding to each of the free living partners, the origin of this clone (fungus or plant) could be determined.

2. The intensity of the hybridization signals on the membranes is proportional to the amount of mRNAs in the different samples (plant, fungus or mycorrhiza). Therefore, the gene expression in the free living partners and in mycorrhiza could be estimated and compared for the different cDNA clones.

3. The relative gene expression of a single cDNA clone in comparison to the other investigated clones.

Among cDNAs of known origin, 53 were plant and 43 fungal clones. The genomic source of the remaining clones was not determined due to either: cross-hybridization yielding identical signals on each set of membranes or a gene expression level in the free living partners that was under the detection limit of the Slot blotting approach. The hybridization intensity which reflects the transcript concentrations of the identified plant clones ranged from barely detectable to intense signals (Table 1).

Northern blot analysis was carried out on selected clones (Fig. 2).

Equal amounts of total RNA isolated from either uninfected roots, mycorrhiza or the fungus were separated on agarose gels and probed with [^{32}P]-labelled cDNAs. The result of the Northern blotting confirmed the data obtained with the Slot blotting approach. Identical rates of the signal intensities in uninfected roots and mycorrhiza were obtained with both approaches. A house-keeping gene (DnaJ, see below) was used as an internal standard to estimate and compare the transcript levels of identified genes in uninfected roots and

Table 1. Signal intensities of plant cDNA clones identified by the differential hybridization procedure in uninfected roots and *E. globulus-P. tinctorius* ectomycorrhiza

	Signal intensity in			Signal intensity in			Signal intensity in	
Clone	Root	Mycorrhiza	Clone	Root	Mycorrhiza	Clone	Root	Mycorrhiza
16	+	+	197	+	0	392	0	–
18	+++	+	208	–	0	431	–	0
32	++++	0	213	0	–	446	0/+	0
38	++	–	229	++	++	447	–	0
42	+	–	235	0/+	0	466	–	0
43	++	++	248	++++	0/+	483	+	–
52	++++	+/++	269	0	–	502	–	+
73	+	0	277	++++	+	517	++	–
78	0	–	286	–	0	523	++	0/+
79	+++++	+++	292	0/+	0	525	–	0
86	+	–	303	++	+	533	0	+/++
98	0	0	307	0	–	534	0/+	–
103	0	–	338	0/+	0	547	–	0
146	+++++	+	340	–	0	564	+	–
171	++++	++	341	0	0	585	0	0
178	+++	+	350	–	0	587	+	0
185	+++	+	351	0	–	588	++	0
194	–	0	368	0	+			

–, no hybridization; 0, very weak hybridization; +,++,+++, ++++ intense hybridization with increasing signal intensity

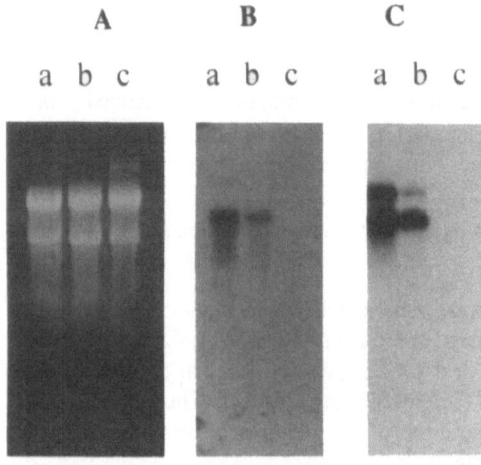

Figure 2. Northern blot analysis of selected plant cDNAs Total RNA of uninfected roots (lane a), mycorrhiza (lane b) and fungus (lane c) were separated on agarose gels transferred to nylon membranes and hybridised with plant cDNAs. (A) Ethidium bromide stained agarose gel; (B) clone 392; (C) clone 447 [auxin regulated protein] together with clone 42 [DnaJ].

mycorrhiza. Ten out of the 53 identified plant genes were differentially regulated during mycorrhiza formation. Their expression was enhanced for 6 and reduced for 4 genes.

Twenty-five plant cDNAs, including SR-cDNAs, were partially sequenced and their deduced protein sequences aligned to the EMBL and GenBank databases, to obtain some insight on their putative biological function. Homology to already known proteins was found for 7 clones (Table 2, Fig. 3).

Cloning of SR-Genes by Substraction-Hybridization

The random isolation procedure yielded 10 differentially regulated plant clones, but this approach was very time consuming due to the occurrence of a large number of fungal cDNAs in the library. Accordingly, an alternate strategy was designed to identify highly expressed plant genes in mycorrhiza (Fig.4).

Table 2. Homology of symbiosis related plant cDNA clones to known proteins and modification of their transcript level in symbiotic tissues during mycorrhiza development

Clone	Homologous to	Genus	Similarity %	Compared amino acids	cDNA length (bp)	Transcript level in mycorrhiza in comparison to uninfected roots
R 32	catalase I	Arabidopsis thaliana	92	50	900	-50%
R 38	cytochrome c I	Solanum tuberosum	96	95	1000	ND[a]
R42	DnaJ	Alium porrum	100	30	1100	identical[b]
R 350	α-tubulin	Zea mays	97	379	1200	+400%
R 447	auxin regulated protein	Nicotiana tabacum	84	67	1000	+300%
R 517	calmodulin I	Arabidopsis thaliana	100	23	650	identical
R 585	cytosolic malate dehydrogenase	Sus scrofa	76	82	800	ND

[a]ND = not determined.
[b]cDNA clone used as internal standard to estimate the level of plant transcripts unaffected by mycorrhiza formation.

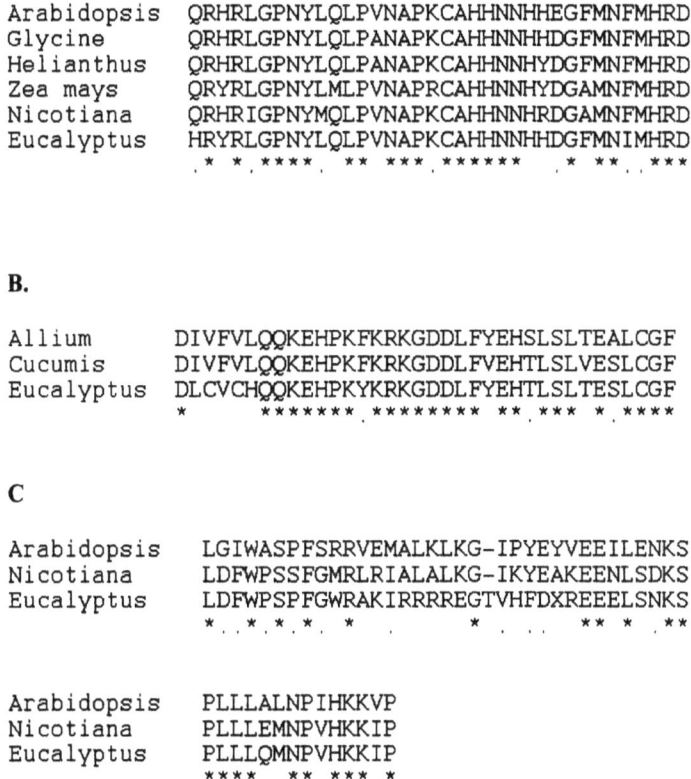

A

Arabidopsis	QRHRLGPNYLQLPVNAPKCAHHNNHHEGFMNFMHRD
Glycine	QRHRLGPNYLQLPANAPKCAHHNNHHDGFMNFMHRD
Helianthus	QRHRLGPNYLQLPANAPKCAHHNNHYDGFMNFMHRD
Zea mays	QRYRLGPNYLMLPVNAPRCAHHNNHYDGAMNFMHRD
Nicotiana	QRHRIGPNYMQLPVNAPKCAHHNNHRDGAMNFMHRD
Eucalyptus	HRYRLGPNYLQLPVNAPKCAHHNNHHDGFMNIMHRD

B.

Allium	DIVFVLQQKEHPKFKRKGDDLFYEHSLSLTEALCGF
Cucumis	DIVFVLQQKEHPKFKRKGDDLFVEHTLSLVESLCGF
Eucalyptus	DLCVCHQQKEHPKYKRKGDDLFYEHTLSLTESLCGF

C

Arabidopsis	LGIWASPFSRRVEMALKLKG-IPYEYVEEILENKS
Nicotiana	LDFWPSSFGMRLRIALALKG-IKYEAKEENLSDKS
Eucalyptus	LDFWPSPFGWRAKIRRRREGTVHFDXREEELSNKS

Arabidopsis	PLLLALNPIHKKVP
Nicotiana	PLLLEMNPVHKKIP
Eucalyptus	PLLLQMNPVHKKIP

Figure 3. Sequence alignments of the deduced protein sequence from cDNA sequences of identified eucalypt clones with known proteins. The deduced protein sequences of partially sequenced eucalyptclones were aligned to homologous proteins using the ClustalV programme. Stars represent identical, points similar aminoacids. (A) catalase I [*Arabidopsis thaliana* (Chevalier et al., 1993), *Glycine max* (Isin and Allen 1992), *Helianthus annuus* (Kleff et al., 1994) *Zea mays* (Abler and Scandalios 1993), *Nicotiana tabacum* (Schultes et al.)]; (B) DnaJ [*Allium porrum* (Bessoule 1993), *Cucumis sativus* (Preisig-Mueller and Kindl 1993)]; (C) auxin-regulated protein [*Arabidopsis thaliana* (Krivitzky et al., 1993), *Nicotiana tabacum* (Takahashi et al., 1989)].

[^{32}P]-labeled first strand mycorrhizal cDNAs were prehybridized with a 100-fold excess of denatured fungal genomic DNA to eliminate the fungal cDNAs from the probe. The resulting cDNA mix, depleted in fungal cDNAs, was used to screen the mycorrhizal cDNA library. From 12,000 investigated clones, approximately 1000 clones hybridized to the "plant-enriched" probe. The hybridization intensities (representing transcript concentrations) varied from barely detectable to strong signals (Fig. 4). Three hundred and eighty of the hybridizing clones were isolated and further investigated by a differential screening procedure using Phage Dot blotting (see Material and Methods) to determine their origin and to determine the influence of mycorrhiza formation on their gene expression (Fig. 5).

Only 21 (approx. 6 %) of the 380 investigated clones were of fungal origin. These fungal clones were highly expressed in mycorrhiza and therefore they were not eliminated by the fungal cDNA-depletion step. Ten of these 21 clones represented only 2 different fungal genes which are highly up-regulated (approx. 10 fold) during mycorrhiza formation. Nev-

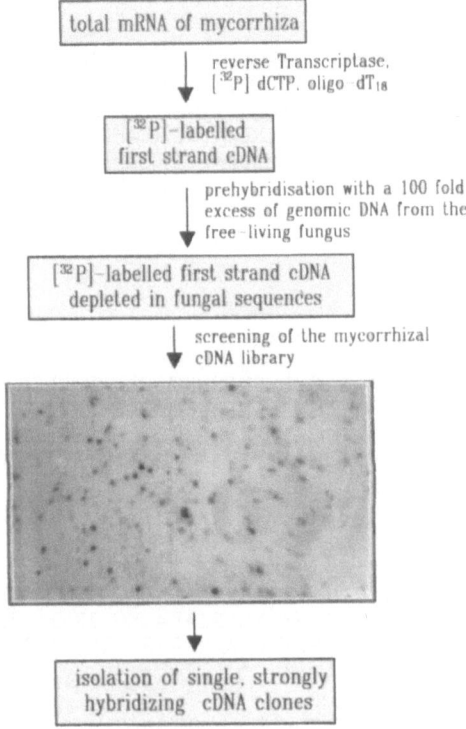

Figure 4. Flow chart of the strategy for cloning of SR-genes by subtractive hybridization.

ertheless, 320 clones hybridized to probes corresponding to uninfected roots and mycorrhiza, revealing their plant origin and indicating the efficiency of this approach. Forty out of these 320 plant clones were differentially regulated (up- or down-regulated) during mycorrhiza formation and are currently characterized.

Estimation of the Plant Transcript Content in Mycorrhiza

To estimate the effect of mycorrhiza formation on the expression of the identified genes, the respective content of plant and fungal mRNAs in mycorrhiza should be determined. This was done by using the fungal 5.8S rRNA (Tagu et al., in preparation) and the DnaJ gene of *Eucalyptus*. The DnaJ protein is involved in protein folding, one of the basic mechanisms of the cell and therefore is thought to be unaffected by mycorrhiza formation.

Fungal RNA, two different amounts of mycorrhizal RNA and a mix of increasing amounts of RNA from uninfected root with decreasing amounts of fungal RNA (artificial mycorrhiza) were separated on agarose gel, transferred to a nylon membrane and hybridized with the DnaJ cDNA (Fig. 6).

Based on the signal intensity, the plant mRNAs represented about 20 % of the mycorrhizal mRNAs.

Based on this plant mRNA content in mycorrhiza, hybridization signals in uninfected roots and mycorrhiza obtained by Northern blotting were then calibrated (Table 2).

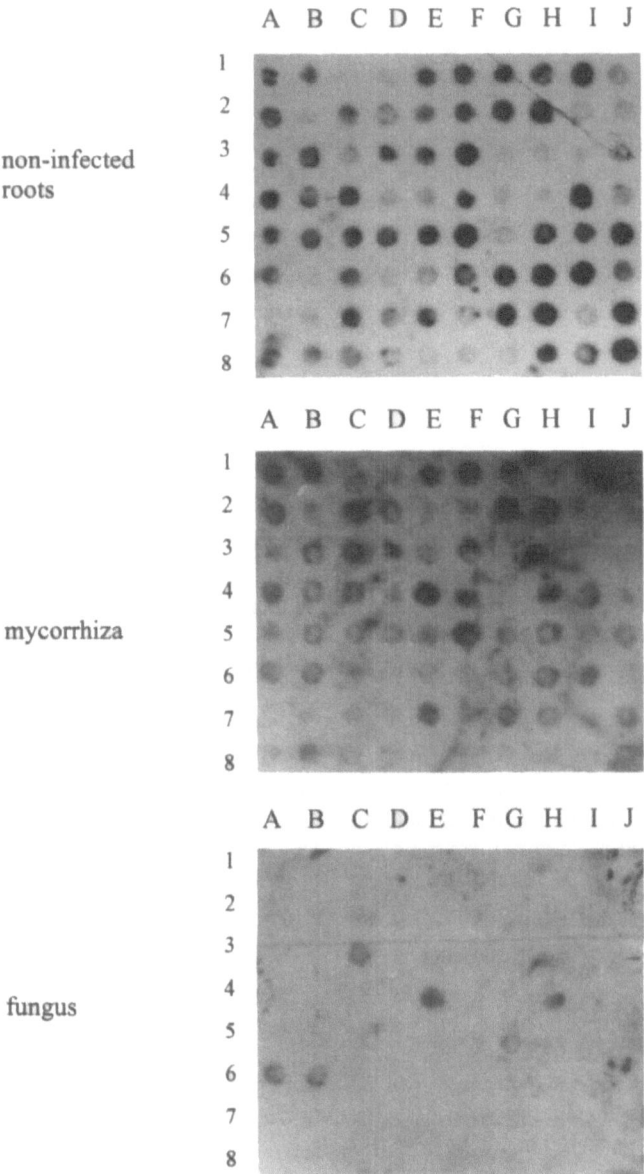

Figure 5. Differential screening of isolated clones using phage Dot blotting Small aliquots of phage suspensions (representing single isolated clones) were applied to a growing bacteria lawn on petri dishes and incubated until lysis plaques were visible Three identical prints were made on nylon membranes and hybridized with [^{32}P]-labeled first strand cDNA of uninfected roots, mycorrhiza or fungus

100 6 14 20 26 % RNA

A B C D E F G

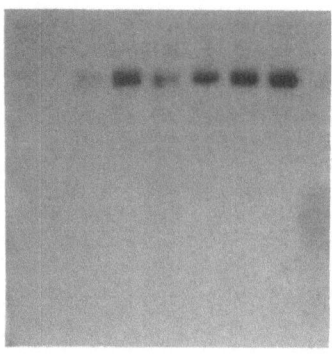

Figure 6. Estimation of the plant mRNA content in ectomycorrhiza The content of plant mRNA in mycorrhiza has been
estimated using the mRNA content of the Eucalypt DnaJ
gene Eight µg fungal RNA (lane A), 3 2 µg (lane B) and
12 5 µg (lane C) mycorrhizal RNA and 0 8 µg (lane D), 1 7
µg (lane E), 2 5 µg (lane F) and 3 3 µg (lane G) root RNA
were separated on an agarose gel, transferred to a nylon
membrane and hybridised with the Eucalyptus DnaJ cDNA.
The numbers on the top represent the percentage of the root
RNA in the lanes in relation to the mycorrhizal RNA in lane C

DISCUSSION

Studies of ectomycorrhiza development at the molecular level were initiated by the
identification of SR-proteins using 2-D-PAGE (Hilbert et al., 1991; Simoneau et al., 1993;
Burgess et al., 1995). These investigations revealed the occurrence of major changes in
protein biosynthesis during symbiosis formation. Using a differential screening procedure
that included PCR amplification of cDNA, Tagu et al. (1993) were able to identify fungal
mycorrhiza-regulated transcripts in *Eucalyptus globulus-Pisolithus tinctorius* ectomycorrhiza.

A major goal of our current research is to obtain a general view of the molecular
events that lead to the development and maintenance of the symbiotic organ in eucalypt
roots. To this end, we cloned plant transcripts representing genes of up- or down-regulated
expression. Here we described different approaches developed to identify plant genes
differentially regulated during early stages of ectomycorrhiza development.

Estimation of the Plant Gene Expression in Mycorrhiza

In mycorrhiza, the fungus and the plant tissues, are tightly connected and the isolated
mRNAs are a mixture from both partners. To estimate whether the expression of identified
plant genes is influenced by the mycorrhiza formation, a critical step was the determination
of the plant RNA content in mycorrhiza. Different criteria were used to estimate the level of
fungal and plant material in *planta*

1. Initial investigations were based on the use of biochemical markers, such as chitin
   for the cell wall or ergosterol for the membrane content (Martin et al., 1990).
   According to these methods, fungal tissues represent about 25-30 % of the
   mycorrhiza (Martin et al., 1990; Burgess et al., 1995), suggesting that plant
   mRNAs account for 70-75 % of the total mRNAs in mycorrhiza. The main
   disadvantage of these biochemical methods is that they do not take into consideration the different transcriptional activities of the two partners which is likely
   to influence their mRNA content in mycorrhiza.
2. To estimate the RNA content of each symbiont in the mycorrhizal tissues we
   therefore used the level of the fungal 5.8S rRNA (Tagu et al., in preparation) and
   the level of transcripts of genes of the basic cellular metabolism which are likely

to be unaffected by mycorrhiza formation. One of the plant clones identified in this work (Table 1) was the DnaJ gene. The DnaJ protein is involved in protein folding (Sadler et al., 1989; Hendric et al., 1993), one of the basic mechanisms within the cell. On the basis of the DnaJ mRNA content in mycorrhiza, we estimated that the plant mRNA represented approx. 20% of the total mycorrhizal mRNA. The discrepancy between the values obtained with biochemical markers (fungal tissue approx. 25-30 % of the mycorrhizal tissue) and molecular markers (fungal mRNAs approx. 70-75 % of mycorrhizal mRNAs) could be explained by:

1. the coenocytic nature and the high cellular activity of the hyphae colonizing the root tissues and by
2. the arrest of the root tip meristems (by far the most active part of the root) during mycorrhiza formation.

Identification of Differentially Regulated Plant Genes

Two approaches were used to identify differentially regulated plant genes. The differential screening procedure was designed to identify both lowly and highly expressed fungal and plant genes. In contrast, the substraction-hybridization strategy was designed for efficient identification of highly expressed plant genes.

Differential Screening Procedure

Randomly selected clones from a mycorrhizal cDNA library were analyzed by a differential screening procedure (Slot blotting). The gene expression in the different tissues was confirmed using Northern blot analysis for selected plant clones.

Hybridization signals from Slot and Northern blotting were comparable in uninfected roots and mycorrhiza. This means that the slot blot analysis could be used as a quick and reliable method to estimate the influence of mycorrhiza formation on gene expression. However, Northern blotting is recommended for accurate quantification of the changes in gene expression, due to its higher sensitivity and low cross artefactual hybridization.

This differential screening procedure has been successfully used to identify mycorrhiza-regulated plant genes. However, it was very time consuming to identify 10 differentially-regulated plant clones out of the 600 which were initially selected.

Substraction-Hybridization Strategy

The key step of this cloning procedure was the prehybridization of a radioactive-labeled mycorrhizal first strand cDNA with an excess of fungal genomic DNA. This prehybridization step removed most fungal cDNAs from the mycorrhizal probe and generated a probe enriched in plant cDNAs. Approximately 1000 (8 %) out of 12,000 investigated cDNA clones hybridized to this probe enriched in plant transcripts.

The isolated clones were further investigated by a differential screening procedure using phage Dot blotting (see Material and methods). The advantage of this phage Dot blot approach was its lower cost and labor requirements in comparison to the differential screening procedure previously used (see above). The substraction-hybridization strategy was mainly designed for the rapid identification of highly expressed plant genes and therefore the sensitivity of the phage Dot blotting was adequate (as indicated by hybridization signals obtained for 361 out of 380 investigated clones).

Only 6 % of the 380 clones analyzed so far were identified as fungal cDNA clones which indicated the efficiency of the developed procedure to eliminate fungal cDNAs. The identified fungal clones were highly expressed in mycorrhiza and therefore the 100-fold excess of fungal genomic DNA used to trap fungal cDNAs was not sufficient to eliminate these cDNAs. Forty plant clones were identified as differentially-regulated during mycorrhiza formation. This cloning technique is therefore well suited for isolating plant SR-genes against a high background of fungal cDNAs.

Putative Biological Function of Identified Plant Clones

The putative cellular function of the protein encoded by the identified plant genes was determined by their sequence homology to known proteins. Identified gene products included DnaJ (protein folding) and α-tubulin (cytoskeleton protein), but also proteins involved in the plant defense reactions, (catalase I, Chevalier et al., 1993), carbon metabolism (cytosolic malate dehydrogenase, Joh et al., 1987), cytochrome C1 (Martinez et al., 1974), signal transduction pathways (calmodulin, Chandra et al., 1993), or modification of transcriptional activity (auxin-regulated protein, Takahashi et al., 1989).

The expression of these identified genes varied during mycorrhiza formation (Table 2). For example, the transcript level of the catalase I gene was decreased by 50 % in mycorrhizal tissues. This presumably leads to an increased H_2O_2 level in eucalypt roots. Active oxygen species and H_2O_2 are components of the plant defense reaction (Degousée et al., 1994).

In contrast, the expression of the α-tubulin gene was enhanced by 400 %. This may reflect the observed changes in the cytoskeleton of the root epidermal cells during Hartig net formation (Peterson and Bonfante, 1994). Furthermore, the expression of the auxin-regulated protein was enhanced 3-fold. This up-regulation underlines the importance of auxin during mycorrhiza formation. A well known feature of auxins produced by ectomycorrhizal fungi is the stimulation of lateral root formation (Burgess et al., 1995). Recently, Gay et al. (1994) have produced mutants of *Hebeloma cylindrosporum* overproducing IAA. These mutants induced pine ectomycorrhiza with an altered morphology (e.g. multiseriate Hartig net) and were able to increase the formation of lateral roots (Gay, personal communication). We will use the identified plant auxin-regulated protein as a molecular marker to further investigate the role of fungal auxins on lateral root formation.

ACKNOWLEDGMENTS

U. Nehls was a recipient of fellowships from the Ministère de la Recherche et de l'Enseignement Supérieure and the Deutsche Forschungsgemeinschaft (DFG). This work was supported by a grant from the Eureka-Eurosilva program to F. Martin. We would like to thank D. Tagu and M. Carnero Diaz for helpful discussion and P. Murphy for a critical reading of the manuscript.

REFERENCES

Abler, M. L. and Scandalios, J. G. 1993. Isolation and characterization of a genomic clone encoding the maize *Cat3* catalase gene. Plant Mol. Biol. 22, 1031-1038.

Altschul, S. F., Gish, W., Miller, W., Myers, E. W., and Lipman, D. J. 1990. Basic local alignment search tool. J. Mol. Biol. 215, 403-410.

Bessoule, J. J. 1993. Occurrence and sequence of a DnaJ protein in plant (*Allium porrum*) cells. FEBS Lett. 323, 51-54.

Burgess, T., Laurent, P., Dell, B., Malajczuk, N., Martin, F. 1995. Symbiosis-related polypeptides in the *Pisolithus tinctorius* (Cocker & Couch) *Eucalyptus grandis* (W. Hill ex Maiden) ectomycorrhiza. Planta, 195, 407-417.

Chandra, A. and Upadhyaya, K. C. 1993. Structure and organization of a novel calmodulin gene of *Arabidopsis thaliana*. Cell. Mol. Biol. Res. 39, 509-516.

Chevalier, C., Yamaguchi, J. and McCourt, P. 1993. Nucleotide sequence of a cDNA for catalase from *Arabidopsis thaliana*. Plant Physiol. 99, 1726-1728.

Degusée, N., Triantaphylidès, C. and Montillet, J.-L. 1994. Involvement of oxidative process in the signaling mechanisms leading to the activation of glyceolin synthesis in soybean (*Glycine max*). Plant Physiol., 104, 945-952.

Dexheimer, J. and Pargney, J.-C. 1991. Comparative anatomy of the host-fungus interface in mycorrhiza. Experientia, 47, 312-320

Gay, G., Marmeisse, B., Sotta, B. and Debaud, J. C. 1994. Auxin overproducer mutants of *Hebeloma cylindrosporum* (Romagnesi) have increased mycorrhizal activity. New Phytol., 128, 645-657.

Hendrick, J. P., Langer, T., Davis, T. A., Hartl, U. and Wiedmann, M. 1993. Control of folding and membrane translocation by binding of the chaperone DnaJ to nascent polypeptides. Proc. Acad. Sci. USA, 90, 10216-10220.

Higgins, D. G. Bleasby, A. J. and Fuchs, R. 1992. CLUSTAL V. improved software for multiple sequence alignment. CABIOS 8, 189-191.

Hilbert, J. L., Costa, G. and Martin, F. 1991 Regulation of the gene expression in ectomycorrhizas. Early ectomycorrhizins and polypeptide cleansing in eucalypt ectomycorrhizas. Plant Physiol. 97: 977-984.

Isin, S. H. & Allen, R. D. 1992. Characterization and expression of a soybean catalase gene. Plant Mol. Biol.

Joh, T. H., Takeshima, H., Tsuzuki, T., Setoyama, C., Shimada, K., Tanase, S., Kuramitsu, S., Kagamiyama, H. and Morino, Y. 1987. Cloning and sequence analysis of cDNAs encoding mammalia cytosolic malate dehydrogenase. J. Biol. Chem. 262, 15127-15131.

Kleff, S., Trelease, R. N. and Eising, R. 1994. Unpublished.

Krivitzky, M., Bonnet, R., Jean-Jacques, I., Kreis, M. and Lecharny, A. 1993. unpublished.

Logeman, J., Schell, J. and Willmitzer L. 1987. Improved method for the isolation of RNA from plant tissues. Analytical Biochemistry, 163, 16-20

Malajczuk, N., Lapeyrie, F. and Garbaye, J. 1990. Infectivity of pine and eucalypt isolates of *Pisolithus tintorius* on roots of *Eucalyptus urophylla in vitro*. New Phytol., 114, 627-631.

Maniatis, T., Sambrock, J. and Fritsch, E. F. 1982. Molecular cloning: A laboratory manual. Cold Spring Harbor Laboratory Press, CSH.

Martin, F., Delaruelle, C., Hilbert, J. L. 1990. An improved ergosterol assay to estimate the fungal biomass in ectomycorrhizas. Mycol. Research 94, 1059-1064.

Martin, F. and Hilbert, J. L. 1991. Morphological biochemical and molecular changes during ectomycorrhizal development Experientia, 47, 321-330.

Martin, F., Burgess, T., Carnero Diaz, M. E., de Carvalho, D., Laurent, P., Murphy, P., Nehls, U. and Tagu, D. 1995. Ectomycohhiza Morphogenesis: Insights from studies of developmentally-regulated genes and proteins. (this volume).

Martinez, G., Rochat, H., Ducet, G. 1974. The amino acid sequence of cytochrome c from *Solanum tuberosum* (potato). FEBS Lett, 47, 212-7.

Peterson, R. L., Bonfante, P. 1994. Comparative structure of vesicular-arbuscular mycorrhizas and ectomycorrhizas. Plant and Soil, 159, 79-88.

Preisig-Mueller, R. and Kindl, H. 1993. Plant DnaJ homologue: molecular cloning, bacterial expression, and expression analysis in tissues of cucumber seedlings. Arch. Biochem. Biophys. 305, 30-37.

Sadler, L., Chiang, A., Kurihara, T., Rothblatt, J., Way, J. and Silver, P. 1989. A yeast gene important for protein assembly into the endoplasmic reticulum and the nucleus has homologie to DnaJ, an *E. coli* heat shock protein. J. Cell. Biol., 109, 2665-2675.

Sanger, F., Nickler, S.,Coulson, A. R. 1977. DNA sequencing with chain terminating inhibitors. Proc. Natl. Acad. Sci. USA 74, 5463-5467.

Simoneau, P., Viemont, J. D., Moreau, J. C., Strullu, D. G. 1993. Symbiosis-related polypeptides associated with theearly stagees of ectomycorrhiza organogenesis in birch (*Betula pendula* Roth). New Phytol. 124, 495-504.

Tagu, D., Phyton, M., Crétin, C., Martin, F. 1993. Cloning symbiosis-related cDNAs from eucalyt ectomycorrhiza by PCR-assisted differential screening. New Phytol. 125, 339-343.

Takahashi, Y., Kuroda, H., Tanaka, T., Machida, Y., Takebe I. and Nagata, T. 1989. Isolation of an auxin-regulated gene cDNA expressed during the transition from G 0 to S phase in tobacco mesophyll protoplasts. Proc. Natl. Acad. Sci. U.S.A. 86, 9279-9283.

12

TRUFFLES

Their Life Cycle and Molecular Characterization

Luisa Lanfranco, Marco Arlorio, Antonella Matteucci, and Paola Bonfante

Centro di Studio sulla Micologia del Terreno del CNR
and Dipartimento di Biologia Vegetale dell'Università
Viale Mattioli 25
10125 Torino, Italy

SUMMARY

Truffles are ascomycetous ectomycorrhizal fungi: the accomplishment of their life cycle relies therefore on the establishment of symbiotic relationships with a narrow range of host plants. Three different phases have been identified in their life cycle: i) growth as filamentous vegetative mycelia, ii) organization of hypogeous fruitbodies with asci and ascospores, and iii) association of the fungal hyphae with the host root to form the fungal mantle and the Hartig net. The identification of truffles is mostly based on the morphological features of the fruitbodies (size and shape of asci, number and wall ornamentations of spores within the asci). However, many of the structures important for taxonomic identification are absent during the phase of filamentous growth and during the contact of hyphae with the host surface in the establishment of mycorrhizae.

The aim of this paper is to illustrate the strategies that were followed to develop molecular probes which would allow truffle identification during their different developmental phases.

DNA extracted from fruitbodies of 10 *Tuber* species, from mycorrhizae and from mycelia growing *in vitro* of *T. magnatum* and of *T. borchii,* was amplified in PCR experiments by using i) ITS and IGS primers designed for ribosomal gene spacers, ii) short arbitrary oligomers in RAPD experiments and iii) specific primers for *T. magnatum* designed on the sequence of a non-polymorphic RAPD band. Amplifications of DNA extracted from fruitbodies with ITS and IGS primers, followed by digestion with restriction enzymes *Hinf* I and *Mbo* I, led to genetic fingerprints which differentiated among the different species in the fruitbodies as well as in the mycorrhizae. RAPD fingerprints revealed a high degree of polymorphism among the different truffle species, but they showed a rather low polymorphism among fruitbodies of the same species, *T. magnatum*. A RAPD band of 1.5 kbp that was found to be specific for *T. magnatum* was cloned and partially sequenced. Two specific primers were designed and allowed us to distinguish unambiguously among the truffle fruitbodies, as they amplified DNA sequences only in *T. magnatum* samples.

Biotechnology of Ectomycorrhizae, Edited by Vilberto Stocchi et al.
Plenum Press, New York, 1995

139

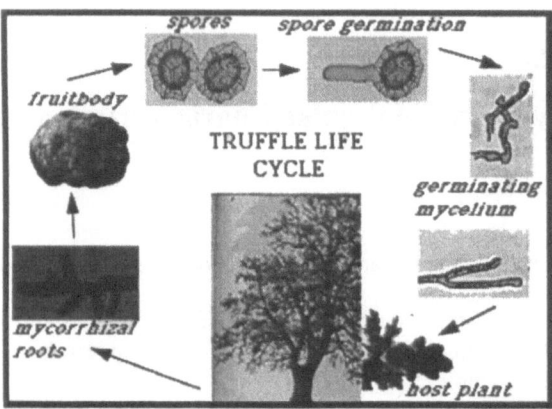

Figure 1. Some steps of the life cycle of a truffle are shown: ascospores are released after the fruitbody breakdown and germinate in the soil; a heterocaryotic mycelium is formed, which makes contact with the host plant, leading to the formation of mycorrhizal roots. The fruitbody is formed only after the establishment of the mycorrhiza.

INTRODUCTION

Truffles are ascomycetous fungi that form hypogeous fruitbodies. Spores are produced inside asci and are then disseminated either through the natural breakdown of the mature fruitbodies or by invertebrates and small mammals (1). As early as 1887, Mattirolo (2) suggested that truffles lived in symbiosis with some trees, but only after the establishment of mycorrhizae under controlled conditions (3) were they universally acknowledged as ectomycorrhizal fungi, associated with the roots of gymnosperm and angiosperm trees (4, 5).

The life cycle of a truffle appears to be very complex, since it requires the succession of many steps: spore dispersion and germination, production of a homonucleate mycelium, establishment of heterocaryotic mycelia, contact with the host root, formation of the fungal mantle and the Hartig net, development of extraradical mycelia and fruitbodies (Fig. 1). Many of these steps will depend on the genetic traits of the fungus (mating-types, compatibility mechanisms, sexual reproduction, fruitbody development), whereas others will depend on the host plant (recognition events, shift of the fungal morphogenesis from a filamentous growth pattern in hyphae towards the formation either of a tissue-like structure in the mantle or of a highly branched pattern in the Hartig net). Finally, steps which are crucial for the ecological success of the truffles (spore dispersion and spore germination), will depend on the dispersal operated by animals and on their interaction in the soil with other soil microrganisms, both fungi and bacteria.

The truffle life cycle can be envisaged therefore as a complex scenario with an interplay among organisms belonging to all the different living kingdoms. Unfortunately, many aspects of this cycle have only been hypothesized. For example, breeding mechanisms, the regulation of sporocarp development as well as the mechanisms of vegetative compatibility have already been partly investigated in some basidiomycetous ectomycorrhizal fungi (6, 7), but they remain obscure for truffles. Truffle biology has in fact been neglected for a long time. In contrast with other ectomycorrhizal fungi like *Pisolithus tinctorius,* which represents a model system for the investigation of plant-fungus relationships (8), truffle mycelia grow very slowly *in vitro* and mycorrhizae are obtained with great difficulties under fully axenic conditions. Understanding how a fungal fruitbody is formed is one of the most challenging fields of research in developmental mycology today: results are limited to some model systems such as *Neurospora* (9) and *Schizophyllum commune* (10): it is not therefore surprising that the development of the truffle fruitbody is a still mysterious event.

Searching for Reliable Identification Parameters

In contrast with the scanty information on the truffle life cycle many efforts have been made, from the Seventies onwards, to increase truffle production, due to the remarkable commercial values of some species, such as *Tuber magnatum* Pico and *T. melanosporum* Vitt.. Agricultural programs for large scale mycorrhizal production have been developed in France, Italy and Spain (11). As with other ectomycorrhizal fungi, such programs require inoculation of the host plants (oak, hazel, willow and poplar are the most commonly used), testing of mycorrhizal formation, planting out of the mycorrhizal seedlings, checking of the presence of the introduced *Tuber* species within the endogenous microbial community and, finally, harvest of fruitbodies. This protocol presents many variables, among which i) heterogeneity in the plant and the fungal material that may lead to mycorrhizal plants with different symbiotic properties (11) and ii) difficulties in the identification of the fungus during the symbiotic phase prior to sporocarp development (Fig.1). The first problem is currently solved by the use of micropropagated plant material and by checking the symbiotic properties of the selected fungal strains, whereas the second aspect opens the question of whether morphological features are sufficient parameters to differentiate among truffles. Sporocarps are usually identified on the basis of size and shape of spores and asci, wall ornamentations, structure of the peridium and gleba, but these features are usually absent during the vegetative phase and the development of the symbiotic structures. The features of the mantle and the Hartig net may be useful for the identification of some closely related fungal symbionts (*T. magnatum* versus *T. borchii*) (12), but their discrimination is a difficult task and requires specialised personnel. The use of molecular techniques in cases where morphological characters are in conflict, ambiguous or missing has been addressed by many authors (13, 14, 15, 16) and already applied to endo- and ectomycorrhizal fungi (17, 18, 19, 20).

Some of these techniques have been recently applied also to the identification of truffles (21, 22, 23). The aim of this paper is to illustrate the strategies we followed to develop molecular probes which would allow us to:

i. identify truffles during their different developmental phases: namely, hypogeous fruitbodies with asci and ascospores, mycorrhizal roots and filamentous vegetative mycelia
ii. evaluate the level of intraspecies polymorphisms shown at the level of the fruitbodies.

In order to meet these goals, primers with different taxonomic value were used in PCR-based techniques: the ITS pair (ITS$_1$ and ITS$_4$) and IGS pair (5SA and CNL12) were used to identify interspecific differences (17, 24), short primers of arbitrary sequences were used in RAPD experiments to reveal intraspecific differences (25, 26) and RAPD-derived specific primers for the identification of *T. magnatum*.

MATERIALS AND METHODS

Fruitbodies of 10 *Tuber* species were collected from different locations in Italy (Table 1) and stored as soon as possible at -20°C; mycorrhizal root tips were selected from the root apparatus of poplar, hazelnut, linden and oak seedlings inoculated with spores from *T. magnatum, T. borchii, T. brumale* and *T. melanosporum*. Mycelium of *T. magnatum* and *T. borchii* originated by the corresponding mycorrhizae was grown on a malt-extract medium.

DNA was extracted according to the protocols described by Henrion et al. (17, 23) and amplified in PCR experiments by using i) ITS_1 and ITS_4 primers, designed to amplify the internal transcribed spacer (5' to 3' sequence: TCCGTAGGTGAACCTGCGG and TCCTCCGCTTATTGATATGC, respectively), ii) CNL12 and 5SA primers, designed to amplify the intergenic spacer IGS_1 (5' to 3' sequence: CTGAACGCCTCTAAGTCAG and CAGAGTCCTATGGCCGTGGAT, respectively), according to 17 and iii) short arbitrary oligonucleotides in RAPD experiments, according to Wyss and Bonfante (19). Specific primers were designed on the sequence of selected RAPD bands according to the protocol described in 27.

RESULTS

Interspecific Discrimination Based on the Internal Transcribed Spacer

Amplification of truffle DNA with the ITS primers resulted in a single band of about 600-650 bp for all the fruitbodies (Fig. 2), with the exception of *T. brumale*, which gave a band of higher molecular weight (1000 bp). Digestion with the restriction enzymes *Hinf* I and *Mbo* I led to an increased number of bands on agarose gels, that many of the truffle species allowed to be distinguished (Fig. 3). After digestion, *Tuber borchii* and *T. maculatum* consistently showed the same fingerprint. The banding pattern of amplified DNA from *T. borchii* mycelium growing *in vitro* matched that of the corresponding fruitbodies, confirming its identity, whereas the fingerprint of *T. magnatum* presented a slightly different pattern after digestion with restriction enzymes. PCR amplification from mycorrhizal roots with ITS primers could differentiate among the truffle symbionts with different degrees of resolution. Amplification of DNA from mycorrhizae of *T. magnatum* on linden roots led to two close bands: the major one had a molecular weight which corresponded to that obtained for the non-inoculated roots or for leaves of the same plants, while the other band corresponded to that shown by the fruitbody (Fig. 4). The amplification of mycorrhizae established between

Table 1. List of truffle species analysed by PCR, PCR-RFLP and RAPD. For each sample, the number of the examined fruitbodies and the origin site are given

Species	Number of fruit-bodies examined	Origin
T magnatum Pico	17	Monta d'Alba (northern Italy)
T magnatum Pico	1	Alba (northern Italy)
T magnatum Pico	4	Urbino (central Italy)
T magnatum Pico	3	Ceva (northern Italy)
T melanosporum Vitt	3	Ceva (northern Italy)
T melanosporum Vitt	3	Val Curone (northern Italy)
T borchii Vitt	3	Urbino (central Italy)
T maculatum Vitt	3	Urbino (central Italy)
T aestivum Vitt	3	Urbino (central Italy)
T aestivum Vitt	1	Monta d'Alba (northern Italy)
T macrosporum Vitt	3	Monta d'Alba (northern Italy)
T rufum Pico	2	Monta d'Alba (northern Italy)

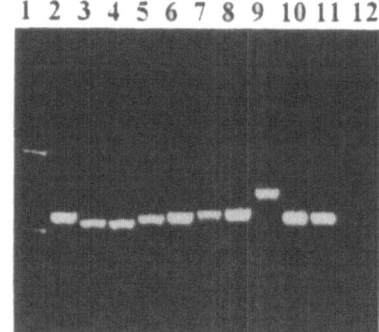

Figure 2. Amplification products with primers ITS1 and ITS4 of DNA extracted from fruitbodies of ten *Tuber* species. Lanes: 1, pBR 322 digested with *Hinf* I; 2, *Tuber magnatum*; 3, *T. borchii*; 4, *T. maculatum*; 5, *T. melanosporum*; 6, *T. macrosporum*; 7, *T. rufum*; 8, *T. aestivum*; 9, *T. brumale*; 10, *T. excavatum*; 11, *T. ferrugineum*; 12, no DNA.

T. borchii and oak roots gave a more intense band that corresponded to the fungal genome, while the plant DNA was only slightly amplified (Fig. 4).

Interspecific Discrimination Based on the Analysis of the Intergenic Spacer

The use of IGS primers allowed the amplification of a DNA fragment characterized by a good level of polymorphism. Many bands resulted from the amplification experiments, giving more complex fingerprints than those obtained by using ITS primers. A good discrimination among truffle species was possible with these primers (Fig. 5), with the exception of *T. borchii* and *T. maculatum*, whose fingerprints were always overlapping. Small differences were also found between *T. melanosporum* and *T. aestivum*. Amplifications from mycorrhizal DNA led to a clear identification of *T. brumale*, as the plant host DNA was not amplified with the IGS primers. Three bands were amplified from the fruitbody, while only the major one was revealed in the mycorrhizal samples (Fig. 6).

RAPD Fingerprints to Reveal Inter- and Intraspecific Polymorphisms

When ITS and IGS primers were used in amplification experiments to distinguish among different fruitbodies of the same truffle species (Table 1), no polymorphism was revealed. Because RAPD is considered a useful diagnostic tool to distinguish between

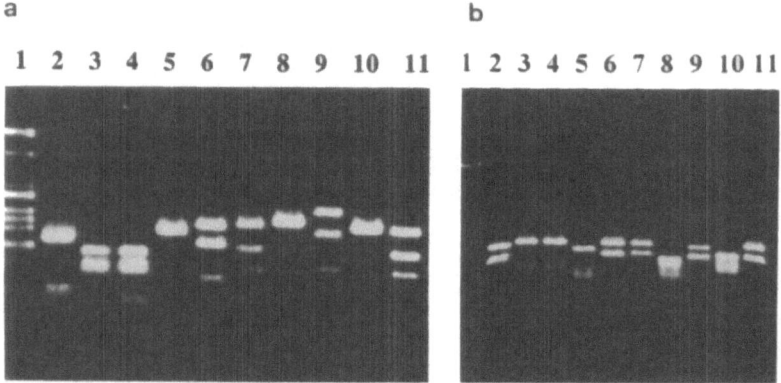

Figure 3. PCR amplified ITS regions digested with *Hinf* I (a) or *Mbo* I (b). Lanes: 1, lambda DNA digested with *Eco* RI-*Hind* III 2, *Tuber magnatum*; 3, *T. borchii*; 4, *T. maculatum*; 5, *T. melanosporum*; 6, *T. macrosporum*; 7, *T. rufum*; 8, *T. aestivum*; 9, *T. brumale*; 10, *T. excavatum*; 11, *T. ferrugineum*.

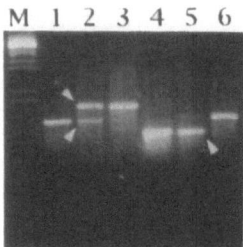

Figure 4. Amplification products with primers ITS1 and ITS4 of DNA extracted from the following samples: fruitbody of *Tuber magnatum* (lane 1); mycorrhizae of *Tuber magnatum* on linden roots (lane 2); linden leaves (lane 3); fruitbody of *Tuber borchii* (lane 4); mycorrhizae of *Tuber borchii* on oak roots (lane 5); oak leaves (lane 6). Lane M: lambda DNA digested with *EcoRI-Hind III*. In lane 2, the upper arrowhead points to the band of host origin, while the lower arrowhead indicates the fungal band of about 600 bp. In lane 5, the arrowhead points to the amplified fungal fragment.

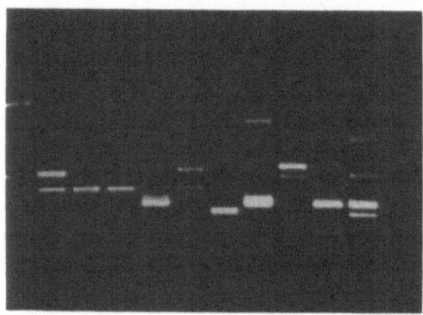

Figure 5. Amplification products with primers CLN12 and 5SA of DNA extracted from fruitbodies of ten *Tuber* species. Lanes: 1, pBR 322 digested with *Hinf* I; 2, *Tuber magnatum*; 3, *T. borchii*; 4, *T. maculatum*; 5, *T. melanosporum*; 6, *T. macrosporum*; 7, *T. rufum*; 8, *T. aestivum*; 9, *T. brumale*; 10, *T. excavatum*; 11, *T. ferrugineum*; 12, no DNA.

isolates, short oligonucleotide primers of arbitrary sequence were tested to investigate the intraspecific polymorphism of truffles. RAPD fingerprints could differentiate well among the different species, and revealed a high degree of interspecific polymorphism (Fig. 7). The level of intraspecific polymorphism was however low, at least for *T. magnatum*. DNA extracted from 25 fruitbodies of *T. magnatum* collected from different sites gave rise to very similar banding patterns after amplification with different primers (Fig. 8) (27). The level of similarity for samples of *T. magnatum* was calculated to be 80-90%, whereas a higher level of polymorphisms was found for samples of *T. borchii* and *T. maculatum* where the level of similarity was 72% and 68%, respectively.

Development of Specific Primers for the Identification of *T. magnatum*

The RAPD technique cannot be used to detect mycorrhizal fungi during their symbiotic phase, because the short oligonucleotide primers will randomly hybridize also to complementary sequences in the host genome, leading to plant amplification products. A different strategy that had already been successful for the detection of the arbuscular

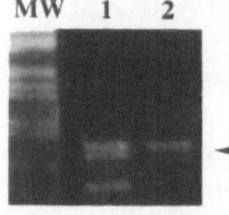

Figure 6. Amplification products with primers CLN12 and 5SA of DNA extracted from *Tuber brumale* fruitbody (lane 1) and oak roots colonized by *T. brumale* (lane 2). The fragment leading to the major band in the fruitbody is present also in the mycorrhizal sample (arrow). MW: lambda DNA digested with *Eco* RI-*Hind* III.

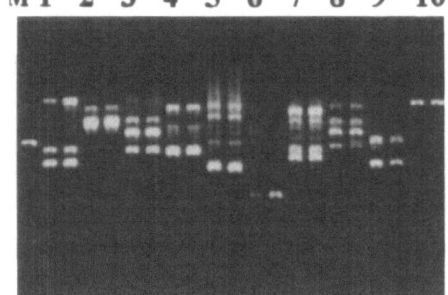

Figure 7. RAPD amplification products with primer OPA-20 (5' GTTGCGATCC 3') of DNA extracted from fruitbodies of ten *Tuber* species. Lanes: M, pBR 322 digested with *Hinf* I; 1, *Tuber magnatum*; 2, *T. borchii*; 3, *T. maculatum*; 4, *T. melanosporum*; 5, *T. macrosporum*; 6, *T. rufum*; 7, *T. aestivum*; 8, *T. brumale*; 9, *T. excavatum*; 10, *T. ferrugineum*. The reactions were repeated twice for each DNA template.

mycorrhizal fungus *Glomus mosseae* (27) was developed to overcome this problem. Briefly, a RAPD-amplified fragment which was consistently found in different samples of *T. magnatum* using the primer OPA-18 was considered as a good marker for this species. This fragment was then subcloned into *Sma* I-digested pBluescript KS⁺ and its specificity for *T. magnatum* was checked by Southern blot experiments (21). The fragment (about 1.5 kbp) was sequenced and a pair of specific primers was designed on the sequence obtained. When used in conventional PCR experiments, these primers (P24 and MO) specifically amplified the *T. magnatum* template (Fig. 9).

DISCUSSION

Experiments based on the use of PCR have allowed us to set up protocols for the reliable identification of truffles during the different steps of their life cycle. Bruns (13), and Bruns and Gardes (28) have clearly stated that identifications based on molecular probes must use a set of oligonucleotide probes and not a single one, because each DNA probe

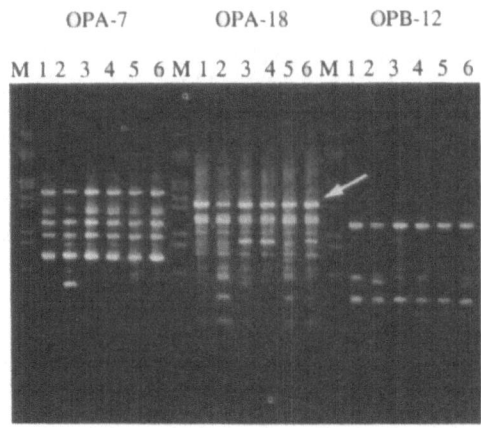

Figure 8. RAPD amplification products with primer OPA-7, OPA-18 and OPB-12 of DNA extracted from *T. magnatum* fruitbodies collected in North (lanes 1-4) and in Central Italy (lanes 5 and 6). Lane M: lambda DNA digested with *Eco* RI- *Hind* III. Arrow points to the 1.5 kbp cloned band. Primers sequences: OPA-7: 5' GAAACGGGTG 3'; OPA-18: 5' AGGTGACCGT 3'; OPB-12: 5' CCTTGACGCA 3'.

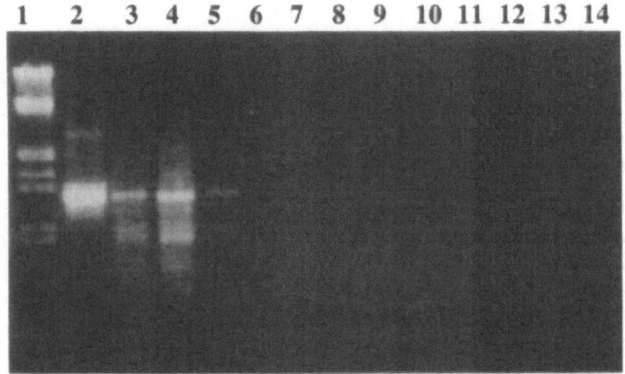

Figure 9. Amplification products with primers P24 (5' AATCTACTAAGATCATGGA 3') and M0 (5' TCTGCAACCCTAAACCTGC 3') of DNA extracted from fruitbodies of ten *Tuber* species. Lanes: 1, lambda DNA digested with *Eco* RI-*Hind* III; 2, recombinant plasmid; 3, 4 and 5 *Tuber magnatum*; 6, *T. borchii*; 7, *T. melanosporum*; 8, *T. macrosporum*; 9, *T. rufum*; 10, *T. aestivum*; 11, *T. brumale*; 12, *T. excavatum*; 13, *T. ferrugineum*; 14, no DNA.

exhibits a range of specificity (isolate, species, genus, family.....). According to these authors, the ideal region for identification of mycorrhizal fungi should have precise characteristics: it should be present in all the fungi of interest, it should be easy to amplify, variable enough to distinguish between taxa but conserved enough to enable the design of probes suitable for several taxonomic hierarchies (28).

Advantages and Disadvantages in the Study of ITS and IGS Regions

The ITS region has recently received much attention because it presents many of the characteristics described above, when applied to the analysis of truffles. In our experiments, amplification with these primers has easily led to consistent banding patterns from mycelia and fruitbodies of all the species sampled, including *T. aestivum*, which could not be previously amplified (23). However, the level of interspecific variability identified with these primers was very low and only *T. brumale* could be easily distinguished, confirming the results by Henrion et al. (23) and Paolocci et al. (29). However, a substantial increase in the resolution among species was obtained after enzyme digestion, allowing a distinction between closely related species (for example *T. magnatum* versus *T. borchii*).

A disadvantage is that ITS does not amplify uniquely fungal DNA when this is mixed with plant DNA, notwithstanding the fact that ITS is supposed to be specific for fungal DNA (24). In certain cases it can also amplify the plant DNA (18, 30). Under our experimental conditions, PCR experiments performed on mycorrhizal roots revealed that DNA from linden roots was easily amplified, although the band was clearly distinguishable from the fungal band, due to its different size. In contrast, the plant DNA from oak mycorrhizal roots was not amplified. In the absence of quantitative results, we cannot claim whether the different amplification pattern depends on the relative quantity of fungal and plant template or whether ITS primers differently bind to different plant genomes.

This disadvantage was partially overcome with the analysis of the IGS region. With the primers used, the plant IGS region was not amplified, probably due to a different organization of these gene regions in plants (17). The exclusive amplification of fungal DNA has therefore led to an easy identification of *T. brumale* during its symbiotic phase.

In conclusion, the ITS and IGS primers are good tools to discriminate truffles in the different phases of their life cycle. Our results show that truffles can be identified in mycorrhizal roots, confirming the results of Paolocci et al. (29) and suggesting in addition that a reduced number of mycorrhizal roots and a quick extraction procedure may be successful.

Advantages and Limitations of the RAPD Method

PCR with single arbitrary short primers relies on the statistical chance that the sites complementary to the primer sequence occur somewhere in the genome as inverted repeats enclosing a relatively short stretch of DNA (30). By using a certain number of arbitrary decamer primers, it was easy to amplify DNA fragments of a given length that produced consistent fingerprints. Amplification products of the same length are considered homologous and the degree of homology mirrors the degree of the genetic polymorphism existing in a population. This makes RAPD the method of choice for many investigations aimed at investigating population dynamics, although some limitations must be taken into account. In our experiments, the use of RAPD revealed only a low degree of intraspecific variability among samples of *T. magnatum* which were harvested in different Italian regions. In contrast, a high degree of intraspecific polymorphism has been reported by Chevalier et al. (31) for many truffle species, with the exception of *T. melanosporum* and *T. brumale*. RAPD analysis was particularly useful in the investigation of *T.borchii* and *T. maculatum*. These two species are closely related from a morphological point of view, so molecular probes are more attractive tools to discriminate them. Their ITS regions, even after enzyme digestion, seemed to be not variable enough to distinguish between them, while by using RAPD, the pattern is much more complex (32). By using 10 primers, 5463 bands were obtained and about 64% of them were shared. This opens the question whether the two truffles are really two separate species or are related taxa with a good level of intraspecies polymorphism.

Another advantage given by RAPD is the possibility of developing specific probes starting from a chosen band. According to the two-step strategy developed for arbuscular mycorrhizal fungi, specific primers can be designed on non-polymorphic bands produced by many isolates of *Glomus mosseae* and, by using this couple of primers, the fungus was identified in mycorrhizal roots (27). Lanfranco et al. (21) and Potenza et al. (22) had already used non-polymorphic RAPD fragments to set up hybridization experiments which allowed the identification of *T. magnatum*. The couple of specific primers here presented led to a reliable and easy identification of *T. magnatum* DNA in a single amplification experiment, result of which is a single major band. The development of such a probe seems of great interest for biotechnological applications: it would allow the fate of the fungus to be monitored during its ectomycorrhizal phase and to check for its presence in food products, like pasta, pate and creams where truffles of high economic value are incorporated.

ACKNOWLEDGMENTS

The authors thank Mr. Francesco Meotto and Mr. Pasquale Carraturo (CMT-CNR) for providing truffle fruitbodies and controlled mycorrhizal samples, and Dr. S. Perotto for a critical reading of the manuscript. Dr. Arlorio's fellowship, as well as the research, was supported by The National Council of Research (CNR), Italy- Strategic Program: Biotecnologia della Micorrizazione. Dr. Matteucci was funded by MURST and Dr. Lanfranco by CNR.

REFERENCES

1 D N Pegler, B M Spooner and T W K Young, British truffles, Whitstable Litho, Kent, UK (1993)
2 O Mattirolo, Sul parassitismo dei tartufi e sulla questione delle micorrize, Atti R Acad Agric Torino, 107 131 (1887)
3 A Fontana, Sintesi micorrizica tra *Pinus strobus* e *Tuber maculatum, Giorn Bot Ital* 101 298 (1967)
4 J L Harley and S E Smith, Mycorrhizal symbiosis Academic Press, London (1993)
5 J M Trappe, The Orders, Families, and Genera of Hypogeus Ascomycotina (truffles and their relatives), *Mycotaxon*, 9 297 (1979)
6 R Marmeisse, J C Debaud and D L Casselton, DNA probes for species and strain identification in the ectomycorrhizal fungus *Hebeloma, Mycol Res* 96 161 (1992)
7 I C Tommerup and N Malajczuk, Genetics and Molecular Genetics of Mycorrhiza *in* "Mycorrhiza synthesis, Advances in Plant Pathology", Academic Press, London (1993)
8 F Martin, P Laurent, D de Carvalho, T Burgess, P Murphy, U Nehls and D Tagu, Fungal gene expression during ectomycorrhiza formation, *Can J Bot* (1995, in press)
9 R L Metzenberger and T A Randall, Mating type and ascus development in *Neurospora*, this volume (1995)
10 J G H Wessels, Gene expression during fruiting in *Schizophyllum commune, Mycol Res* 96 609 (1992)
11 G Chevalier, Evolution des recherches sur les plants mycorhizes par la truffe et perspectives de development, *Giorn Bot Ital* 128 7 (1994)
12 A Fontana, A Ceruti and F Meotto, Criteri istologici per il riconoscimento delle micorrize di *Tuber albidum*, Convegno Internazionale sul tartufo, L'Aquila, 5-8 Marzo 1992, *Micol Veg Medit* 2 121 (1992)
13 T D Bruns, Fungal molecular systematics, *Ann Rev Ecol Syst* 22 525 (1991)
14 L M Kohn, Developing new characters for fungal systematics an experimental approach for determining the rank of resolution, *Mycologia* 84 139 (1992)
15 E C Swann and J M Taylor, Higher taxa of basidiomycetes an 18S RNA gene perspective, *Mycologia*, 85 923 (1993)
16 L M Foster, K R Kozak, M G Loftus, J J Stevens and I K Ross, The polymerase chain reaction and its application to filamentous fungi, *Mycol Res* 97 (7) 769 (1993)
17 B Henrion, F Le Tacon and F Martin, Rapid identification of genetic variation of ectomycorrhizal fungi by amplification of ribosomal RNA genes, *New Phytol* 122 289 (1992)
18 M Gardes and T D Bruns, ITS primers with enhanced specificity for Basidiomycetes application to identification of mycorrhizae and rusts, *Mol Ecol* 2 113 (1993)
19 P Wyss and P Bonfante, Amplification of genomic DNA of arbuscular-mycorrhizal (AM) fungi by PCR using short arbitrary primers, *Mycol Res* 97 1351 (1993)
20 L Simon, J Bousquet, R C Levesque and M Lalonde, Origin of endomycorrhizal fungi and coincidence with vascular land plants, *Nature*, 363 67 (1993)
21 L Lanfranco, P Wyss, C Marzachi and P Bonfante, DNA probes for identification of the ectomycorrhizal fungus *Tuber magnatum* Pico, *FEMS Microbiol Lett* 114 245 (1993)
22 L Potenza, A Amicucci, I Rossi, F Palma, R De Bellis, P Cardono and V Stocchi, Identification of *Tuber magnatum* Pico DNA markers by RAPD markers, *Biotechnol Techn* 8 93 (1994)
23 B Henrion, G Chevalier and F Martin, Typing truffle species by PCR amplification of the ribosomal DNA spacers, *Mycol Res* 98(1) 37 (1994)
24 T J White, T Bruns, S Lee and J Taylor, Amplification and direct sequencing of fungal ribosomal RNA genes for phylogenetics, *in* "PCR Protocols A guide to methods and Applications" M A Innis, D H Geland, J J Sninsky and T J White, Academic Press, San Diego
25 J G K Williams, A R Kubelik, K J Livak, J A Rafalski, S V Tingey, DNA polymorphisms amplified by arbitrary primers are useful as genetic markers, *Nucl Acids Res* 18 (22) 6531 (1990)
26 H Hadrys, M Balick and B Schierwater, Applications of random amplified polymorphic DNA (RAPD) in molecular ecology, *Mol Ecol* 1 55 (1992)
27 L Lanfranco, P Wyss, C Marzachi and P Bonfante, Generation of RAPD-PCR primers for the identification of isolates of *Glomus mosseae*, an arbuscular mycorrhizal fungus, *Mol Ecol* in press
28 T D Bruns and M Gardes, Molecular tools for the identification of ectomycorrhizal fungi-taxon-specific oligonucleotide probes for suilloid fungi, *Mol Ecol* 2 233 (1993)
29 F Paolocci, P Angelini, E Cristofari, B Granetti and S Arcioni, Characterization of some Tuber spp and relative mycorrhizae by molecular markers, this volume (1995)
30 K Bachmann, Molecular markers in plant ecology, *New Phytol* 126 403 (1994)

31. G. Chevalier, F. Martin, P. Nicolas, D. Gandebouef, B. Henrion, C. Dupre, P. Drevet, V. Coehlo, L. Gentzbittel, Characterization and identification of *Tuber* species using molecular criteria, this volume (1995).
32. M. Arlorio, L. Lanfranco, P. Bonfante, A. Fontana, L. Potenza, I. Rossi and V. Stocchi, Molecular analyses to discriminate among the "Bianchetti" truffles. International Symposium on: *Biotechnology of ectomy-corrhizae: Molecular approaches*, Urbino, November 9-11, 1994, Abstracts, 49 (1994).

13

MOLECULAR IDENTIFICATION OF *TUBER* SPECIES AND ISOLATES BY PCR-BASED TECHNIQUES

D. Gandeboeuf,[1] B. Henrion,[2] C. Dupré,[1] P. Drevet,[3] P. Nicolas,[3] G. Chevalier,[1] and F. Martin[2]

[1] INRA, Station d'Agronomie et de Mycologie
  12 avenue du Brézet
  63039 Clermont-Ferrand Cedex, France
[2] INRA, Equipe de Microbiologie Forestière
  Champenoux 54280 Seichamps, France
[3] Université Blaise Pascal
  Unité associée à l'INRA "Organisation et Variabilité des Génomes Végétaux"
  63000 Clermont-Ferrand, France

SUMMARY[*]

Morphological features of fruit bodies and ectomycorrhizas allow the identification of most truffle species. However, economically important truffles (e.g. *Tuber aestivum / T. uncinatum; T. borchii/ T. magnatum*) are difficult to distinguish using these criteria. It is therefore required, when morphological traits are unsufficient for discriminating, to call for additional markers (e.g. isozymes and DNA). DNA polymorphisms in truffle genomes have therefore been used to generate molecular markers by PCR amplification of specific regions and by random amplification of polymorphic DNA (RAPD). Amplification of the internal transcribed spacer (ITS) and intergenic spacer (IGS) of the ribosomal DNA combined with endonuclease digestion yielded unique diagnostic patterns for most European truffle species. However, rDNA polymorphism was not high enough to distinguish *T. aestivum* and *T. uncinatum*. This approach revealed a low intraspecific variation within *T. melanosporum*. The PCR amplification of the rDNA spacers can be combined with RFLP, allowing truffle species to be typed in less than a day and is most immediately applicable to molecular epidemiology. DNA polymorphism has been further assessed by RAPD technique. The high interspecific genomic variability already observed for most truffle species using PCR-RFLP

Biotechnology of Ectomycorrhizae, Edited by Vilberto Stocchi et al.
Plenum Press, New York, 1995

was confirmed. In addition, large intraspecific variations have been found in all species, except *T. melanosporum* and *T. brumale*. Molecular techniques provide an efficient tool, on the one hand for typing fruiting bodies for the food industry, mycelium for patenting purposes as well as mycorrhiza for checking the inoculated seedlings and on the other hand for tracking the fate of mycorrhizal seedlings within truffières and forest ecosystems.

INTRODUCTION

Commercial inoculation of oak and hazel seedlings with ecologically-adapted ectomycorrhizal truffles has been applied for a long time in France and Italy (Chevalier and Grente, 1979; Bencivenga, 1982). This controlled inoculation of seedlings is currently expanding, but formation of truffle mycorrhizas and fruit bodies is still difficult to manage. After inoculation, the production of truffles depends on the rate of initial mycorrhiza formation, persistence and dissemination of the inoculated fungus, and poorly defined biotic and abiotic factors specific to each site (Chevalier, 1990; Grove and Le Tacon, 1993). In natural sites where indigenous ectomycorrhizal communities can be large and competitive, the local *Tuber* species can outcompete the artificially introduced fungi and lead to the production of unwanted fruiting bodies (Chevalier *et al.*, 1982; Le Tacon *et al.*, 1992). Later during the production stage, precise regulations for collecting and marketing truffles are requested to avoid frauds. For example, some truffle species, such as *T. uncinatum* Chat. and *T. brumale* Vitt. var. *moschatum* Ferry mentioned in the present French and Italian legislations, are doubtful.

Tuber species can generally be identified by the morphological features of fruit bodies and mycorrhizas. However, there are some exceptions within economically important species. For example, fruit bodies of *T. aestivum* (summer truffle) and those of *T. uncinatum* (Burgundy truffle) are closely similar. The ectomycorrhizal status of root systems after inoculation can be evaluated by morphological identification of the fruit bodies of the fungal symbiont or by detailed microscopic examination of the mycorrhizas. However, using this approach it is difficult to distinguish some types of mycorrhizas, such as those of *T. albidum* (bianchetto) and *T. magnatum* (Piedmont white truffle) as well as *T. aestivum* and *T. uncinatum*. Moreover, it is impossible to ascertain whether the identified fungal symbiont is the isolate initially inoculated on seedlings. Thus, production of truffles could possibly be unrelated to the selected fungal inoculum initially introduced in sites where the natural propagules are abundant. Therefore, techniques other than morphological ones are necessary to identify truffle species and isolates.

Biophysical, biochemical, and immunochemical approaches have been used to identify truffle species (Mouches *et al.*, 1981; Papa *et al.*, 1987; Pacioni and Pomponi, 1989, 1991; Palenzona *et al.*, 1990; Papa and Polimeni, 1990; Dupré and Chevalier, 1991; Dupré *et al.*, 1992; Corocher *et al.*, 1992; Dupré *et al.*, 1993; Zambonelli, 1993). However, these techniques are not sensitive enough to identify fungal symbionts from a single mycorrhiza and molecular methods can substantially facilitate the characterization of truffles.

Restriction fragment length polymorphisms (RFLPs) of polymerase chain reaction (PCR) generated fragments (Gardes *et al.*, 1991; Henrion *et al.*, 1992, 1994; Mehmann *et al.*, 1994) have been used to identify ectomycorrhizal fungi, including truffles (Henrion *et al.*, 1994; Mehmann *et al.*, 1994). The latter approach has rapidly gained favour because of its conceptual simplicity and potential to rapidly screen a large number of individuals on single root tips (Gardes *et al.* 1991; Henrion *et al.*, 1994). Another approach to identification of truffles involves random amplification of polymorphic DNA (RAPD) (Williams *et al.* 1990). This technique can discriminate *Tuber* species and provides markers for population studies (Lanfranco *et al.*, 1993; Potenza *et al.*, 1994). In the present paper, we will discuss the use of molecular markers for identifying European truffle species and analyzing truffle populations. We will first outline the DNA techniques used in our programme, then discuss the properties of the different markers, and finally review the recent data obtained on the DNA polymorphism in truffles.

THE EXPERIMENTAL PROCEDURES

Fungal Isolates and Culture Conditions

Most isolates are maintained in the collection of *Tuber* cultures at the INRA Mycology Unit (Clermont-Ferrand, France) (Table 1). The mycelial cultures were obtained as described by Henrion *et al.* (1994). Fruit bodies of eleven taxa [*T. aestivum, T. borchii* (=*T. albidum*), *T. brumale, T. brumale* var *moschatum, T. excavatum, T. macrosporum, T. magnatum, T. melanosporum, T. mesentericum, T. rufum, T. uncinatum*] were chosen according to their economical importance and their frequent presence in truffières. They were collected in different locations in France and in Italy and stored at -20°C (Table 2).

DNA Extraction, PCR Amplification of rDNA and RFLP Analysis

Total DNA extraction and PCR amplification of ITS and IGS of the rRNA genes were carried out according to Henrion *et al.* (1994).

Genomic Fingerprinting by RAPD

RAPD was performed according to Williams *et al.* (1990) using Operon Technologies primers. The amplification reactions were carried out in a final volume of 25 µl containing: 1X reaction buffer (Appligene), 200 mM dNTPs (Pharmacia), 0.2 mM primers , 0.5 units *Taq* polymerase (Appligene), and 50 ng DNA. The thermal cycler was programmed for an initial 5 min denaturation at 93°C, followed by 40 cycles of denaturation (1 min, 91°C), annealing (1 min, 36°C), and extension (2 min, 70°C) with the fastest ramping time between all temperature points. A final 10 min extension followed the 40 cycles. Amplification

Table 1. Mycelial isolates and ascocarps of *Tuber* used to amplify the ribosomal DNA spacers

Mycelial isolates		Geographical origin	Biological origin	Ascocarp	Origin
T. aestivum	Aes 1	France	Ascocarp	*T. aestivum 1*	France
T. borchii	A1	Italy	Ascocarp	*T. aestivum 2*	France
T. borchii	A'	Italy	Mycorrhiza	*T. borchii*	France
T. borchii	Gre7	Italy	Mycorrhiza	*T. brumale*	France
T. brumale	Mos Pey	France	Ascocarp	*T. magnatum*	Italy
T. melanosporum	Bal3	Italy	Ascocarp	*T. melanosporum*	France
T. melanosporum	1015	France	Ascocarp	*T. uncinatum 1*	France
T. melanosporum	1017	France	Mycorrhiza	*T. uncinatum 2*	France
T. melanosporum	Mel 3	Italy	Ascocarp		
T. melanosporum	Mel 10	France	Mycorrhiza		
T. melanosporum	Mel24	France	Mycorrhiza		
T. melanosporum	Mel28	France	Mycorrhiza		
T. melanosporum	Mel T	Italy	Ascocarp		
T. uncinatum	U11	France	Ascocarp		

Table 2. List of ascocarps of truffle analysed by RAPD

Origin		Origin	
T. melanosporum		*T. uncinatum*	
1 to 2	Lot (France)	1	Yonne (France)
3 to 6	Dordogne (France)	2	Haute-Saône (France)
7 to 13	Charentes (France)	3	Meuse (France)
14 to 17	Meuse (France)	4	Puy-de-Dôme (France)
T. brumale		5	The Marches (Italy)
1	Dordogne (France)	*T. mesentericum*	
2	canning factory	1	Dordogne (France)
3	Charentes (France)	2	Meuse (France)
4	Yonne (France)	3	Yonne (France)
5	Meuse (France)	4	Avelino (Italy)
T. moschatum		5	Saou (France)
1	Dordogne (France)	*T. excavatum*	
2	Dordogne (France)	1	Puy-de-Dôme (France)
3	Vaucluse (France)	2	Haute-Saône (France)
4	canning factory	3	Yonne (France)
5	Ardèche (France)	4	Allier (France)
T. rufum		5	Meuse (France)
1	Puy-de-Dôme (France)		
2	Meuse (France)	*T. magnatum*	
3	Puy-de-Dôme (France)	1	Piedmont (Italy)
4	Yonne (France)	2	The Marches (Italy)
5	Puy-de-Dôme (France)	3	The Marches (Italy)
T. aestivum		4	canning factory
1	Lot (France)	5	canning factory
2	Hérault (France)	*T. borchii*	
3	Bouches du Rhône (France)	1	Puy-de-Dôme (France)
4	Soria (Spain)	2	Puy-de-Dôme (France)
5	Berga (Spain)	3	Puy-de-Dôme (France)
T. macrosporum		4	Puy-de-Dôme (France)
1	Meuse (France)	5	The Marches (Italy)
2	Yonne (France)		
3	Côte d'Or (France)		
4	Piedmont (Italy)		
5	The Marches (Italy)		

products were analyzed by electrophoresis in 1.4% agarose gels in tris-borate buffer, stained with ethidium bromide and visualized by illumination with UV light.

POLYMORPHISM IN RIBOSOMAL RNA GENES

If a technique is to be useful for analyzing DNA polymorphism in closely related fungal species, it must meet two criteria: 1) it should provide sufficient numbers of easily-scored markers and 2) it should allow efficient typing of large numbers of samples. PCR amplification of ribosomal DNA (rDNA) meets these two conditions. The tandemly-repeated rDNA is one of the most frequently analyzed nuclear DNA in fungi (White *et al.*, 1990). The transcribed region contains four genic sequences that code for the ribosomal RNA subunits (17S, 5.8S, 25S, and 5S), and intergenic spacers (ITS and IGS). In ectomycorrhizal basidiomycetes, interspecific variation was found in the ITS using PCR amplification and RFLPs (Gardes *et al.*, 1991; Henrion *et al.*, 1992). In these species, most intraspecific

Table 3. The amplified rDNA ITS of various species of *Tuber* from mycelium, ascocarp and ectomycorrhiza

		ITS (bp)		
		1000	630	600
Mycelium				
T. brumale	(Mos Pey)	1	0	0
T. borchii	(A1)	0	0	1
T. borchii	(A')	0	0	1
T. melanosporum	(Mel24)	0	1	0
Ascocarp				
T. melanosporum		0	1	0
T. borchii		0	0	1
T. magnatum		0	0	1
T. brumale		1	0	0
Mycorrhiza				
T. melanosporum		0	1	0

variation in the rDNA occurs in the number of repeat units and the length of IGS with variations occuring both between and within individuals (Henrion *et al.*, 1992, 1994).

Use of PCR for DNA allowed identification of several truffle species and characterization of the symbionts on ectomycorrhizal tips by amplification of the ITS and IGS (Henrion *et al.*, 1993; Paolocci *et al.*, this volume).

Polymorphism in the rDNA ITS

PCR amplifications of the ITS were succesfully performed with total DNA extracted from mycelium, fruit bodies, and ectomycorrhizas of most truffle species (Table 1), except *T. aestivum* and *T. uncinatum* (Table 3). Amplifications yielded a single ITS fragment with an interspecific length variation [600-650 bp (most species) to 1000 bp (*T. brumale*)]. RFLPs obtained with the endonucleases *AluI/HinfI* yielded unique diagnostic patterns for all *Tuber* species (Table 4). The ITS of *T. melanosporum* was amplified from single *Corylus avellana* mycorrhizal tips and the host-plant material did not interfere with the DNA extraction and amplification procedures. Therefore, RFLP of PCR-amplified ITS should prove useful for rapidly identifying any suspicious fruit bodies and in examining the structure of communities of truffle species in microbial ecology. In contrast, this technique is poorly adapted to the analysis of population heterogeneity. Limited variation was found between the different RFLP patterns obtained from eight *T. melanosporum* isolates of various geographical origins (Henrion *et al.*, 1994).

Table 4. Hinf/Alu 1 - RFLPs of the amplified rDNA ITS of *Tuber*. Presence of the fragments is noted by 1 and absence by 0

	Ascocarp			
	T. melanosporum	*T. borchii*	*T. magnatum*	*T. brumale*
ITS (bp) restriction fragments				
320	0	1	0	0
250	0	0	1	0
200	0	1	0	0
190	1	0	0	1
170	1	0	0	0
150	0	0	0	1
140	1	0	0	0
120	0	1	0	0
100	0	0	1	0
80	0	0	1	0

Table 5. Amplification of the rDNA IGS and RFLP analysis of the amplified rDNA ITS from various species of *Tuber*. DNAs were extracted from mycelium and ascocarp. Presence of fragments is noted by 1 and absence by 0

	IGS					IGS restriction fragments (bp)											
Mycelium	850	430	390	310	150	600	550	450	400	350	300	250	200	150	130	120	100
T. melanosporum (MelT)	0	0	0	1	1	0	0	0	1	0	1	0	0	1	0	1	1
T. borchii (A1)	0	1	1	0	0	1	1	1	0	0	0	0	0	0	0	0	0
T. borchii (A')	0	1	1	0	0	1	1	1	0	0	0	1	0	0	0	0	0
T. borchii (Gre7)	1	1	1	0	0	1	1	1	0	1	0	1	0	0	1	0	1
Ascocarp	700	300	270	100							280	250	200	100			
T. aestivum 1	1	1	1	1							1	1	1	1			
T. aestivum 2	1	1	1	1							1	1	1	1			
T. uncinatum 1	1	1	1	1							1	1	1	1			
T. uncinatum 2	1	1	1	1							1	1	1	1			

Polymorphism in the rDNA IGS

The IGS-PCR technique uses the fact that fungi frequently have repetitive DNA sequences scattered at various locations throughout their IGS. The length of DNA between the coding regions (25S –> 5S, 5S –> 17S) thus varies among different species and isolates, and can serve as the basis for identifying a specific fungal isolate. Primers specific for the fungal IGS have been designed and used to amplify these polymorphic intervening sequences in several ectomycorrhizal species (Henrion *et al.*, 1992, 1994). Amplification of the IGS sequence of *T. aestivum*, *T. uncinatum*, *T. albidum* and *T. melanosporum* produced an incremental set of characteristic DNA band patterns that uniquely identify a particular species (Table 5) (Henrion *et al.*, 1994). RFLP analysis of the amplified products using *Rsa*I provided an additional approach to discriminate the investigated *Tuber* species. However, IGS patterns and scored RFLP markers were not sufficient to distinguish *T.aestivum* and *T. uncinatum*. The low genetic variability between these two species had already been documented using allozyme analysis (Dupré *et al.*, 1991; Gandeboeuf *et al.*, 1993; Pacioni and Pomponi, 1991).

Intraspecific variation can be revealed by the PCR/RFLP of IGS as shown by the analysis of three isolates of *T. albidum* of different geographical origins (Henrion *et al.*, 1994). In contrast, a limited intraspecific variation was detected within *T. melanosporum* (Henrion *et al.*, 1994), supporting the low level of enzymatic polymorphism already observed (Gandeboeuf, 1993; Pacioni and Pomponi, 1989, 1991).

In the past, the rate of identification of truffle fruiting bodies and ectomycorrhizas was limited by the length of time required to carry out the full gamut of differential morphological and biochemical tests. This process can take days, and be quite involved for these slow-growing ectomycorrhizal fungi. PCR amplification of the rDNA spacers can be combined with RFLP to let truffle species be typed in less than a day, and is most immediately applicable to molecular epidemiology. But many other applications, including identification of harvested fruit bodies, food quality testing, and typing of fungal symbionts occurring on inoculated seedlings, can be envisioned for the future. It is possible that, if a computer database of DNA patterns is developed and maintained, this method could be used for fungal identification in the food industry and microbial ecology.

T. melanosporum

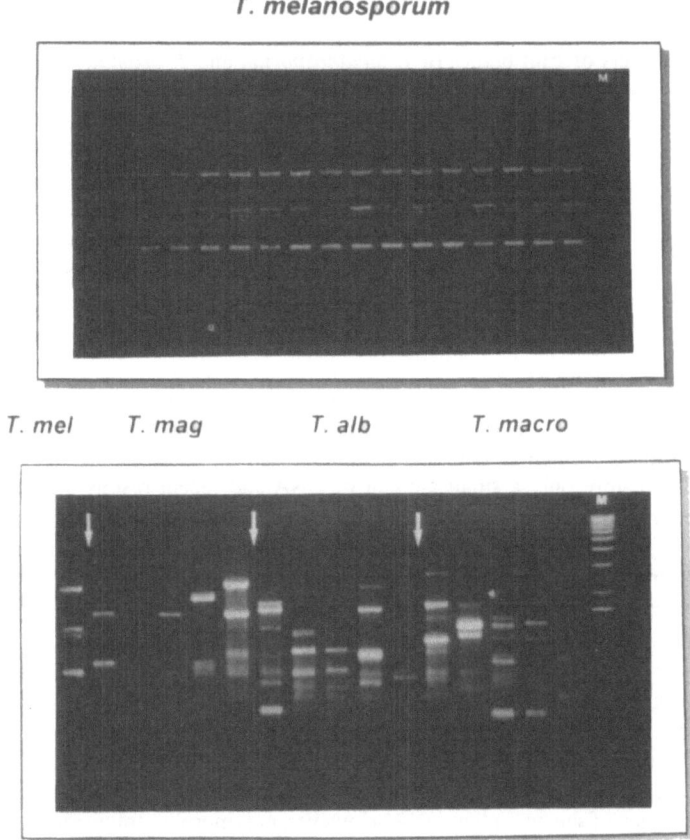

Figure 1. Amplification products with primer OPE20 of DNA extracted from *T. melanosporum* (*T. mel*), *T. borchii* (*T. alb*), *T. magnatum* (*T. mag*) and *T. macrosporum* (*T. macro*) fruit bodies.

FINGERPRINTING TRUFFLES WITH RAPD PRIMERS

As suggested by the low polymorphism of rDNA spacers of *T. melanosporum*, the PCR amplification of specific regions of the genome has limitations in its ability to distinguish between populations and truffle isolates of different geographic sources. RAPD primers have instead been successfully used to fingerprint isolates of several species of fungi, including truffles (Lanfranco *et al.*, 1993; Potenza *et al.*, 1994). This technique is precise, fast, and can provide markers for population studies. Therefore, we used this approach to discriminate different *Tuber* species and assess the isolate DNA polymorphism within *T. melanosporum* (Gandeboeuf *et al.*, unpublished results).

RAPD amplifications were carried out on DNA extracted from *T. melanosporum* fruit bodies of different geographic origin. Among 150 Operon primers tested, only 50 yielded amplification products. Many of these primers were subsequently eliminated because they failed to produce any amplified DNA fragments during the screening of additional *Tuber* taxa (*T. aestivum*, *T. borchii*, *T. brumale*, *T. brumale* var. *moschatum*, *T. excavatum*, *T. macrosporum*, *T. magnatum*, *T. melanosporum*, *T. mesentericum*, *T. rufum*, *T.uncinatum*).

Successful amplifications were obtained for this set of species with only 10 primers. These primers yielded specific diagnostic patterns of reliably amplified fragments for all the tested species.

DNA extracts of fruit bodies of *T. melanosporum* and *T. brumale s.l.* collected on different sites yielded highly similar fingerprints (Fig.1). The level of similarity was higher than 90% indicating a low intraspecific genomic polymorphism within these two species. These data are in agreement with the results obtained by isoenzyme analysis (Pacioni and Pomponi, 1989, 1991). In contrast, RAPD of isolates belonging to other species (e.g. *T. albidum, T. macrosporum*) generated highly polymorphic fingerprints (Fig. 1) suggesting a high intraspecific variability. In our case, DNA extracts of the various isolates of *T. magnatum* gave rise to very different fingerprints. This contrasts with data reported by Lanfranco *et al.* (1993) showing a very low level of RAPD polymorphism within this species. The reason for the discrepancy between these RAPD data is not known, but is currently being explored.

The high similarity between RAPD patterns of taxa which are morphologically very close (e.g. *T. aestivum - T. uncinatum* and *T. brumale - T. brumale* var. *moschatum*) casts doubt on the validity of their taxonomic attribution.

RAPDs require only a small amount of DNA and recent results indicate that the differences among truffle species are readily detected by a wide array of primers (this study, Lanfranco *et al.*, 1993; Potenza *et al.*, 1994). When looking at the variability within a species, some populations (e.g. *T. albidum, T. macrosporum*) appeared to be easily differentiated by the selected primers. In contrast, the variability within populations of *T. melanosporum* was harder to detect. These results suggest that the genetics of the different species of European truffles may be very different. Intraspecific differences should therefore be related to the geographic origin of the isolates. The importance of recognizing the geographic origin of isolates of commercially important truffles lies in the possibility of determining the provenance of distributed fruit bodies.

RAPD cannot be used directly on ectomycorrhizas to type the associated symbiont, due to the presence of the host-plant DNA. However, polymorphic fragments generated by RAPD, and present in all isolates within a species, can be used as specific probes in Southern and dot blot procedures (Lanfranco *et al.*, 1993; Potenza *et al.*, 1994). This approach has been used to type truffle species on mycorrhizal roots (Potenza *et al.*, this volume). Occurrence of bacteria and fungi within truffle fruit bodies may be a problem in RAPD typing due to the lack of specificity of the primers used.

CONCLUSION AND FUTURE DEVELOPMENTS

Up to now the identification of ectomycorrhizal truffles has been based largely on morphological and biochemical features. However, taxonomy remains questionable for several species and the application of these approaches to single ectomycorrhizal tips of inoculated seedlings is a difficult task. Now that the market of seedlings inoculated with selected edible truffles is expanding, new, more objective criteria are needed. This need has led to the development of a large array of molecular techniques that can be used to determine the identity of the fruit body, mycelium and mycorrhiza. The most powerful procedures are based on genetic characterization. We and others have used PCR-based techniques employing either specific primers or single primers of arbitrary sequence to examine differences in amplified fragment patterns. These PCR-based procedures have led to the molecular identification of most European truffle species. Many species showed extreme diversity in PCR-amplified fragments. In contrast, *T. melanosporum* isolates showed a high similarity. Why does *T. albidum* show such wide-ranging variation while *T. melanosporum* does not?

T melanosporum shows genetic evidence for a strongly clonal origin for the various strains found in France The clonal nature of this species suggests a common origin and perhaps, a recent evolutionary history, in comparison to *T albidum*

The results demonstrate that molecular biology techniques provide an efficient tool, not only in typing fruiting bodies for the food industry, mycelium for patenting purposes, mycorrhizae for checking the inoculated seedlings, but also for tracking the fate of mycorrhizal seedlings within truffieres and ecosystems However, our ability to type fungal material at the isolate level is still limited for species such as *T melanosporum* and the polymorphism of additional genomic regions should be explored

REFERENCES

Armstrong J L , Fowles N L and Rygiewicz P T (1989) Restriction fragment length polymorphisms distinguish ectomycorrhizal fungi *Plant and Soil* 116, 1-7

Bencivenga M (1982) Alcune metodiche di micorrizazione di piante forestali con il tartufo nero pregiato di Norcia e di Spoleto (*Tuber melanosporum* Vitt) *L Informatore Agrario* 38, 21155-21163

Chevalier G (1990) Recherche et experimentation sur la culture de la truffe en France *P H M* 312, 17-23

Chevalier G , Giraud M and Bardet M C (1982) Interactions entre les mycorhizes de *Tuber melanosporum* et celles d'autres champignons ectomycorhiziens en sols favorables a la truffe In *Les mycorhizes partie integrante de la plante biologie et perspectives d utilisation* (eds Giannazzi S and Giannazzi-Pearson V) INRA Paris, 313-321

Chevalier G and Grente J (1979) Application pratique de la symbiose ectomycorhizienne production a grande echelle de plants mycorhizes par la truffe *Mushroom Science* 10, 483-505

Corocher N , Polimeni C , Giraudi G and Papa G (1992) Sviluppo di un metodo immunoenzimatico (ELISA) per la caratterizzazione di ectomicorrize di *Tuber magnatum* e *T albidum* *Micologia e Vegetazione Mediterranea* 7 (1), 151-158

Duchesne L C and Anderson J B (1990) Location and direction of transcription of the 5S rRNA gene in *Armillaria* *Mycological Research* 94, 266-269

Dupre C and Chevalier G (1991) Analyse electrophoretique des proteines fongiques de differents *Tuber* en association ou non avec *Corylus avellana* *Cryptogamie Mycologie* 12 (4), 243-250

Dupre C Chevalier G and Gandeboeuf D (1991) Analyse electrophoretique des proteines totales des ascocarpes, du mycelium en culture et des mycorhizes de differents *Tuber* *C R Groupe de travail Application des mycorhizes* , Bordeaux, 10-12 octobre 1991

Dupre C , Chevalier G , Palenzona M , Ferrara A M , Nascetti G , Mattiucci S , D'Amelio S , La Rosa G and Biocca E (1992) Differenzazione genetica di ascocarpi, miceli e micorrize di differenti specie di *Tuber* *Micologia e Vegetazione mediterranea* 7 (1), 139-144

Dupre C Chevalier G , Palenzona M and Biocca E (1993) Caracterisation des mycorhizes de differents *Tuber* par l'etude du polymorphisme enzymatique *Cryptogamie Mycologie* 14 (4), 163-170

Gandeboeuf D (1993) Differenciation des truffes europeennes d'interet commercial par l'analyse des isoenzymes *D E A* Sciences de l'alimentation, Universite de Clermont-Ferrand II, 23 p

Gardes M , White T J , Fortin J A , Bruns T D and Taylor J W (1991) Identification of indigenous and introduced symbiotic fungi in ectomycorrhizae by amplification of nuclear and mitochondrial ribosomal DNA *Canadian Journal of Botany* 69, 180-190

Grove T S and Le Tacon F (1993) Mycorrhiza in plantation forestry *Advances in plant Pathology* 23, 191-227

Henrion B , Le Tacon F and Martin F (1992) Rapid identification of genetic variation of ectomycorrhizal fungi by amplification of ribosomal RNA genes *New Phytologist* 122, 289-298

Henrion B , Chevalier G and Martin F (1994) Typing truffle species by PCR amplification of the ribosomal DNA spacers *Mycological Research* 98 (1), 37-43

Lanfranco L , Wyss P , Marzachi C and Bonfante P (1993) DNA probes for identification of the ectomycorrhizal fungus *Tuber magnatum* Pico *FEMS Microbiology Letters* 114, 245-252

Le Tacon F , Alvarez I , Bouchard D *et al* (1992) Variations in field response of forest trees to nursery ectomycorrhizal inoculation in Europe In *Mycorrhizas in Ecosystems* (eds Read D J , Lewis D , Fitter A , Alexander I), CAB International, Wallington, UK, 119-134

Martin F , Zaiou M , Le Tacon F and Rygiewicz P (1991) Strain specific differences in ribosomal DNA from the ectomycorrhizal fungi *Laccaria bicolor* (Maire) Orton and *Laccaria laccata* (Scop ex Fr) Br *Annales des Sciences forestieres* 48, 297-305

Mehmann, B , Brunner, I Braus, G H (1994) Nucleotide sequence variation of chitin synthase genes among ectomycorrhizal fungi and its potential use in taxonomy *Appl Environ Microbiol* 60, 3105-3111

Mouches C , Duthil P , Poitou N , Delmas J and Bove J M (1981) Caracterisation des especes truffieres par analyse de leurs proteines en gel de polyacrylamide et application de ces techniques a la taxonomie des champignons *Mushroom Science* 11, 819-831

Pacioni G and Pomponi G (1989) Chemotaxonomy of some italian species of *Tuber Micologia e Vegetazione Mediterranea* 4 (1), 63-72

Pacioni G and Pomponi G (1991) Genotypic patterns of some italian populations of the *Tuber aestivum T mesentericum* complex *Mycotaxon* 62, 171-179

Palenzona M , Biocca E , Nascetti G , Ferrara A M , Mattiucci S , D'Amelio S and Balbo T (1990) Studi preliminari sulla tipizzazione genetica (sistema gene-enzyma) di specie del genere Tuber In *Atti del II Congresso Internazionale sul Tartufo* (ed B Granetti and M Bencivenga), Spoleto, November 25-28, 53-58

Papa G , Balbi P and Ausidio G (1987) Preliminary study of fungal spores by pyrolysis-gas chromatography *Journal of Analytical and Applied Pyrolysis* 11, 539-548

Papa G and Polimeni C (1990) Caratterizzazione delle proteine nelle *Tuberaceae* I Studio preliminare di una banda elettroforetica di *Tuber magantum* Pico *Micologia italiana* 2, 37-43

Potenza L , Amicucci A , Rossi I , Palma F , De Bellis R , Cardoni P and Stocchi V (1994) Identification of *Tuber magnatum* Pico DNA markers by RAPD analysis *Biotechnology Techniques* 8, 93-98

Sen R (1990) Isozymic identification of individual ectomycorrhizae synthetized between Scots pine (*Pinus sylvestris* L) and isolates of two species of *Suillus New Phytologist* 114, 613-622

White T J , Bruns T , Lee S and Taylor J (1990) Amplification and direct sequencing of fungal ribosomal RNA genes for phylogenetics In *PCR Protocols A guide to Methods and Applications* (eds Innis M A , Gelfand D H , Sninsky J J and White T J), Academic Press, San Diego, 315-322

Williams J G K , Kubelik A R , Livak K J , Rafalaski J A , Tingey S V (1990) DNA polymorphisms amplified by arbitrary primers are useful as genetic markers *Nucleic Acids Research* 18, 6531-6535

Zambonelli A (1993) An enzyme-linked immunosorbent assay (ELISA) for the detection of *Tuber albidum* ectomycorrhiza *Symbiosis* 15, 71-76

14

MOLECULAR CHARACTERIZATION OF SOME TRUFFLE SPECIES

Barbara Lazzari,[1] Elisabetta Gianazza,[2] and Angelo Viotti[1]

[1] Istituto Biosintesi Vegetali
CNR
Via Bassini 15, 20133 Milano, Italy
[2] Istituto di Scienze Farmacologiche
Università degli Studi di Milano
Via Balzaretti 9, Milano, Italy

SUMMARY

In order to attempt the molecular characterization of ectomycorrhizal Ascomycetous fungi eight *Tuber* species (*Tuber magnatum* , *Tuber melanosporum, Tuber aestivum, Tuber rufum, Tuber ferrugineum, Tuber macrosporum, Tuber brumale* and *Tuber borchii*) were analyzed at the DNA and protein levels. DNA was extracted from truffle fruitbodies and digested with EcoRI to perform Southern analysis. Ribosomal DNA was used as a probe, and gave rise to highly polymorphic patterns. Small differences were noted even between two samples of *T. magnatum* harvested in different geographical areas. A cDNA library of *T. magnatum* was prepared in order to obtain truffle homologous probes. SDS-PAGE analysis was performed on total protein extracts leading to similar conclusions: patterns are specific for each truffle, with the exception of *T. magnatum* for which small differences between the two samples previously described were observed, and no difference was noticed among the patterns of proteins extracted from morphologically different parts of the same fruitbody. Isoelectric focusing (IEF) was carried out on six protein extracts (from *T. borchii, T. melanosporum, T. magnatum, T. uncinatum* and two samples of *T. aestivum,*), giving rise to specific patterns for each truffle. Two-dimensional analysis (IEF followed by SDS-PAGE) of truffle total proteins could be very helpful in differentiating between similar species.

These molecular approaches have proved to be particularly useful in discriminating among the different truffle species and it is worth verifying whether the same techniques can also be adopted to identify these symbiotic fungi in ectomycorrhizae.

Biotechnology of Ectomycorrhizae, Edited by Vilberto Stocchi et al.
Plenum Press, New York, 1995

INTRODUCTION

Many truffle species are known to form mycorrhizae both with angiosperm and gymnosperm roots, but only some of these species are commercially important. *T. magnatum* Pico and *T. melanosporum* Vitt. are the most valuable truffles and for this reason many attempts were made to inoculate host plant roots with these fungi in order to obtain large quantities of fruitbodies (1, 2, 3, 4). The development of mycelia of the different species in the soil is affected by many different factors - many of which are still unknown - and endogenous species often get the upper hand of the inoculated ones. This implies that, after planting out mycorrhized seedlings, it is necessary to verify that the desired *Tuber* species are present in the soil. Identification of different truffles has always been difficult, in part due to their particular life cycle. The most common parameters for truffle identification refer to the morphological features of the sporocarps, but these are absent in the first stages of the life cycle, when the mycelium is formed and the symbiosis established. In any case, discriminating among different fruitbodies is also often a difficult task because of the high degree of similarity among closely related species. Morphological features such as the structure of the peridium and gleba or the size and shape of asci and spores are sometimes insufficient to identify truffles and even the most accurate analysis can be misleading.

Finding a reliable and rapid method for the identification of *Tuber* species would also be very important for quality control of commercial truffles: *T. magnatum* and *T. borchii*, for example, are morphologically and biochemically very similar (5), but the first is much more valuable than the latter and is often subject to fraud (6).

Molecular biology techniques could be helpful for this purpose, both to obtain a reliable classification of fruitbodies and to discriminate among truffles in different stages of their life cycle: for example, at the mycelium stage.

PCR-based techniques have already been adopted with success and allowed the identification of inter- and intraspecific differences among symbiotic fungi (7,8,9,6). Specific probes have also been cloned by PCR, and used to assess for the presence or absence of *T. magnatum* (10).

In our lab, we concentrated on the analysis of DNA and proteins extracted from fruitbodies of different *Tuber* species in order to find out differences among species that could lead to the identification and classification of different truffles. A cDNA expression library from *T. magnatum* has also been prepared, in order to isolate homologous probes.

MATERIALS AND METHODS

DNA Extraction and Southern Analysis

Total DNA was extracted from approximately 0.5 gr of frozen *Tuber* fruitbodies. Samples were crushed in liquid nitrogen with alcoa and the DNA was extracted using 8 M urea, 0.35 M NaCl, 50 mM Tris/HCl pH 7.6, 20 mM EDTA pH 8 and 2% Sodium Lauril Sarcosine (SLSar), followed by phenol-chloroform extraction and isopropanol precipitation. DNA was then purified by CsCl gradient.

EcoRI digests of DNAs extracted from different truffle species were separated by agarose gel electrophoresis, transferred to Hybond-N+ nylon membranes (Amersham) and hybridized to ribosomal DNA. Probes were prepared via the random primer method, using a-^{32}P-dCTP (3000 Ci/mmol, Amersham) and Promega's Prime-a-Gene Labeling System, obtaining a specific activity of about 10^8 cpm/µg. Hybridization was carried out overnight at 53°C using 1 ng of labeled DNA/ml hybridization solution (10% Dextran sulphate, 2%

SLSarcosine, Heparin 0.5 mg/ml, 0.6 M NaCl, 0.18 M Na_2HPO_4 pH 6.5, 6 mM EDTA). Filter washes were performed as follows: 5 min at room temperature in 2x SSC, 0.1% SDS; 30 min in 2x SSC, 0.1% SDS at 53°C; 30 min in 0.2x SSC, 0.1% SDS at 53°C. Autoradiography was carried out for 24 hours at -80°C using Kodak-XAR films and with an intensifying screen. Probe removal was obtained by pouring boiling 0.5% SDS on the filter and allowing it to cool to room temperature.

Protein Extraction and Electrophoresis

Total truffle proteins were extracted in sample buffer (SB = 60 mM Tris/HCl pH 6.8, 1.5% SDS, 4% b-mercaptoethanol, 10% sucrose) from frozen homogenized fruitbodies. Truffle protein extracts seem to be quite susceptible to degradation even during storage in SB at -20°C, so best results were obtained adding specific inhibitors of proteolytic activities to the samples immediately after the extraction (0.1-0.2% v/v antiproteinase mix) (11).

To perform SDS-PAGE analysis, extracts were dialyzed overnight against SB at room temperature in a System 100 microdyalizer apparatus (Pierce), while to perform Isoelectric Focusing (IEF) analysis, samples were precipitated in 93% acetone (30' at -20°C), resuspended in 8M urea 4% β-mercaptoethanol and dialyzed overnight at room temperature against the same solution.

SDS-PAGE was performed on a 15 or 17% polyacrylamide separating gel (4% crosslinker) with a 5% stacking gel on a Protean apparatus (Biorad). Samples were heated before loading for 5' at 95°C when in SB. In a few experiments 2 M urea was added to SB and consequently samples were heated at 37°C instead of 95°C. Gels were fixed for 1 h in 30% methanol, 10% acetic acid and stained with Coomassie Brilliant Blue (0.25% Coomassie R-250 in 30% methanol, 10% acetic acid) for 1 h for 1 mm thick gels (1 h 30 min for 1.5 mm thick gels). Destaining was performed in 30% methanol, 10% acetic acid.

IEF analysis was carried out both on ampholine gels and on immobilized pH gradient (IPG) gels. Ampholine gels were prepared as follows: 6 M urea, 4.5% acrylamide/bisacrylamide (28.8:1.2), 2% ampholines pH 3.5-10 (Pharmacia). Alternatively, IEF analysis was performed on non-linear 4-10 immobilized pH gradients (12). In both cases samples were loaded at the cathode.

Preparation of the cDNA Library

Two gr of frozen *Tuber magnatum* Pico fruitbody were crushed in liquid nitrogen; nucleic acids were extracted in Tris/HCl 100 mM pH 8.5, 100 mM NaCl, 20 mM EDTA, 1% SLSar and purified by phenol and phenol-chloroform extraction. Following isopropanol precipitation a second precipitation step was performed in 2 M LiCl to remove RNA from DNA. After a few washing steps, total RNA (about 700 μg) was resuspended in sterile distilled water and passed twice through an oligo-dT cellulose column to obtain about 4 μg of poly-A+ RNA.

The cDNA library was prepared according to Promega's protocols for the Riboclone cDNA Synthesis System using the λ-gt11/EcoRI vector, obtaining a total amount of recombinant plaques of about 10^5.

RESULTS

Southern Analysis

DNA from eight different *Tuber* species (as reported in Fig. 1) was digested with EcoRI to perform Southern analysis. Two different accessions both of *T. magnatum* and of

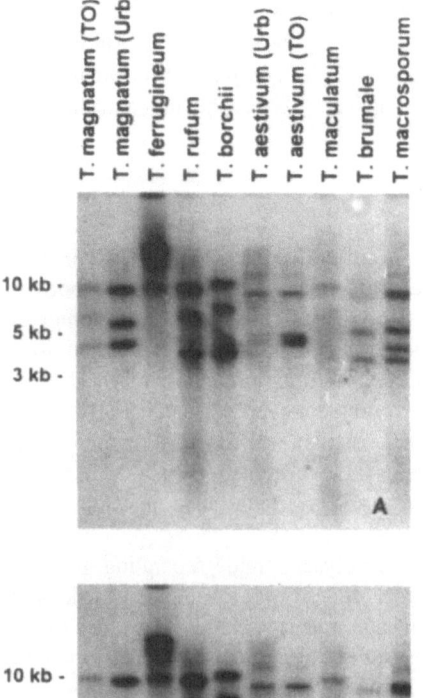

Figure 1. Southern analysis of DNA extracted from fruitbodies of different *Tuber* species, as indicated at the top 1 μg of EcoRI digested *Tuber* DNA was loaded in each lane Two accessions both of *T magnatum* and of *T aestivum* were analyzed, coming from different geographical areas (TO = Torino and Urb = Urbino) Panel A hybridization with ribosomal DNA from flax (*L usitatissimum*) Panel B same filter, hybridized with ribosomal DNA from *N crassa* after probe melting

T. aestivum were analyzed, coming from different geographical areas. Two methods were compared to extract DNA from fruitbodies: SLSar-urea and guanidine isothiocyanate. The first allowed higher DNA recoveries. Further purification of DNA after extraction proved to be indispensable. Purification by CsCl gradient was demonstrated to be the most reliable method to obtain complete removal of RNA and good restrictions of genomic DNA (data not shown).

Digests were probed with heterologous ribosomal DNA from flax (*Linum usitatissimum*) (Fig. 1a) and from *Neurospora crassa* (Fig. 1b).

The resulting patterns are highly polymorphic: even if some bands are conserved in different truffles, the complete pattern is specific to each *T.* taxon. The two samples of different origin of *T. aestivum* share the same pattern and only some difference in band intensity can be noted, while the two samples of *T. magnatum* have slightly different patterns and only one of the three bands remains exactly the same for the two accessions.

No difference was observed probing the same digests with the two different probes, thus indicating a high degree of interspecific conservation of rDNA.

Some other Southern experiment was carried out using actin and γ-tubulin genes from *Aspergillus nidulans* as probes but results were unsatisfactory, even under low stringency hybridization conditions, probably because of a too low homology between templates and probes (data not shown).

Some preliminary experiment towards the analysis of mycorrhizal root tips was carried out. The problem of DNA purification seems to be of relevant importance in this case as the quantity of DNA extracted from root tips is always very poor, and frequently under the minimum amount required to perform CsCl purification. This can affect Southern analysis results, as unpurified DNA does not undergo good restriction. In any case, DNA extracted from the most common host plants (*Corylus avellana, Quercus robur* and *Tilia platiphillum*) and digested with EcoRI gave specific patterns, clearly distinguishable from those of the analyzed truffles (data not shown), so that we may assume that the presence/absence of the fungus in root tips could be easily detected by Southern analysis. Digestion with methylation sensitive endonucleases (e.g. PvuII or PstI) could also be useful for this purpose, as fungal DNA is usually methylated to a much lower extent than plant DNA, and this difference could lead to easy identification of the fungal pattern, if present.

SDS-PAGE analysis

Several truffle samples were analyzed by SDS-PAGE under many different conditions and specific patterns were obtained for *T. magnatum, T. borchii, T. melanosporum, T. aestivum, T. brumale, T. ferrugineum, T. macrosporum* and *T. rufum*, showing a high level of heterogeneity (Fig. 2).

Predominant electrophoretic bands, in fact, are typical to each truffle and there is usually no correspondence even between patterns of species that are morphologically

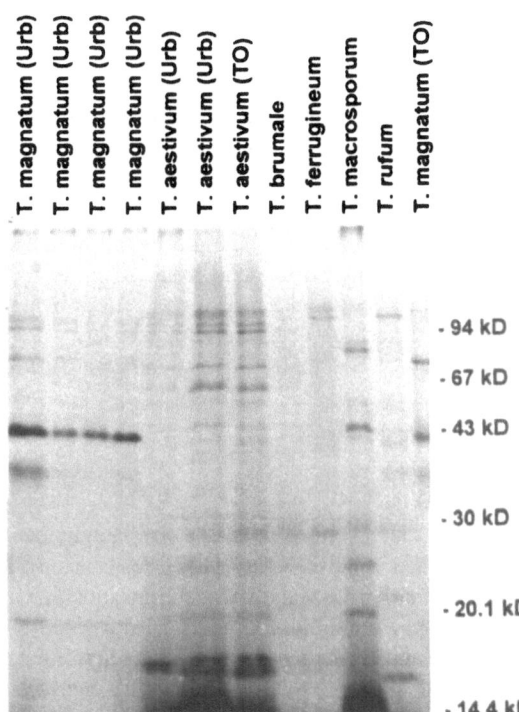

Figure 2. Fruitbody total protein analysis by SDS-PAGE on a 17% polyacrylamide gel Proteins loaded in lanes 2, 3 and 4 were extracted from morphologically different parts of the same fruitbody (Urb) = from Urbino (TO) = from Torino *T* species are indicated at the top

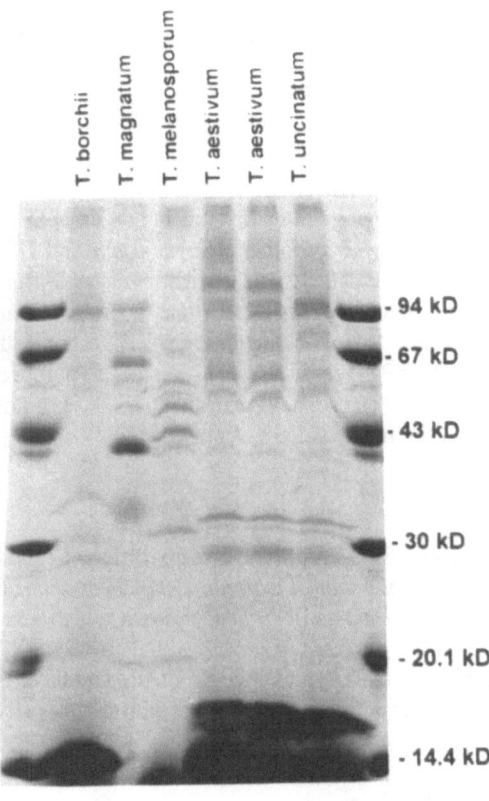

Figure 3. Fruitbody total protein analysis by SDS-PAGE on a 15% polyacrylamide gel. All samples come from Torino, except for T. *magnatum* which comes from central Italy T. species are indicated at the top

very close to one another (for example *T. magnatum* and *T. borchii*). The protein profile seems to be the same for the whole fruitbody, as no difference was seen among patterns of extracts obtained from morphologically different parts of the same fruitbody (*T. magnatum*).

Furthermore, SDS-PAGE profiles allowed discrimination between *T. aestivum* and *T. uncinatum*, which differ one from the other for the presence or absence of a high molecular mass band, the rest of the pattern being only slightly different (Fig. 3). To further confirm these data, the same extracts were analyzed under different conditions (both changing SB composition or sample purification procedures), obtaining always the same results.

IEF Analysis

The data obtained both from IEF-ampholine gels (data not shown) and -IPG gels (Fig. 4) confirm SDS-PAGE results. Isoelectric profiles of total protein extracts are in fact typical to each truffle and allow discrimination among species and varieties. In particular, referring to the small differences revealed by SDS-PAGE analysis between the patterns of *T. aestivum* and *T. uncinatum*, we can state that IEF patterns as well reveal some difference, and conclude that both SDS-PAGE and IEF analyses can be considered reliable techniques towards the identification of *Tuber* fruitbodies.

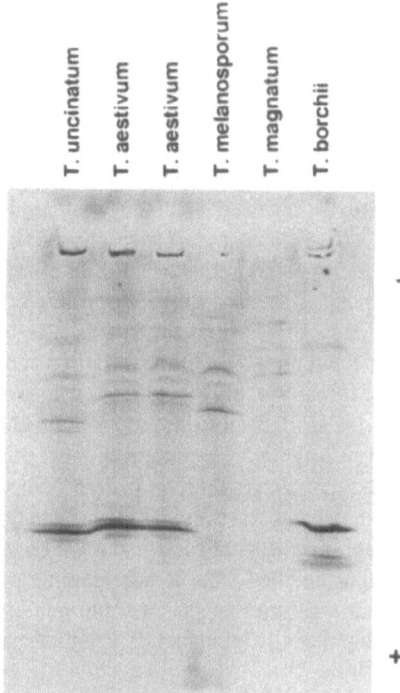

Figure 4. IEF analysis on a non-linear immobilized pH gradient. The same protein extracts have been differently processed to obtain samples loaded in this fig. and in fig. 3. T. species are indicated at the top.

DISCUSSION

Molecular biology techniques can provide methods for fast and reliable identification of different truffle species. This is supported by the complete accordance with data obtained by morphological classification (mostly based on the analysis of fruitbodies and spores). Several strategies can be adopted, depending on the available material and on the purposes of the analyses.

PCR with ITS and IGS primers has proved to be a reliable tool towards the identification of interspecific differences, while the RAPD technique can be adopted in order to reveal inter- and intraspecific differences (13,14). As all PCR-based techniques require very small sample quantities, they can also be useful for the analysis of truffles during different phases of their life cycle. The choice of oligonucleotides is obviously crucial for the success of the method, as the characteristics of the amplified regions have a strong influence on the final conclusions (13): in this respect, the use of a mixture of oligonucleotides instead of a couple certainly leads to more secure results.

As ITS and IGS primers are representative of interspecific conserved regions both the fungus and the host DNA are amplified when analyzing mycorrhizae. This implies that mycorrhizal roots analysis must be supported by the fungus and the analysis of non-mycorrhizal roots and electrophoretic patterns of amplified fragments must be sufficiently different to allow discrimination between the truffle and the plant.

Southern analysis can also be efficient to discriminate among different truffle species or varieties, but it requires higher quantities of DNA, and this could be a disadvantage when working with mycorrhized root tips. Apart from this consideration, Southern analysis can be successfully performed to assess for the presence of the fungus in the mycorrhizal root, digesting with either methylation-sensitive or insensitive enzymes. Ribosomal DNA can be considered a good probe as it gives rise to specific patterns on EcoRI *Tuber* digests; in any case, it would be useful to have homologous probes. Establishing specific cDNA libraries can be useful for this purpose and can also be considered the first step towards a deeper investigation of truffle genetic organization.

These DNA-based techniques can be supported by the analysis of total protein extracts, that can also be useful at the level of monodimensional analysis (SDS-PAGE and IEF) to distinguish even among very similar truffles (15).

Two dimensional electrophoresis of non-mycorrhizal roots and of the fungus proteins can be compared to the 2-D profile of mycorrhizal roots, in order to discriminate between the spots of plant or fungal origin (16). Furthermore, 2-D analysis can be considered an important tool to assess for the presence of symbiosis-specific polypeptides (mycorrhizins) or, on the contrary, for the absence of some polypeptides specific to one of the two organisms in the symbiosis (polypeptide cleansing) (17,18).

Unfortunately, this kind of approach could be misleading. By applying microelectrophoresis techniques to protein extracts from single root tips of the same plant, in fact, it was proved that electrophoretic patterns can vary from one tip to the other, depending on the developmental stage and on the growth conditions (16). This happens both in mycorrhizal and non-mycorrhizal roots and must be taken into consideration expecially when conventional techniques - that require pooling of a large number of root tips to extract sufficient protein quantities to perform the analysis - are used in order to discriminate between developmentally regulated polypeptides and mycorrhiza-specific polypeptides.

ACKNOWLEDGMENTS

We thank Prof. V. Stocchi, Prof. P. Bonfante and Mr. F. Meotto for providing us *Tuber* fruitbodies and Prof. P. Macino for providing us the rDNA probe from *N. crassa*. B. L. as well as the research were supported by The National Council of Research (CNR), Italy - Strategic Program: Biotecnologia della Micorrizazione.

REFERENCES

1. Fontana, A.. Spoleto,.24-25 maggio 1968, Miceli di funghi ipogei in coltura pura, *Atti del Congresso Internazionale sul Tartufo*.
2. Fontana, A.. 1971, Il micelio di *Tuber melanosporum* Vitt. in coltura pura, *Allionia* 17:19-23.
3. Zambonelli, A., Govi, G. and Previati, A.. 1989, Micorrizazione *in vitro* di piantine micropropagate di *Populus alba* con micelio di *Tuber albidum* in coltura pura, *Mic. Ital.* 3:105-111.
4. Chevalier, G.. 1994, Evolution des recherches sur les plantes mycorhizés par la truffe et perspectives de développement, *Giorn. Bot. Ital.* 128:7.
5. Fontana, A., Ceruti, A., and Meotto, F.. 1992, Criteri istologici per il riconoscimento delle micorrize di *Tuber albidum*, Convegno Internazionale sul tartufo, L' Aquila, 5-8 marzo 1992, *Micol. Veg. Medit.* 2:121.
6. Henrion, B., Chevalier G., and Martin, F.. 1994, Typing truffle species by PCR amplification of the ribosomal DNA spacers, *Mycol. Res.* 98(1):37.
7. Henrion, B., Le Tacon, F., and Martin, F.. 1992, Rapid identification of genetic variation of ectomycorrhizal fungi by amplification of ribosomal RNA genes, *New Phytol.* 122:289.
8. Wyss, P., and Bonfante, P.. 1993, Amplification of genomic DNA of arbuscular mycorrhizal (AM) fungi by PCR using short arbitrary primers, *Mycol. Res.* 97:1351.
9. Gardes, M., and Bruns, T. D.. 1993, ITS primers with enhanced specificity for Basidiomycetes: application to identification of mycorrhizae and rusts, *Mol. Ecol.* 2:113.
10. Lanfranco, L., Wyss, P., Marzachì, C., and Bonfante, P.. 1993, DNA probes for identification of the ectomycorrhizal fungus *Tuber magnatum* Pico, *F.E.M.S. Microbiol. Lett.* 114:245.

11. Faust, P. L., Wall, D. A., Perara, E., Lingappa, V. R., and Kornfeld, S.. 1987, Expression of human cathepsin D in *Xenopus* oocytes: phosphorylation and intracellular targeting, *J. Cell Biol.* 105:1937-1945.

12. Gianazza, E., Giacon, P., Sahlin B., and Righetti, P. G.. 1985, Non-linear pH courses with immobilized pH gradients, *Electrophoresis* 6:53-56.

13. Lanfranco, L., Wyss, P., Marzachì, C., and Bonfante, P.. Generation of RAPD-PCR primers for the identification of isolates of *Glomus mosseae*, an arbuscular mycorrhizal fungus, *Mol. Ecol.* in press.

14. Potenza, L., Amicucci, A., Rossi, I., Palma, F., De Bellis, R., Cardono, P., and Stocchi, V.. 1994, Identification of *Tuber magnatum* Pico DNA markers by RAPD markers, *Biotechnol. Techn.* 8:93.

15. Stocchi, V., De Bellis, R., Vallorani, L., Piccoli, G., and Dachà, M.. 1994, Characterization and purification of protein markers from various species of white truffle, International Symposium on: *Biotechnology of ectomycorrhizae: Molecular approaches*, Urbino, November 9-11, 1994, Abstracts, 41.

16. Guttenberger, M., and Hampp, R.. 1992, Ectomycorrhizins - symbiosis-specific or artifactual polypeptides from ectomycorrhizae?, *Planta* 188:129-136.

17. Hilbert, J. L., and Martin, F.. 1988, Regulation of gene expression in ectomycorrhizae I. Protein changes and the presence of ectomycorrhiza-specific polypeptides in the *Pisolithus-Eucalyptus* symbiosis, *New Phytol.* 110:339-346.

18. Martin, F., and Hilbert, J. L.. 1991, Morphological, biochemical and molecular changes during ectomycorrhiza development, *Experientia* 47:321-331.

15

THE POLYMORPHISM OF THE rDNA REGION IN TYPING ASCOCARPS AND ECTOMYCORRHIZAE OF TRUFFLE SPECIES

F. Paolocci,[1] E. Cristofari,[2] P. Angelini,[3] B. Granetti,[3] and S. Arcioni[1]

[1] Istituto di Miglioramento Genetico delle Piante Foraggere del CNR
via Madonna Alta 130, 06128 Perugia, Italy
[2] A.R.S.-Parco Tecnologico Agroalimentare dell'Umbria
06050 Pantalla, Todi-Perugia, Italy
[3] Dipartimento di Biologia Vegetale dell'Università di Perugia
Borgo XX giugno 74, 06121 Perugia, Italy

ABSTRACT

Methods were developed to type some of the most economically important *Tuber* spp. (*Tuber aestivum* f. *uncinatum, Tuber albidum, Tuber magnatum, Tuber brumale, Tuber melanosporum*) and some of their mycorrhizae by the use of molecular markers. The polymorphism of the rDNA region was investigated by polymerase chain reaction (PCR) and Southern experiments on DNA isolated from ascocarps, uninoculated host roots and mycorrhizae. Using specific primer pairs, PCR amplification of the internal transcribed spacers (ITS) and intergenic spacer (IGS) of DNA isolated from fruitbodies allowed fungus discrimination only by the length or the number of the ITS and IGS amplified fragments. These results were confirmed by Southern analysis carried out on fungus DNA restricted with EcoRI and probed with the ITS fragment of *T. brumale*: a specific banding pattern for each species tested was evident. The same analyses were then performed on mycorrhizal root tips and inoculated host plant roots. Each fungus-host plant interaction showed a specific pattern due to the combination of traits of the two partners. It is concluded that rDNA region is a suitable molecular marker to characterize both ascocarps and mycorrhizae of truffle species by PCR and/or Southern analysis.

INTRODUCTION

Truffle cultivation and production represent a significant agricultural income in marginal areas of those regions (such as the inland areas of central Italy) where economically important endogenous species are present and where artificial mycorrhization of forest trees to increase the production of fruibodies is widely practised. The knowledge of biotic and

Biotechnology of Ectomycorrhizae, Edited by Vilberto Stocchi et al.
Plenum Press, New York, 1995

abiotic factors characteristic of the habitats of the most marketable *Tuber* spp., and above all the development and optimization of the techniques for the inoculation and growth of artificially mycorrhized plants (Chevalier 1994), make the production and subsequent planting of a wide range of symbiotic host plants possible. However, little is known about the interaction between endogenous flora and the exogenous flora associated with the newly introduced mycorrhized plants. Also, monitoring the development of desirable symbiotic agents in host plant root systems over the years is hardly possible because of the persisting difficulty in characterizing the ectomycorrhizae of truffle species. The distinctiveness of *Tuber* spp. is in fact one of the major topics in truffle research, and in this context several studies have been carried out to develop more sophisticated methodologies in the last few years. In fact, it has been realized that although the morphological and structural traits can be suitable to classify ascocarps of all the *Tuber* spp., even fine biochemical approaches such as isoenzyme and protein assays or immunoenzymatic analyses might not help in typing mycorrhizae of morphologically very similar *Tuber* spp. (Corocher *et al.* 1992; Dupré & Chevalier 1991; Dupré *et al.* 1993; Frizzi *et al.* 1992; Mouches *et al.* 1981; Palenzona *et al.* 1990; Papa & Polimeni 1990; Papa *et al.* 1992; Plattner *et al.* 1991, Potenza *et al.* 1994; Lanfranco *et al.* 1993, Zambonelli *et al.* 1992). However, mycorrhizal identification is still largely based on the analysis of morphological traits (Bencivenga & Granetti 1990; Ceruti 1960; Fontana *et al.* 1990, Fontana *et al.* 1992; Giraud 1988; Granetti 1990; Granetti *et al.* 1990; Zambonelli *et al.* 1993) which may evolve and vary with the season, so that in winter it may well be impossible to distinguish mycorrhizae of *Tuber brumale* from those of *Tuber melanosporum*. Moreover, in the summer, as a consequence of soil drought, mycorrhizae of *Tuber albidum* are not very different from those of *Tuber magnatum*. The development of a reliable, and relatively simple and inexpensive method to check for the presence of symbiotic fungi in a root apparatus could help regulate the market of mycorrhized plants. This paper reports a method based on the analysis (through PCR and Southern experiments) of the structure of rDNA in fungus and host plant genomes. The rDNA is a genomic region often used in taxonomic and phylogenetic studies (Lane *et al.* 1985; Sogin & Gunderson 1987; Sogin 1990, John & Miklos 1988; White *et al.* 1990; Egger *et al.* 1991) on account of the following characteristics: it is present in high copy number (Rogers & Bendich 1987); it is arranged as a tandem repeated array with both variable and highly conserved regions; the sequence of the conserved regions is known for many species, so that it is possible to select probes and primers to be used for Southern analysis and PCR amplification (Gardes *et al.* 1991); the spacer regions evolve rapidly, even among closely related taxa or species (Jorgensen & Cluster 1988), so that polymorphism among species is easily detectable. In the experiments reported here, the successful typing of truffle species is based on the polymorphism of the ribosomal spacers: IGS (intergenic spacer) and ITS (internal transcribed spacer). The proposed method allows a rapid identification of 1) fruitbodies of *Tuber aestivum* Vitt. f. *uncinatum* (Chatin) Fischer, *T. albidum* Pico, *T. magnatum* Pico, *T. brumale* Vitt. and *T. melanosporum* Vitt., 2) the ectomycorrhizae of the last four *Tuber* spp. on *Corylus avellana* L. and 3) the mycorrhizae of *T. melanosporum* on *Quercus pubescens* Willd. The use of this method in quantifying the degree of mycorrhization of host plants is discussed.

MATERIALS AND METHODS

Preparation of Mycorrhized Plants

        In order to have a sufficient quantity of mycorrhizae for morphological and molecular characterization, hazelnut seedlings were mycorrhized with *T. brumale*, *T. melanosporum*, *T. albidum*, *T. magnatum*. In October, hazelnuts, cultivar "Tonda Romana", were gathered from a plantation located near Perugia (Italy). The hazelnuts were immersed in a 1% H_gCl_2 solution for 10 min and then washed three times with sterile water. They were placed in containers (sterilezed with absolute ethanol) and maintained at +1°C. In January, the hazelnuts were sown in a substrate made up of 50 parts vermiculite and 50 parts agriperlite which had been steam sterilized for 2h at 120°C (Bencivenga, 1982). They were placed in sterilized plastic trays with some holes in the bottom. The trays were placed in a small greenhouse which was also carefully disinfected with sodium hypoclorite. The day temperature was 14-16°C and the night temperature about 13-14°C to favor germination. When irrigation was necessary sterile water was used. In April, the average seedlings height was 12 cm and the plants had well-developed root systems. The roots were inoculated using a spore-suspension method (Bencivenga, 1982). For this purpose, the sporocarps needed were collected when they were perfectly mature and healthy as resulting by peridium and gleba colour and by morphology and colour of spores (Ceruti, 1960); they were checked to ascertain that they belonged to one of the four species being examined (Ceruti 1960, Chevalier *et al*. 1990, Gross 1987). After careful washing, the truffles were sterilized externally by immersing them in absolute ethanol for 2 minutes. They were placed in sterile plastic trays containing sterile moist vermiculite. The trays were closed with plastic wrap and refrigerated (+1°C) until the truffle were used for inoculation.

For the inoculation, a spore suspension was prepared by finely grinding the truffles, in the presence of sterile water, in a Osterizer homogenizer. One kg of well-cleaned truffles was used to make up a two-liter spore and gleba suspension. This liquid was absorbed by the sterile vermiculite. Separately, soil taken from a field adjacent to a natural truffle area was steam sterilized for 4 hours at 120°C in plastic bags, and then was mixed with the previously prepared spore-loaded vermiculite so that every liter of substrate contained 24 g of truffle. At the end of April, hazel seedlings were removed from the trays, transplanted into sterile rigid plastic pots and filled with spore-containing substrate. Eight grams of truffles were placed in each pot. The transplanted seedlings were grown in a disinfected greenhouse, without heating and covered with a black shade net that reduces the sunlight by 50%. Sterile water was used to irrigate the plants when necessary. The air temperature was slightly higher than the outside temperature because a small fan was used to extract the air when the temperature exceeded 24°C. A filter was used at the air intake to remove dust and reduce the risk of contamination. By the following October, the plants had an average height of 25 cm and had formed abundant root systems. A preliminary examination of these showed an adequate level of mycorrhization in the plants inoculated with *T. magnatum*, and a high percentage of mycorrhization in those inoculated with the other *Tuber* spp.

Description of Mycorrhizae

The species of the genus *Tuber* form ectomycorrhizae on the roots of various trees and are characterized by a multi-layered mycoclena and Hartig net that involves the rhizoderm and the cells underlying the primary bark. The descriptions of the mycorrhizae of the four species in the present work are based on research carried out by the Authors, considering those structural features able to differentiate these mycorrhizae and taking into account the results already published (Bencivenga & Granetti 1990, Chevalier & Desmas 1977, Fontana *et al*. 1990, Fontana *et al*. 1992, Granetti 1990, Guinberteau *et al*. 1990, Palenzona 1969, Palenzona *et al*. 1972, Parguey *et al*. 1990, Zambonelli *et al*. 1993).

Mycorrhizae of T. albidum. The mycorrhizae of *T. albidum* are prevalently oblong and club-shaped with very rounded ends. They can be simple or branched, depending on the monopodial arrangement. In the resting stage, these mycorrhizae are uniformly amber-colored from the base to the apex; in the active growing phase they are very clear amber with slightly darker spots and a whitish apex. The surface cells of the mycoclena have a slightly wavy edge with poorly pronounced lobes. During growth, numerous spinules form in the mycorrhizae. The spinules are hyaline or a very pale yellow; the base is enlarged where it attaches to the mycoclena; there are 1 or 2 septa in the proximal part and they have a well-thickened wall that almost completely occludes the cell lumen in the upper half of the spinule. The apex of the spinule is rounded. The length of the spinule varies between 72 and 108 µm (average 106 µm). The basal part of the spinule, from 2.4 to 6 µm (average 4.5 µm) is wider than that of *T. magnatum*. The peritrophic hyphae, which are of little diagnostic importance, are pale yellow, simple or slightly branched, winding with tranverse septa throughout the entire length. The Hartig net has chained hyphae that develop prevalently on the rhizoderm cells (Fig. 1 A, B, C).

Mycorrhizae of T. Magnatum. These mycorrhizae are club-shaped or oblong with rounded ends. They can be simple or branched, depending on the monopodial arrangement. In the resting phase the mycorrhizae have uniform dark amber color from the base to the apex; during active growth the apex is whitish while the remaining part of the mycorrhiza is bright amber; this is much more marked then the mycorrhizae of *T. albidum*.

The surface cells of the mycoclena have a wavy edge with definite lobes that are more pronounced than those observed in *T. albidum* and are generally a slightly smaller. These mycorrhizae produce abundant spinules during the growth phase. The spinules are hyaline, have one or two transverse septa and have a slightly dilated base; the apex is narrow and rounded. The spinule wall is thin. The spinules are between 2.4 and 4.8 µm (average 3.0 µm) at the base, and are between 55 and 108 µm (average 78 µm) long. The peritrophic hyphae are hyaline, simple or slightly branched, winding with tranverse septa throughout the entire length. The Hartig net has chained hyphae that also develop between the cortical cells located under the rhizoderm (Fig. 1 D , E, F).

Mycorrhizae of T. Brumale. These mycorrhizae are club-shaped with well-rounded apices; they are only slightly branched. The mycorrhizae are brownish from the base to the apex (they are darker than *T. melanosporum*). During the active growing stage, they are amber colored in the first two-thirds with a whitish apex. The surface cells of the mycoclena have quite wavy edges (a little less than in *T. melanosporum*) but also have cells that are only slightly lobed. The spinules are thick and robust, from 36-98 µm long (average 60 µm), and widen noticeably towards the base where the width varies from 2.3 to 7.1 µm (average 5.1 µm); they have 1 or 2 septa in the lower half; the extremity is rounded. The wall of the spinule is well-thickened up to the apex; it is a distinct yellow. The peritrophic hyphae, which are of little diagnostic interest, are simple or branched, with multiple, winding septa and pale yellow. The Hartig net is made up of hyphae of neighboring septa which also extend below the rhizoderm (Fig.1 G, H, I).

Mycorrhizae of T. Melanosporum. The mycorrhizae of *T. melanosporum* are club-shaped with well-rounded apices, often extensively branched depending on the monopodial arrangement. During the resting stage they are a uniform dark amber color from the base to the apex; in the growing phase the apex is whitish and the rest of the mycorrhiza is bright amber. The surface cells of the mycoclena have a wavy edge with well-pronounced lobes. In some periods of the year, these mycorrhizae produce hyaline spinules (cystids) that are between 66,9 and 126,7 µm long (avarage 102,4), slender, with the base slightly enlarged near the point of attachment to the mycoclena (average of the width at this point is 2,7 µm) and a narrow, rounded apex; the spinules have one or two tranverse septa located on the lower half. The wall of the spinule is thin throughout its length. At the apex the spinule can

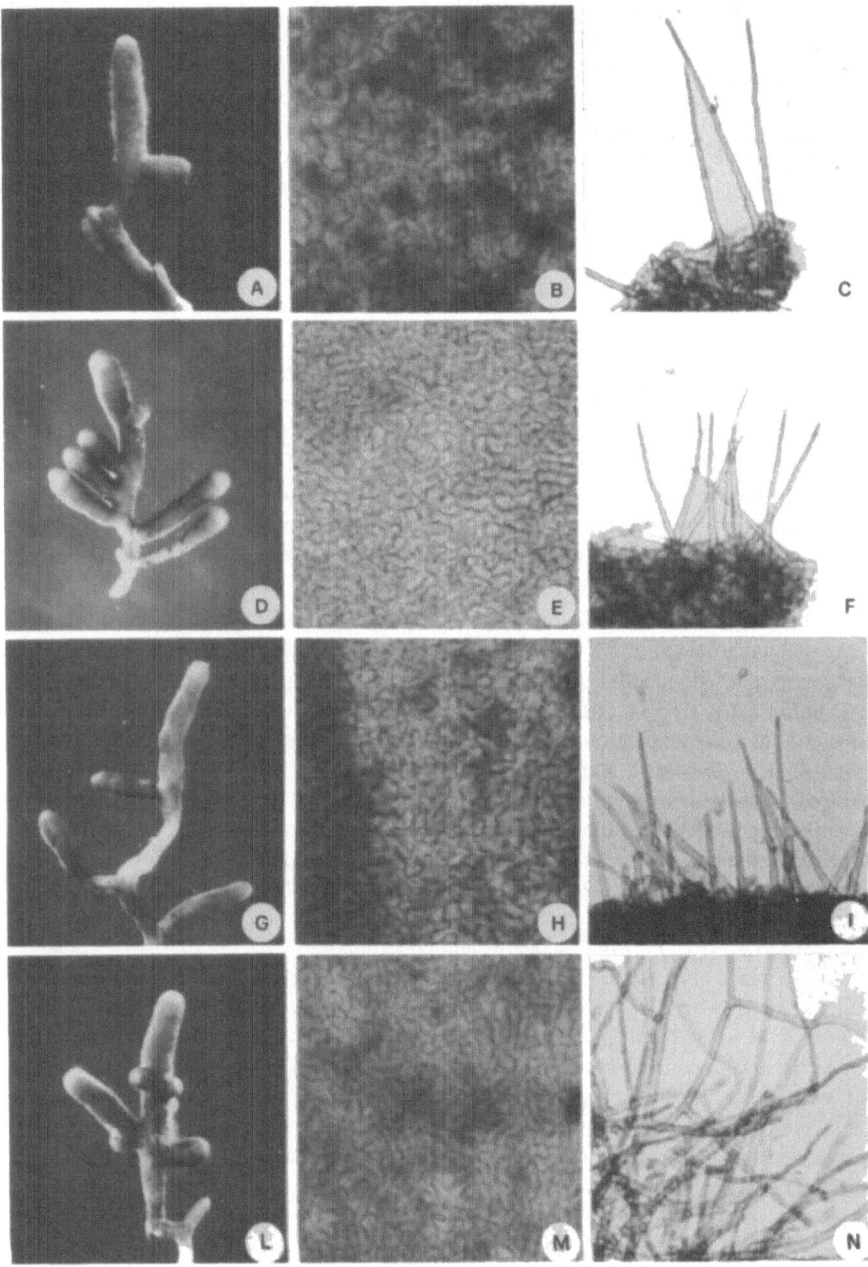

Figure 1. Ectomycorrhizae, surface cells of the mycoclena and emanating hyphae of *T albidum* (A, B, C), *T magnatum* (D, E, F), *T brumale* (G, H, I) and *T melanosporum* (L, M, N)

form a typically vegetative hypha with a constant diameter and a well-rounded apex that can also be branched (this phenomenon can also be observed in other *Tuber* species). During the growing stage, the mycorrhize of *T. melanosporum* preferentially produce peritrophic hyphae that attach to the cells of the mycoclena with parallel walls (not enlarged as in the spinules). These hyphae are straight, rigid and form some right-angled branches in the proximal part, followed by other straight branches which are at about 45° angles to the principal hypha. The peritrophic hyphae have mainly a light yellow-amber wall in the basal parts while they are light yellowish or even hyaline in the young ramifications. This particular type of peritrophic hyphae is very characteristic of *T. melanosporum*. The Hartig net made up of hyphae in neighboring septa developed between the cells of the rhizoderm and between the cortical cells (Fig. 1 L, M, N).

Sample Collection and DNA Isolation

For each species considered, fresh ascocarps were collected in central Italy, washed in water, peeled to remove superficial contaminants and analyzed microscopically according to the protocols routinely followed for species characterization. Ectomycorrhizal root tips were collected under stereo-microscope and characterized by morphological analysis according to the method described in the previous sections, frozen in liquid nitrogen and immediately used for DNA extraction or stored at -70°C. Genomic DNA was extracted from about 2 grams of ascocarps, ectomycorrhizae and roots of uninoculated plants according to Rogers *et al.* (1989) with slight modifications: tissues were ground in the presence of liquid nitrogen and resuspended in a buffer containing 2% CTAB, 100 mM Tris-HCl pH8.0, 20 mM EDTA pH 8.0, 1.4M NaCl, 1% polyvinylpyrrolidone (PVP, MW 360000) and 0.2% mercaptoethanol, heated at 65°C. The tissues were maintained at 65°C for 30'-60'. After chloroform-isoamylic alcohol extraction followed by centrifugation, 1/10 volume of 10% CTAB buffer (10% (w/v) CTAB, 0.7 M NaCl) heated at 65°C was added to the top phase recovered. Then a second chloroform-isoamylic alcohol extraction was performed and 1/10 volume of 3M Na-acetate and 2 volumes of ethanol was added to the aqueous phase recovered. Nucleic acids were cold-precipitated and pelleted by centrifugation. The pellet was washed in 70% ethanol and subsequently resuspended in a TE buffer (10mM Tris-HCl, 10mM EDTA), a CsCl gradient ultracentrifugation was then performed according to Maniatis *et al* (1982). After CsCl gradient centrifugation, the DNA isolated from fruitbodies showed two bands in test tubes examined under UV light (Fig. 2), while only one band was visible for DNA extracted from ascocarps and host roots.

Two bands obtained in CsCl gradient from ectomycorrhizae generally gave the same Southern pattern when collected separately and hybridized with the same probe, although the DNA of the lower band appeared less pure than that of the upper band. The experiments reported here were carried out on DNA from the two mixed bands or from just the upper band, when it was distinctly separated from the lower one in the centrifuge tube. Once the bands were collected, they were dialyzed against TE buffer and then ethanol-precipitated. The recovery of DNA was evaluated by optical density at 260 nm, or by agarose gel electrophoresis.

PCR Amplification and Probe Preparation

In order to amplify the ITS region two primers were chosen from the conserved sequences of 17S and 25S of rDNA, (primer ITS1: 5'-TCCGTAGGTGAACCTGCGG and primer ITS4: 5'-TCCTCCGCTTATTGATATGC), while to amplify the IGS two primers were selected from the conserved region of 25S rDNA and 5S rDNA (primer CLN12: 5'-CTGAACGCCTCTAAGTCAG and primer 5 SA: 5'- CAGAGTCCTATGGCCGTGGAT)

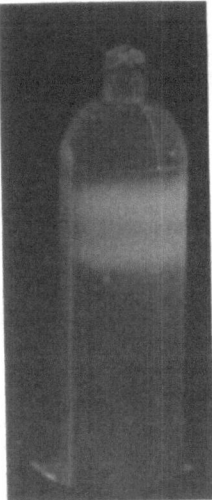

Figure 2. DNA bands of ectomycorrhizae of *T.brumale* on hazel resulting from CsCl-gradient.

(Henrion *et al.* 1992). ITS from both fungi and host plants and IGS from fungi were amplified by PCR with a 3-step thermal cycle, repeated 25 times (120 s at 95°C, 25 s at 55°C, and 120 s at 72°C, plus a final extension for 10 min at 72°C). An aliquot of the PCR products was submitted to electrophoresis in 1.5% and 2% agarose gels for IGS and ITS, respectively. For amplification of DNA isolated from ectomycorrhizae, an aliquot (1-1.5 μl) of the products of the first round of amplification was used to set a second PCR amplification cycle, in order to have more easily detectable bands on agarose gels. ITS fragments obtained from amplification of *T. brumale* and *T. albidum* were electroeluted from agarose gels, and after two phenol-chloroform extractions and ethanol precipitation they were radiolabelled with ^{32} P dCTP (Promega kit) and used as probes in Southern experiments (Feinberg & Vogelstein 1984).

Southern Analyses

Genomic DNA isolated from ascocarps, roots of host plants and ectomycorrhizae, was digested with EcoRI, run on 0.8 % agarose gel and alkaline blotted on Hybond N$^+$ membrane (Amersham). Pre-hybridization and hybridization were performed at 65°C according to the filter supplier's instructions. Two or three high stringency washes were performed at 65°C in 0.1x SSC and 0.1% (w/v) SDS. The membrane was exposed for 24-48 hrs to Kodak films with two intensifyng screens and maintained at -70°C.

RESULTS AND DISCUSSION

PCR Amplification of rDNA Spacers of Fungi and Host Plants

Each species tested showed only one ITS band, with exception of *T. aestivum* which showed two bands, suggesting the possibility of a polymorphism within its rDNA gene family (Fig. 3). Each single band has a specific length that allows a first discrimination

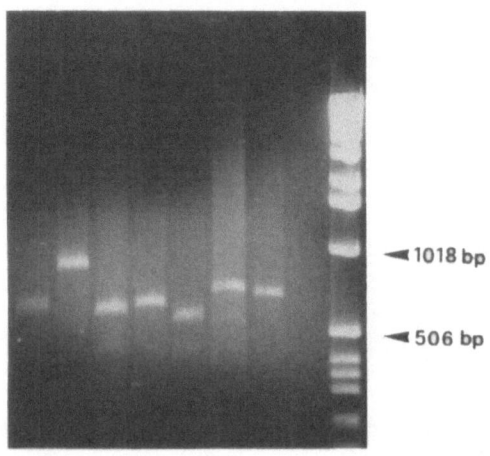

◀ 1018 bp

◀ 506 bp

Figure 3. PCR amplification of the ITS region of DNA isolated from ascocarps of *Tuber* spp. and roots of host plants. Lane 1: oak; lane 2: *T.brumale*; lane 3: *T.melanosporum*; lane 4: *T.magnatum*; lane 5: *T. albidum*; lane 6: *T.aestivum*; lane 7: hazel; lane 8: control (no template); lane 9: size markers (1 Kb DNA ladder, GIBCO BRL).

among all the truffle spp. considered; in particular such differentiation is easily detectable also between morphologically similar spp. such as *T. melanosporum* vs. *T. brumale*, or *T. albibum* vs. *T. magnatum*. Contrary to results reported by Henrion *et al* (1994) host plants showed amplification products, suggesting that the selected ITS primers are also homologous to the plant DNA sequences. This means that at least another fragment can be expected to be amplified when PCR is carried out on DNA isolated from mycorrhized apical root tips. The presence of this supplementary band might not allow the typing of fungus species by RFLP analysis of the fragments produced by PCR amplification of DNA isolated from ectomycorrhizae, as suggested by Henrion *et al.* (1994). To confirm the ITS amplification observed in oak and hazel, DNA was isolated from a number of plant species, including wheat and petunia, and all these spp. showed a band after amplification by using the ITS1/ITS4 primer pairs and the same cycle parameters as reported above. PCR amplification carried out on DNA from *Tuber* spp. and host plants by utilizing CLN12/5SA primer pairs, showed the possibility to also use IGS region in truffle typing. In fact, the IGS amplification of truffle spp. produced distinct and complex banding patterns due to the presence of multiple (major and minor) bands for all the spp. except *T. magnatum* which showed a single well-defined fragment of approx 500 bp (data not shown). Due to the complexity of the patterns produced, and, above all, to their high rates of variation even within species (Jorgensen & Cluster 1988), the use of the ITS rather than the IGS region should be considered for diagnostic purposes. Furthermore, the primers selected to amplify the IGS of fungi gave rise to bands even when the DNA of host plants was amplified. The IGS primers were not supposed to amplify the plant DNA, because CLN12 and 5SA are homologous to the 25S and 5S genes, respectively, and in plants, unlike fungi, the two genes are situated in different gene families. These results suggest that in the higher plants either the 5S gene is still partially located close to the other rDNA genes, or that the 25S gene partially migrated with the 5S gene family elsewhere in the genome.

PCR Amplification of ITS from Ecomycorrhizae

PCR amplification carried out on DNA isolated from mycorrhized apical root tips, showed the possibility of typing the truffle spp. examined even at the symbiotic stage. Fig. 4 shows the products of PCR amplification of ITS region performed using as template the

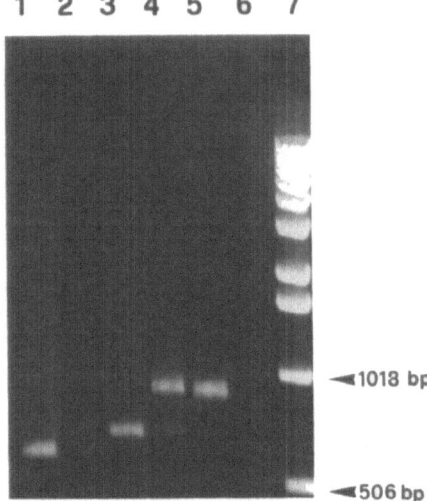

Figure 4. PCR amplification of the ITS region of DNA isolated from ascocarps and ectomycorrhizae of two *Tuber* spp. and from roots of the host plants. Lane 1: *T. melanosporum*; lane 2: ectomycorrhizae of *T. melanosporum* on hazel; lane 3: hazel; lane 4: ectomycorrhizae of *T. brumale* on hazel; lane 5: *T. brumale*; lane 6: control (no template); lane 7: size markers (1 Kb DNA ladder, GIBCO BRL).

DNA isolated from ascocarps of *T. melanosporum* (lane 1) and *T. brumale* (lane 5), from uninoculated host plant (lane 3) and from ectomycorrhizae of *T. melanosporum* (lane 2) and *T. brumale* (lane 4), both on hazel. The two ITS bands derived from the amplification of each ectomycorrhiza, correspond exactly to the single bands specific to the two symbiotic agents. Therefore, the ITS pattern of each mycorrhiza was the sum of the patterns characteristic of fungus and host plants, and the variability of the ITS region permitted the typing of fungi and their mycorrhizae.

Southern Analyses

In order to confirm and extend the results obtained by PCR amplification, experiments on DNA-DNA hybridization were carried out. Such analyses were first performed on genomic DNA isolated from ascocarps of all the truffle species considered and from the two host plants, using as a probe the ITS fragment of *T. brumale* amplified by PCR. Each truffle species showed a different banding pattern, but very similar patterns were obtained when utilizing as probes the ITS region of *T. brumale* or *T. albidum*, suggesting high sequence homology between the two probes.

The possibility of typing fungi by Southern hybridization was based on the different number and/or molecular weight of bands displayed by each species considered (Fig. 5). *T. melanosporum* showed at least three specific bands not displayed by other fungi; *T. brumale* and *T. albidum* showed characteristic fragments of around 600 bp and 300 bp, respectively. *T. magnatum* and *T. aestivum* both presented two bands and the distinction between these two spp. was based only on the different size of the hybridizing fragments. Hazel and oak (see Fig. 7) gave only one band of around 3-3.5 kbp, in agreement with the restriction analysis of the fragments produced by PCR amplification with ITS1/ITS4 primers. As a matter of fact the digestion with EcoRI did not produce any additional fragment indicating the absence of restriction sites within the PCR-amplified fragments. These results indicate that Southern analysis can be used in addition to PCR for typing *Tuber* spp.. Southern analyses were also performed on DNA isolated from mycorrhized apical root-tips and, as expected, the pattern was a combination of those of the fungus and host plant responsible for symbiosis. A representative example is reported in Fig. 6. Therefore, this analysis allowed the differentiation and identification of all the ectomycorrhizae investigated. As reported above, when

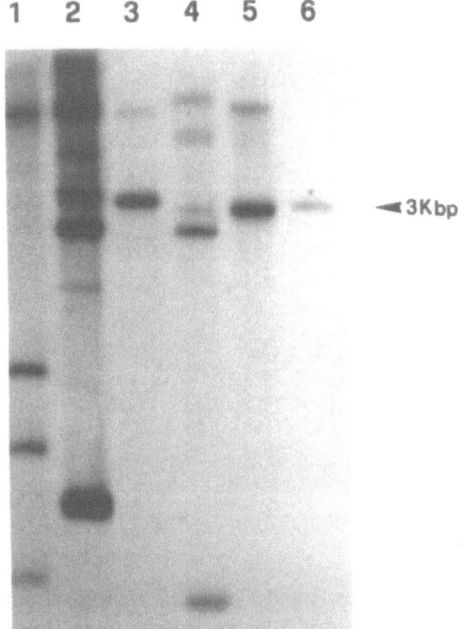

Figure 5. Southern analysis of DNA isolated from ascocarps and roots of hazel, probed with the ITS fragment of *T albidum* Lane 1 *T melano sporum*, lane 2 *T brumale*, lane 3 *T aestivum*, lane 4 *T albidum*, lane 5 *T magnatum*, lane 6 hazel

the two ectomycorrhizal bands produced by centrifugation on CsCl gradient were separately hybridized with the same probe, they showed the same pattern (Fig 7, lanes 2 and 3) even if, in some instances, the upper band, more clearly than the lower one, gave rise to the expected combined pattern

DISCUSSION

The methods developed here appear to be another powerful tool to investigate and identify *Tuber* spp Molecular approaches are now widely utilized for species identification

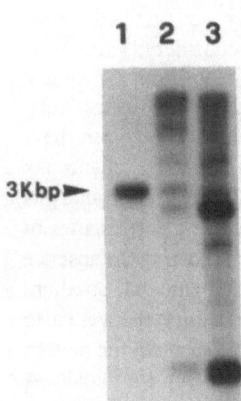

Figure 6. Southern analysis of DNA isolated from ascocarps of *T brumale*, its ectomycorrhizae on hazel and roots of hazel, probed with the ITS fragment of *T brumale* Lane 1 roots of hazel, lane 2 mycorrhizae, lane 3 ascocarps

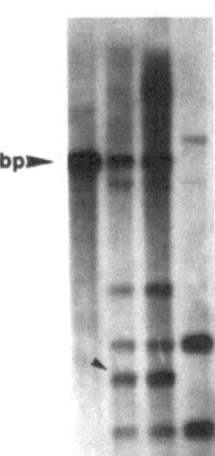

Figure 7. Southern analysis of DNA isolated from ascocarps of *T. melano-sporum*, its ectomycorrhizae on oak, and roots of oak, probed with the ITS fragment of *T. brumale*. Lane 1: roots of oak; lanes 2 and 3: lower and upper CsCl-gradient bands, respectively, of mycorrhizae of *T. melanosporum* on oak; lane 4: ascocarps of *T. melanosporum*. Arrow indicates the band never found in the Southern analysis of *T. melanosporum*.

(many examples in this volume) and this work adds the use of both Southern and PCR for positive typing of ascocarps and ectomycorrhizae. Applications of these two methods have allowed a double check of truffle species and provided new information about the genomic structure of both fungi and host plants. Moreover, the protocol described here is open to a wide range of applications: in fact, rather than performing species-typing on a single mycorrhized apical root tip, as suggested by other Authors (Henrion *et al.* 1994), our analyses were carried out on a large number of mycorrhized apical root tips per plant, so as to achieve a more accurate evaluation of the *Tuber* spp. actually present on the root apparatus. This methodology allowed the identification of more than one truffle species on a plant root system.

Fig. 7 shows the Southern analysis performed on ectomycorrhizae harvested from the root tips of an oak plant supposedly mycorrhized with *T. melanosporum*, using the ITS fragment of *T. brumale* as a probe. The pattern obtained showed at least one band more than would have been expected on the basis of the individual patterns shown by the two symbiotic agents; moreover, the additional band had a molecular weight similar to a specific band of *T. brumale*. The possibility that *T. brumale* had contaminated the oak roots mycorrhized with *T. melanosporum* was later confirmed by an accurate morphological analysis (data not shown). These results emphasize the possibility of achieving positive truffle characterization with a single discriminating probe; moreover, once the initial inoculum has been charac-terized, molecular analyses of randomly chosen mycorrhized plants, can be carried out to check for the presence of the desired symbiotic fungus.

Whatever methodology is used to type *Tuber* spp., a method to assess both the degree of inoculation of the root tips and the presence of contaminating truffle species needs to be developed. In this context preliminary Southern analyses were carried out on mixtures, in different ratios, of DNA isolated from ascocarps of two truffle species, in order to determine the lowest level of foreign DNA that can be revealed by this method.

The smallest amount of contaminating DNA that could be detected in a radiograph and subsequently measured with the aid of a densitometer was estimated at about 5% of the total DNA examined (Fig. 8). If this method could be applied on DNA isolated from ectomycorrhizae, it could give fully exhaustive information on the presence of a symbiotic

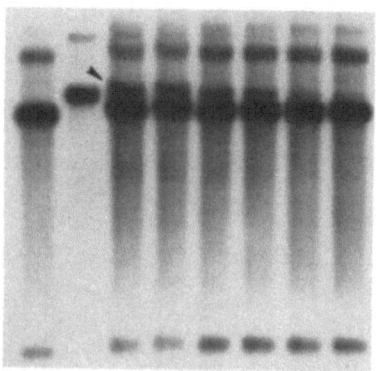

Figure 8. Southern analysis of DNA isolated from ascocarps of *T. albidum* and *T. aestivum* and of their DNA mixtures, probed with the ITS fragments of *T. albidum*. Lane 1: *T. albidum* (2 µg); lane 2: *T. aestivum* (2 µg); lane 3: *T. albidum* (2 µg) and *T. aestivum* (2 µg); lane 4: *T. albidum* (2 µg) and *T. aestivum* (1.5 µg); lane 5: *T. albidum* (2 µg) and *T. aestivum* (1 µg); lane 6: *T. albidum* (2 µg) and *T. aestivum* (0.5 µg); lane 7: *T. albidum* (2 µg) and *T. aestivum* (0.2 µg); lane 8: *T. albidum* (2 µg) and *T. aestivum* (0.1 µg). Arrow indicates the *T. aestivum* fragment considered as the diagnostic band.

fungus on a host root apparatus, thus contributing to more transparency in the market of mycorrhized plants.

ACKNOWLEDGMENTS

The authors wish to thank Dr. F. Damiani and Dr. P.D. Cluster for their critical reading of the manuscript and Mr. A. Bolletta for photographic plates. This work was supported by a grant from the Regione Umbria.

REFERENCES

Bencivenga, M., 1982, Alcune metodiche di micorrizazione di piante forestali con il tartufo nero pregiato di Norcia o di Spoleto (*Tuber melanosporum* Vitt.), *Inf. Agr.* 38: 21, 21155-21163.

Bencivenga, M., and Granetti, B., 1990, Valutazione biometrica di *Ostrya carpinifolia* Scop. prodotte da tartufi di varie specie, In: Comunità Montana dei Monti Martani e del Serano (ed.), *Atti del secondo Congresso Internazionale sul Tartufo*, Spoleto (Italy) 24-27 November 1988, pp 265-270.

Ceruti, A., 1960, *Elaphomycetales* et *Tuberales*, Tridenti (ed.), *Iconografia Mycologica*, Bresadola J, 28 suppl.

Corocher, N., Polimeni, C., Giraudi, G., and Papa, G., 1992, Sviluppo di un metodo immunoenzimatico (ELISA) per la caratterizzazione di ectomicorrize di *Tuber magnatum* e *T. albidum*, Convegno internazionale sul Tartufo, L'Aquila 5-8 marzo 1992, *Micol. Veget. mediter.*, 7: 151-158.

Chevalier, G., and Desmas, C., 1977, Synthèse des mycorhizes de *Tuber melanosporum* avec *Corylus avellana* sur gélose a partir de spores, *Ann. Phytopathol.* 9: 5-31.

Chevalier, G., Riousset, G, Riousset, L., and Dupré, C., 1990, Taxonomie des truffes européennes, In: Comunita montana dei Monti Martani e del Serano (ed.), *Atti del secondo Congresso internazionale sul Tartufo*, Spoleto (Italy), 24-27 November 1988, pp 37-44.

Chevalier, G., 1994, Evolution des recherches sur les plants mycorrhizés par la truffe et perspectives de développement, *Giorn. Bot. Ital.* 128: 7-18.

Dupré, C., and Chevalier, G., 1991, Analyse électrophorétique des protéines fongiques de différents *Tuber* en association ou non avec *Corylus avellana*, *Crypt. Mycol.* 12: 243-250.

Dupré, C., Chevalier,G., Palenzona, M., and Biocca, M., 1993, Caractérisation des mycorhizes de différents *Tuber* par l'étude du polymorphisme enzymatique, *Crypt. Mycol.* 14: 163-170.

Egger, K.,N., Danielson R.,M., and Fortin, J.A, 1991, Taxonomy and population structure of E-strain mycorrhizal fungi inferred from ribosomal and mitochondrial DNA polymorphism, *Myc. Res.* 95: 866-872.

Feinberg, A.,P., and Vogelstein, B., 1984, A technique for radiolabelling DNA restriction endonuclease fragments to high specific activity, *Anal. Biochem.* 137: 266-267.

Fontana, A , Ceruti A , and Meotto, F ,1990, Criteri istologici per il riconoscimento delle micorrize di *Tuber magnatum* Pico, In Comunita montana dei Monti Martani e del Serano (ed), *Atti del secondo Congresso internazionale sul Tartufo*, Spoleto (Italy), 24-27 November 1988, pp 141-154

Fontana, A , Ceruti , and Meotto F , 1992, Criteri istologici per il riconoscimento delle micorrize di *Tuber albidum*, In Pacioni G (ed), *Atti del Congresso Internazionale sul Tartufo*, L'Aquila (Italy), pp 121-136

Frizzi, G , Pacioni, G , and Visca, C , 1992, Chiavi biochimiche per il riconoscimento dei Tartufi di interesse commerciale, *Micol Veget mediter* 7 145-150

Gardes, M , White, T ,J , Fortin J ,A , Bruns, T ,D , and Taylor J ,W , 1991, Identification of indigenous and introduced symbiotic fungi in ectomycorrhizae by amplification of nuclear and mitochondrial ribosomal DNA, *Can J Bot* 69 180-190

Giraud, M , 1988, Prelevement et analyse de mycorhizes, *Bull FNPT*, 10 46-63

Granetti, B , 1990 Caratteri morfologici, biometrici e strutturali delle micorrize di *Tuber magnatum* Pico con *Pinus pinea* L , In Comunita Montana dei Monti Martani e del Serano (ed), *Atti del secondo Congresso Internazionale sul Tartufo*, Spoleto (Italy) 24-27 November 1988, pp 273-281

Granetti, B , Mincigrucci, G , and Bricchi, E , 1990, Analisi biometrica e morfologica delle ascospore di alcune specie del genere *Tuber*, In Comunita Montana dei Monti Martani e del Serano (ed), *Atti del secondo Congresso Internazionale sul Tartufo*, Spoleto (Italy) 24-27 November 1988, pp 59-100

Gross, G , 1987, Zu den europaischen Sippen der Gattung *Tuber*, In Derbsch H et Schmitt J A *Atlas der Pilze des Saarlandes, 2 Nachweise, Okologie, Vorkommen und Beschreibungen* 79-99 Delattinia (ed), Saarbrucken, pp 816

Guinberteau, J , Salesses, G , Olivier, J M , and Poitou, N , 1990, Mycorhization de vitroplants de noicetiers clones *Atti del secondo Congresso Internazionale sul Tartufo*, Spoleto (Italy) 24-27 November 1988, pp 205-2159-100

Henrion, B , Le Tacon, F , and Martin F , 1992 Rapid identification of genetic variation of ectomycorrhizal fungi by amplification of ribosomal RNA genes, *New Phytol* 122 289-298

Henrion, B , Chevalier, G , and Martin, F , 1994, Typing truffle species by PCR amplification of the ribosomal DNA spacer, *Mycol Res* 98 37-43

John, B , and Miklos, G , 1988, The eukaryote genome in development and evolution, Allen & Unwin, Inc , Winchester, MA pp 177-182

Jorgensen, R ,A , and Cluster P ,D , 1988 Modes and tempos in the evolution of nuclear ribosomal DNA new characters for evolutionary studies and new markers for genetic and population studies, *Ann Mo Bot Gar* 75 1238-1247

Lane, D ,J , Pace, B , Olsen, G ,J , Sthal D ,A , Sogin M ,L , and Pace N ,R , 1985, Rapid determination of 16S ribosomal RNA sequences for phylogenetic analyses, *Proc Natl Acad Sci* USA 82 6955-6959

Lanfranco, L , Wyss, P , Marzachi, C ,and Bofante P , 1993, DNA probes for identification of the ectomycorrhizal fungus *T magnatum* Pico, *FEMS Microb Lett* 114 245-252

Maniatis, T , Fritsch, E F , and Sambrook, J , 1982, Molecular cloning A laboratory manual, Cold Spring Harbor Laboratory, Cold Spring Harbor, NY

Mouches, C , Duthil, P , Poitou, N , Delmas, J , and Bove, J M , 1981, Caracterisation des especes truffieres par analyse de leurs proteines en gel de polyacrylamide et application de ces techniques a la taxonomie des champignons, *Musch Sci* 11 819-831

Palenzona, M , 1969, Sintesi micorrizica tra *Tuber aestivum, T brumale, T melanosporum* e semenzali di *Corylus avellana*, *Allionia* 15 121-131

Palenzona, M , Chevalier, G , and Fontana, A , 1972, Sintesi micorrizica tra i miceli in coltura di *T brumale, T melanosporum, T rufum* e semenzali di conifere e latifoglie, *Allionia* 18 41-52

Palenzona, M , Biocca, E , Nascetti, G , Ferrara, A , M , Mattiucci, S , D'Amelio, S , and Balbo, T , 1990, Studi preliminari sulla tipicizzazione genetica (sistema gene-enzima) di specie del genere *Tuber*, In G Pacioni (ed), *Atti del secondo Convegno Internazionale sul Tartufo*, L'Aquila (Italy) pp 53-58

Papa, G , and Balbi, P , 1988, Micorrize identificazione mediante pirolisi, Riassunti del secondo congresso internazionale sul Tartufo Spoleto, 24-27 novembre 1988 17

Papa, G , and Polimeni, C , 1990, Caratterizzazione delle proteine nelle *Tuberacee* I Studio preliminare di una banda elettroforetica di *Tuber magnatum* Pico, *Micol Ital* 2 37-43

Papa, G , Polimeni, C , Mischiati, P , and Cantini Cortellazzi, G , 1992, Immunological aspects of the characterization of *Tuber magnatum* and *Tuber albidum*, In Read D J *et al* (eds) *Mycorrhizas in Ecosystem* 395, CAB International, Cambridge

Parguey, J C , Leduc, J P , Dexheimer J , and Chevalier G , 1990, Etude ultrastructurale et cytochimique de l'association mycorhizienne *Tuber melanosporum/Corylus avellana*, In Comunita Montana dei

Montı Martanı e del Serano (ed), *Attı del secondo Congresso Internazıonale sul Tartufo*, Spoleto (Italy) 24-27 November 1988, pp 129-133

Plattner, I, Grabher, T , Hall, I, Haselwandter, K, and Stoffler, G , 1991, Identıfıcatıon of ectomycorrhızal fungı by use of ımmunologıcal tecnıques Abstract of Thırd European Symposıum of Mycorrhızas - Mycorrhızas ın ecosystem - Structure and functıon Scheffıeld, August 19-23

Potenza, L , Amıcuccı, A , Rossı, I , Palma, F, De Bellıs, R , Cardonı, P, and Stocchı, V, 1994, Identıfıcatıon of *T magnatum* Pıco DNA markers by rapd analysıs, *Bıotec Tech* 8 93-98

Rogers, S , O , Rehner, S , Bledsoe, C , Mueller, G ,J , and Ammıratı, J ,F, 1989, Extractıon of DNA from Basıdıomycetes for rıbosomal DNA hybrıdızatıon, *Can J Bot* 67 1235-1243

Rogers, O , and Bendıch, A , J , 1987, Rıbosomal RNA genes ın plants varıabılıty ın copy number and ın the ıntergenıc spacer, *Plant Mol Bıol* 9 509-520

Sogın, M ,L , and Gunderson, J , H , 1987, Structural dıversıty of eukaryotıc small subunıt rıbosomal RNAs evolutıonary ımplıcatıons Endocytobıology III, *Ann N Y Acad Scı* 503 125-139

Sogın, M ,L , 1990, Amplıfıcatıon of rıbosomal RNA genes for molecular evolutıon studıes, In Innıs M A , Gelfand, D ,H , Snınsky, J ,J , Whıte, T ,J (eds), *PCR protocols A guıde to methods and applıcatıons*, Academıc Press, San Dıego, pp 307-314

Whıte, T ,J , Bruns, T , Lee, S , and Taylor, J , 1990, Amplıfıcatıon and dırect sequencıng of fungal rıbosomal RNA genes for phylogenetıcs, In Innıs, M ,A , Gelfand, D ,H , Snınsky,J ,J , Whıte, T ,J , (eds), *PCR protocols A guıde to methods and applıcatıos*, Academıc Press, San Dıego,pp 315-322

Zambonellı, A , Gıunchedı, L , and Poggı Pollınі, C , 1992, An Enzyme-Lınked Immunosorbent Assay (ELISA) for the Detectıon of *Tuber albıdum* Ectomycorrhıza, *Symbıosıs*, 15 71-76

Zambonellı, A , Salomonı, S , and Pısı, A , 1993, Caratterızzazıone antomorfologıca delle mıcorrıze dı *Tuber* spp su *Quercus pubescens* Wılld , *Mıcol Ital* 22 73-79

16

BIOCHEMICAL CHARACTERIZATION OF VARIOUS SPECIES OF WHITE TRUFFLE

A Preliminary Study

P. Cardoni, L. Vallorani, L. Cucchiarini, M. Betti, C. Pierotti, and
V. Stocchi

Istituto di Chimica Biologica "Giorgio Fornaini"
Università degli Studi di Urbino
Via Saffi, 2
61029 Urbino
Italy

SUMMARY

In this paper we report a preliminary biochemical characterization of different species of white truffle: *Tuber magnatum* Pico, *Tuber borchii* Vitt. and *Tuber maculatum* Vitt.. In particular we have evaluated the level of enzymes involved in the glycolytic and pentose phosphate pathways, along with the levels of compounds involved in maintaining the energetic state of the cell, such as the adenine nucleotides and their degradation products, inosine and hypoxanthine. Finally, we examined the distribution of free amino acids in the different species of *Tuber* studied. A discussion concerning the results obtained will be presented.

INTRODUCTION

Almost nothing is known about the truffle metabolism, although these fungi are a very interesting genera of Ascomycetes due to their organoleptic properties and economic value. Many authors have studied the maturation process of the truffle but, until recently, their interest has been focused primarily on the morphological aspects. In fact, it is well known that *Tuber* sporocarps have a very slow maturation process (1-3) characterized by a succession of various events (3-6) resulting in the constitution of the spores (7) required for the production of new primary mycelium. Current knowledge concerning the physiology of *Tuber* spp. is very incomplete for all the phases of development namely the germination of spores, the symbiotic interaction between mycelium and plant roots and the production of carpophores. In order to obtain more information on the biochemical characteristics of the hyphae and spores which constitute the fruitbody, we investigated the level of enzymes of

Biotechnology of Ectomycorrhizae, Edited by Vilberto Stocchi et al.
Plenum Press, New York, 1995

185

the glycolytic and pentose phosphate pathways, as well as the presence of compounds involved in maintaining the energetic state of the cells, in three different species of white truffle: *Tuber magnatum* Pico, *Tuber borchii* Vitt. and *Tuber maculatum* Vitt.. In particular, we evaluated the activity levels of both the glycolytic and pentose phosphate pathways in homogenates of freshly collected truffle samples. These pathways are involved in the production of energy, using glucose as substrate, and in supplying reducing power, with formation of reduced pyridine coenzymes. We focused our attention on hexokinase, the first enzyme of the glycolytic pathway, which catalyzes the phosphorylation of glucose to glucose-6-phosphate. In mammalian tissues, many authors have demonstrated that hexokinase is present as distinct enzymatic forms (8-12). In order to characterize the different species of *Tuber*, we evaluated the chromatographic pattern of hexokinase using hydrophobic interaction (HIC) and ion exchange chromatography (IEC).

We also evaluated the levels of compounds involved in maintaining the energetic state of the cell, such as the adenine nucleotides (ATP, ADP, AMP) and their degradation products (inosine and hypoxanthine). The ATP, ADP, AMP levels influence cell functionality and allow determination of the energy charge of the cell, an indicator of cellular integrity (13-15).

Furthermore, since very little information is available in the literature about the amino acid content in truffles (16), we analyzed the levels of free amino acids in freshly collected white truffles. Moreover, we investigated the distribution of free amino acids in carpophores maintained for several days at room temperature in order to assess the presence of possible proteolytic activity during the degradation of fruitbody.

The metabolic properties studied could be of interest in the characterization of *Tuber* fruitbodies and for the identification of the possible molecular mechanisms involved in the degradation process.

MATERIALS AND METHODS

Homogenization Procedures

The fruitbodies were washed and deprived of peridium, then minced and homogenized (1g/6ml) in 5 mM sodium-potassium phosphate buffer, pH 8.1, containing 3 mM KF, 3 mM β–MSH, 1 mM DTT and 5 mM glucose (Homogenization Buffer), using a glass pestle potter. The homogenate consisted mainly of broken hyphae, while the majority of the spores remained intact. The suspension obtained was then centrifuged at 14,000 rpm for 30 minutes to remove unbroken hyphae and spores and the supernatant was used for the analyses.

Evaluation of Glycolytic and Pentose Phosphate Pathway Enzymes

The supernatants obtained as above were diluted and used for the enzymatic assays as described in (17) with slight modifications. The activities of hexokinase (HK, EC 2.7.1.1), glucose phosphate isomerase (GPI, EC 5.3.1.9), phosphofructokinase (PFK, EC 2.7.1.11), aldolase (EC 4.1.2.13), glyceraldehyde phosphate dehydrogenase (GAPD, EC 1.2.1.13), triose phosphate isomerase (TPI, EC 5.3.1.1), phosphoglycerate kinase (PGK, EC 2.7.2.3), monophosphoglyceromutase (MPGM, EC 5.4.2.1), enolase (EC 4.2.1.11), pyruvate kinase (PK, EC 2.7.1.40), lactate dehydrogenase (LDH, EC 1.1.1.27), glucose-6-phosphate dehydrogenase (G-6-PD, EC 1.1.1.49), 6-phosphogluconate dehydrogenase (6-PGDH, EC 1.1.1.44) and phosphoglucomutase (PGluM, EC 5.4.2.2) were the means of seven independent experiments.

Data are expressed as I.U./mg of proteins; the protein concentration was determined spectrophotometrically at 700 nm according to the method of Lowry (18).

Chromatographic Analysis of Hexokinase Activity

The hexokinase pattern of truffle homogenates was studied by hydrophobic interaction (HIC) and ion exchange chromatography (IEC). The supports used were Toyopearl Phenyl 650S (for HIC) and Toyopearl DEAE 650S (for IEC), obtained from Tosohaas Technical Center (Woburn, MA, USA). The columns were prepared as described in (19) and (20).

The Toyopearl Phenyl 650S columns were equilibrated in Homogenization Buffer with the addition of 30% (w/v) ammonium sulphate. After charging the sample in cold room, the hexokinase activity was eluted, by a Beckman HPLC system (Berkeley, CA, USA), operating at a flow rate of 1 ml/min using a two-step descending gradient of ammonium sulphate: from 30 to 10% in 8 minutes and from 10 to 0% in 45 minutes. The Toyopearl DEAE 650S columns were equilibrated in Homogenization Buffer. The elution of hexokinase was obtained at a flow rate of 1 ml/min using a 200-200 ml linear gradient of KCl from 40 to 200 mM in the equilibrating buffer. Fractions of 1 ml were collected and assayed for hexokinase activity and the protein concentration was then determined spectrophotometrically at 280 nm in the course of elution from the columns.

Analysis of Nucleotides

Procedure for Extract Preparation. 1 gram of minced truffle was added to 6 ml of perchloric acid and homogenized using a glass pestle potter. After centrifugation at 3,000 rpm for 10 minutes, the supernatant was separated and neutralized with 3 M K_2CO_3. The solution was again centrifuged at 3,000 rpm for 10 minutes and supernatant was removed and used for chromatographic analysis.

Chromatographic Apparatus and Conditions. The HPLC system used was from Varian (Palo Alto, CA, U.S.A.) and consisted of two Model 2010 pumps, a Model 2020 solvent programmer and a model 2050 variable-wavelength detector. Integration of peak areas was obtained by means of an HP 3390A electronic integrator (Hewlett-Packard, Avondale, PA, U.S.A.). A 5-μm Supelcosil LC-18 column (25 cm x 4.6 mm I.D.) (Supelco, Bellefonte, PA, U.S.A.) protected by a guard column (Pelliguard LC-18, 20 mm x 4.6 mm I.D., 40-μm particles) was used throughout these studies. The injection volume was 50 μl. The mobile phase used for the separation of nucleotides consisted of two eluents: 0.1 M potassium dihydrogen phosphate, pH 6.0, containing 8 mM tetrabutylammonium hydrogen sulfate (Buffer A) and 0.1 M KH_2PO_4 solution, pH 6.0, containing 8 mM tetrabutylammonium hydrogen sulfate and 20% (v/v) of CH_3OH (Buffer B) (21).

Determination of Free Amino Acids

10 μl of truffle homogenate were pipetted into an eppendorf tube, dried under vacuum and resuspended in 50 μl of 0.2 M $NaHCO_3$, pH 9.0, then treated with 100 μl of DABS-Cl solution (4 nmol/μl). The mixture was heated at 70°C for 10 minutes; the sample was dried under vacuum and the residue was redissolved in 500 μl of 70% (v/v) ethanol (HPLC grade). 5 μl of this preparation were used for RP-HPLC analysis.

Separation of free amino acids was performed using a 3 μm Supelcosil LC-18 T column (15 cm x 4.6 mm I.D.) (Supelco, Bellefonte, PA, U.S.A.) protected with a 5 μm

Supelcosil LC-18 T guard column (2.0 cm x 4.6 mm I.D.). Solvent A was 25 mM potassium dihydrogen phosphate buffer, pH 7.05, and solvent B was acetonitrile and methanol (70:30). Separation of DABS-amino acid derivatives was obtained at room temperature at a flow rate of 1.5 ml/min. The detection was performed at 436 nm (22,23).

RESULTS

Enzymatic Analyses

The sporocarps of *Tuber magnatum*, *Tuber borchii* and *Tuber maculatum* were initially studied in order to evaluate the activity levels of both the glycolytic and the pentose phosphate pathway enzymes. It is known that these processes supply the cells with both energy and reducing power.

The results obtained, shown in Table I, revealed different levels of specific activity in the three species examined. In particular it is possible to observe that, in the fresh truffle homogenates, the main differences among the three species studied reside in the key enzymes of the glycolytic pathway: hexokinase, phosphofructokinase and pyruvate kinase. This fact is of great importance because these enzymes have a pivotal role in the regulation of glycolysis and could influence the efficiency of glucose utilization significantly. Further-more, the glucose-6-phosphate dehydrogenase (G-6-PD) and 6-phosphogluconate dehydro-genase (6-PGDH), pentose phosphate pathway enzymes, also show very low activity levels especially in *T. maculatum*. In particular, we focused our attention on the hexokinase activity, studying the chromatographic profile of the enzyme in the different species of white truffle. In fact, this enzyme, as reported in the literature (8-12,24), is present in multiple forms which are the product of different genes (25), the result of post-translational modification(s) or even due to a different intracellular distribution (26).

Figure 1 shows the HIC profiles of the fruitbody homogenates of *T. magnatum*, *T. borchii* and *T. maculatum* obtained using Toyopearl Phenyl 650S columns. The hexokinase

Table I. Activity of glycolytic and pentose phosphate pathway enzymes in some white truffle species

	T. magnatum	*T. borchii*	*T. maculatum*
Hexokinase	0.018 ± 0.002	0.039 ± 0.014	0.012 ± 0.003
Glucose phosphate isomerase	1.5 ± 0.1	1.08 ± 0.27	0.53 ± 0.02
Phosphofructokinase	0.15 ± 0.02	0.037 ± 0.02	0.013 ± 0.004
Aldolase	0.019 ± 0.004	0.11 ± 0.04	0.076 ± 0.003
Triose phosphate isomerase	0.41 ± 0.06	0.25 ± 0.08	0.165 ± 0.02
Glyceraldehyde phosphate dehydrogenase	0.13 ± 0.01	0.082 ± 0.018	0.04 ± 0.003
Phosphoglycerate kinase	0.29 ± 0.08	0.23 ± 0.05	0.11 ± 0.01
Monophosphoglycero mutase	0.2 ± 0.002	0.34 ± 0.15	0.155 ± 0.007
Enolase	0.23 ± 0.001	0.64 ± 0.17	0.35 ± 0.05
Pyruvate Kinase	0.1 ± 0.01	0.31 ± 0.1	0.12 ± 0.02
Lactate Dehydrogenase	0.016 ± 0.004	0.09 ± 0.03	0.013 ± 0.0004
Glucose-6-phosphate dehydrogenase	0.05 ± 0.008	0.01 ± 0.003	0.0017 ± 0.0003
6-Phosphogluconate dehydrogenase	0.017 ± 0.002	0.004 ± 0.001	0.0011 ± 0.0002
Phosphoglucomutase	0.34 ± 0.03	0.32 ± 0.09	0.035 ± 0.02

The enzymatic activities are expressed as U.I./mg proteins. The values reported represent the means of 7 different experiments for each species of truffle. The supernatants, obtained as described in Materials and Methods, were used for the enzymatic assay as described in (17) with slight modifications.

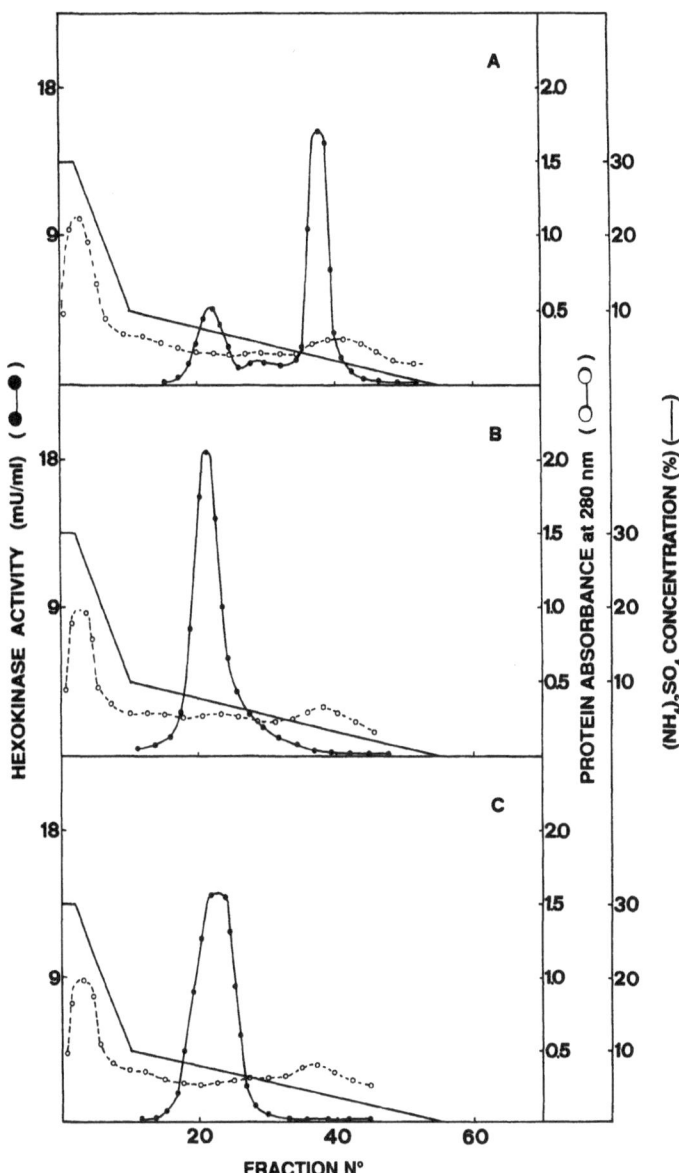

Figure 1. Hydrophobic interaction chromatographic profiles of hexokinase in different species of white truffle. In each experiment, 500 µl of homogenate, prepared as described in Materials and Methods, were charged onto the Toyopearl Phenyl 650S columns (5.0 cm x 1.2 cm I.D.) equilibrated using 5 mM sodium-potassium phosphate buffer, pH 8.1, containing 3 mM KF, 3 mM β-MSH, 1 mM DTT, 5 mM glucose and 30% (w/v) ammonium sulphate. The hexokinase activity was eluted using a two-step descending gradient of ammonium sulphate from 30 to 10% in 8 min and from 10 to 0% in 45 min. Fractions of 1 ml were collected and assayed for hexokinase activity.

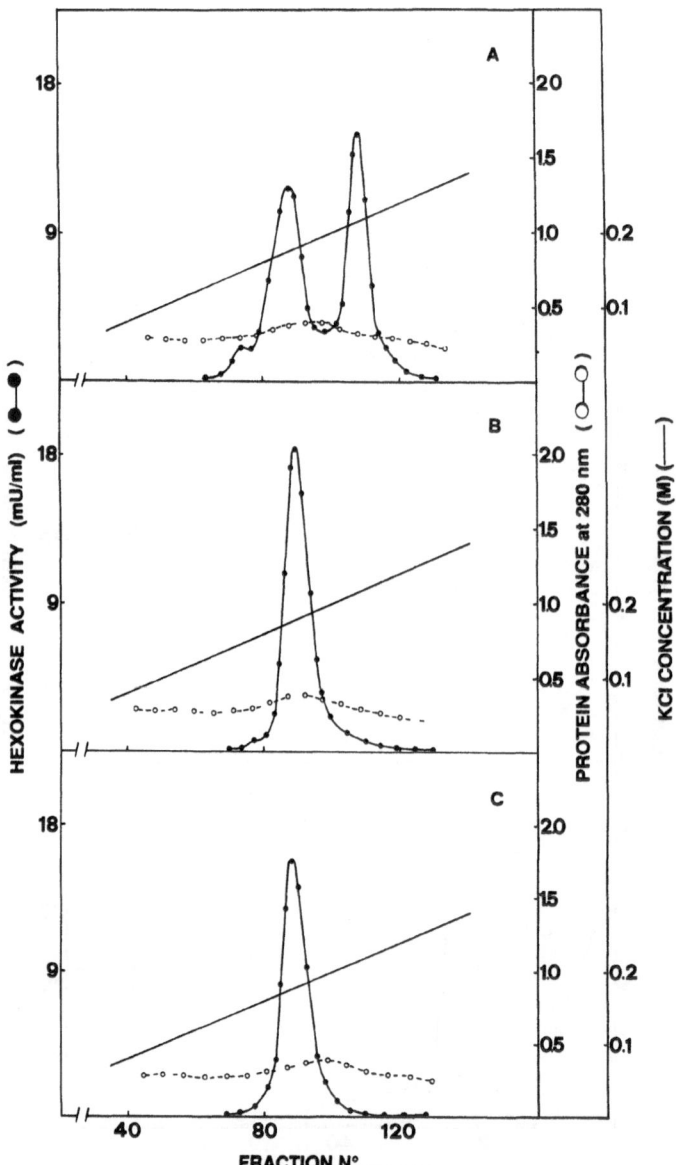

Figure 2. Anion exchange chromatographic profiles of hexokinase in different species of white truffle In each experiment, 500 μl of homogenate, prepared as described in Materials and Methods, were charged onto the Toyopearl DEAE 650S columns (5 0 cm x 1 2 cm I.D) equilibrated using 5 mM sodium-potassium phosphate buffer, pH 8 1, containing 3 mM KF, 3 mM β-MSH, 1 mM DTT and 5 mM glucose The elution of hexokinase activity was obtained using a 200-200 ml linear gradient of KCl from 40 to 200 mM in the same equilibrating buffer Fractions of 1 ml were collected and assayed for hexokinase activity

pattern of *T. magnatum* (Fig. 1A) shows the presence of two different forms of the enzyme while, in *T. borchii* (Fig. 1B) and *T. maculatum* (Fig. 1C) a single peak of hexokinase is present. It is interesting to note that this activity corresponds to the first peak eluted in *T. magnatum*. A comparable situation was also observed by analyzing the same homogenates by IEC.

Figure 2 shows the hexokinase profiles obtained using the Toyopearl DEAE 650S columns. In this case as well, it was possible to separate two forms of the enzyme only in *T. magnatum* (Fig. 2A) the first of which presents the same elution time as the *T. borchii* (Fig. 2B) and *T. maculatum* (Fig. 2C) hexokinase. The results obtained with the different chromatographic supports indicate that the latter two enzymes and the first enzyme eluted for *T. magnatum* have similar hydrophobic characteristics and similar isoelectric points.

Nucleotide Determinations

Adenine nucleotides, ATP, ADP, AMP, are involved in maintaining the energetic state of the cell. The evaluation of these compounds and their degradation products (hypoxanthine, inosine) allows us to determine the functionality and integrity of cells. The results obtained, using a Supelcosil LC-18 column (25 cm x 4.6 mm I.D.) showed the presence of significant amounts of ATP, ADP and AMP in the different species of freshly collected truffle. In particular, we observed a significant variability in the levels of these compounds among different samples of *Tuber* belonging to the same species. This could be related to the fact that the fresh truffle samples may have been collected at different stages of maturation. This hypothesis is also supported by the evidence that perchloric acid extracts of fresh truffle samples left at 4°C showed a significant decrease in ATP and ADP with a concomitant increase of AMP. From this point of view, the level of ATP could be considered a useful index of the freshness of the fruitbody collected. This aspect should be investigated more carefully in trying to understand the molecular mechanism(s) involved in the degradation of the fruitbody.

Furthermore, the chromatograms reported in Figure 3 show the presence of distinct peaks that need to be identified. Under the experimental conditions used we were able to exclude that these peaks, the chromatographic profile of which differs significantly among the species examined, were not adenine nucleotides, their degradation products, pyridine coenzymes, guanine nucleotides, peptides, or flavonoids.

Evaluation of Free Amino Acids

Finally, we evaluated the distribution of free amino acids in the three different species of white truffle: *T. magnatum*, *T. borchii* and *T. maculatum*. Amino acids were determined in the homogenates from freshly collected fruitbodies. The results showed a high content of alanine, glutamic acid, glutamine in all three species, while arginine is present in high amounts only in *T. magnatum* (Table II).

In particular, we focused our attention on *T. magnatum*, the most valuable species. The analysis of a large number of fruitbodies revealed that the amount of free amino acids varied significantly among the different samples. We evaluated the percentage of essential amino acids in all samples in the search for a common parameter for fruitbody samples of the same species. These values were similar and ranged from 13.25±1.37, suggesting that in samples from the *T. magnatum* species, the relative distribution of free amino acids is very close.

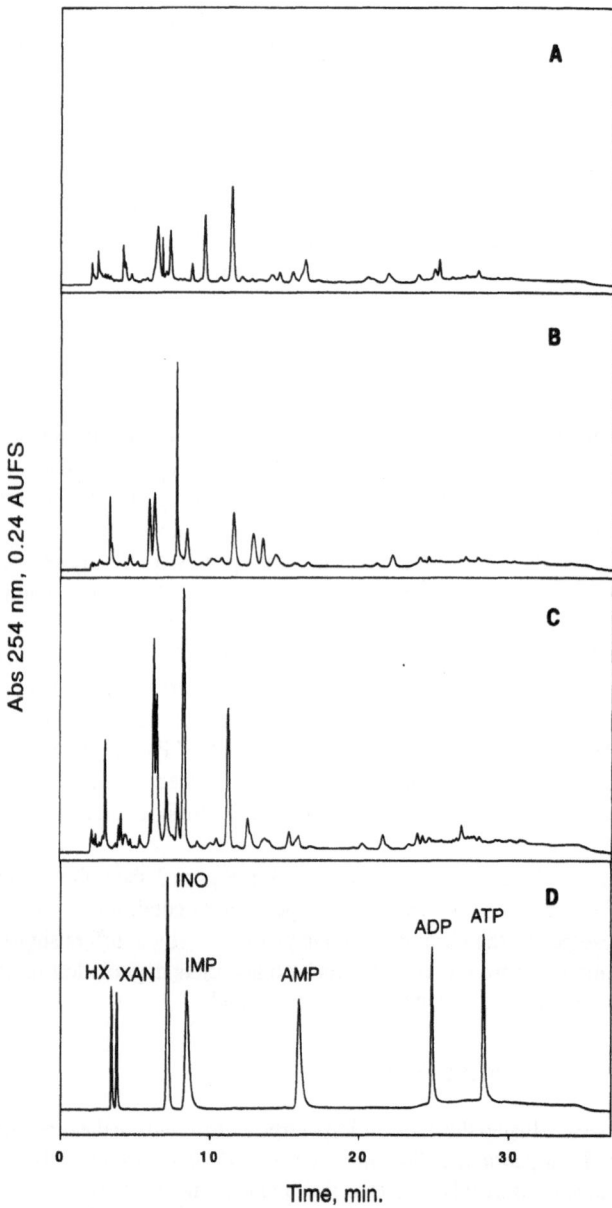

Figure 3. Separation of adenine nucleotides and their degradation products present in *T. magnatum*, *T. borchii* and *T. maculatum* by ion-pair reversed-phase HPLC. A) *Tuber magnatum*; B) *Tuber borchii*; C) *Tuber maculatum*; D) Standard nucleotides: HX = hypoxantine, XAN = xantine, INO = inosine, IMP = inosinemonophosphate, AMP = adenosinemonophosphate, ADP = adenosinediphosphate, ATP = adenosinetriphosphate.

Table II. Composition of free amino acids from homogenates of *Tuber magnatum*, *Tuber borchii* and *Tuber maculatum*

	Tuber magnatum Pico	*Tuber borchii* Vitt.	*Tuber maculatum* Vitt.
Aspartic acid	47	31.9	28.3
Glutamic acid	262	276.6	100
Asparagine	74.8	31.9	40.2
Glutamine	182	212.7	110.13
Serine	157	10.6	30.2
Threonine	38	37.22	12.6
Glycine	87.7	95.7	78.6
Alanine	214	404	138
Arginine	154	not detectable	not detectable
Proline	56	21.3	43.9
Valine	71	53.2	21.4
Methionine	3.8	8.5	not detectable
Isoleucine	47	39.1	8.79
Leucine	24	17	not detectable
Phenylalanine	25	27.6	3.7
Cystine	35	25.5	39.2
Lysine	17.5	21.3	not detectable
Histidine	43.8	24.2	not detectable
Tyrosine	4.14	4.2	not detectable

Values are presented as nmol/mg of proteins. Amino acids were evaluated as DABS-derivatives using a RP-HPLC system with a Supelcosil LC-18 T column (15 cm x 4.6 mm I.D.) protected with a LC-18 T guard column (2.0 cm x 4.6 mm I.D.). Detection was performed at 436 nm.

In order to verify that the free amino acid content depends on the maturity of the truffle, we maintained a *T. magnatum* fruitbody at room temperature for several days and regularly (at days 0, 3, 5, 8) evaluated the amounts of these amino acids.

The results reported in Table III show that only alanine increased while the other amino acids decreased or were stable. At the present time, we are not able to explain which metabolic processes are responsible for these data or if the activities of any proteolytic enzymes are involved.

CONCLUSIONS

The different species of *Tuber* analyzed showed different levels of activity for the enzymes examined and, in particular, for the regulatory enzymes of the glycolytic pathway: hexokinase, phosphofructokinase and pyruvate kinase. Furthermore, we observed the very low activity levels of pentose phosphate pathway enzymes (glucose-6-phosphate dehydrogenase and 6-phosphogluconate dehydrogenase) especially for *T. maculatum*.

Hydrophobic interaction or anion exange chromatographic analyses of fruitbody homogenates showed the presence of two different forms of hexokinase only in *T. magnatum*. The single form of the enzyme present in *T. borchii* and *T. maculatum* corresponds to the first peak eluted in *T. magnatum* using either hydrophobic interaction or anion exchange chromatography. This fact pointed out that these forms of hexokinase have similar isoelectric points and hydrophobic properties.

In freshly collected samples, ion-pair reversed-phase analyses of fruitbody extracts revealed the presence of numerous unidentified peaks, in the first part of chromatograms, and ATP, ADP and AMP. After maintaining the fruitbody samples at 4°C for few days, the

Table III. Comparison of free amino acids content in a *Tuber magnatum* fruitbody maintained at room temperature for several days

	I day	III day	V day	VIII day
Aspartic acid	15	ND	ND	10
Glutamic acid	239	101	255	208
Asparagine	53	30	55	30
Glutamine	224	91	117	107
Serine	127	128	157	25
Threonine	24	44	70	54
Glycine	140	143	202	149
Alanine	1,692	1,947	2,646	2,306
Arginine	21	16	34	28
Proline	41	109	152	104
Valine	119	117	192	113
Methionine	8	5	26	6
Isoleucine	43	51	108	52
Leucine	43	46	104	39
Tryptophane	10	14	43	12
Phenylalanine	23	28	48	25
Cystine	217	170	291	74
Lysine	6	30	80	17
Histidine	38	13	25	ND
Tyrosine	4	3	1	11

The *Tuber magnatum* fruitbody was divided in four pieces of 0 2 g, maintained at room temperature and homogenized at days 0,3,5,8
Values are presented as nmol/mg of total proteins

adenine nucleotides considerably decreased, in fact, the analyses of 15 days-old samples show only the presence of the unidentified compounds, that are not pyridine nucleotides, pyridine coenzymes, peptides or flavonoids.

The evaluation of free amino acids in the homogenates obtained from freshly collected *Tuber* fruitbodies, revealed a wide variability among samples of different species and samples of the same species too. However, in *T. magnatum* we observed that the percentage of essential amino acids was similar, even in samples which had a very different content of free amino acids. Among the three species, *T. maculatum* presents the lowest content of free amino acids, some of which, such as leucine, lysine, histidine, were in fact not detectable.

Our studies represent a preliminary characterization of *Tuber* fruitbody that could be useful for the comprehension of the complex processes involved in the metabolism and biological cycle of the truffles.

ACKNOWLEDGMENTS

This work was supported by CNR, P.S. "Biotecnologia della micorrizazione".

REFERENCES

1 Montant C., Kulifaj M. and Gleize R., 1983, Note sur la recolte de jeunes ascocarpes du *Tuber melanosporum* Vitt. (truffe noir du Perigord) et leur evolution, *Hebd Séances Acad Sci.*, III 296, 463-468

2 Montant C and Kulıfaj M , 1985, Donnees nouvelles sur la bıologıe du *Tuber melanosporum* Vıtt (truffe du Perıgord) ıncıdence sur la productıon, *Resau Mycologıe* (Reunıon 1985, 11-12 octobre) Toulouse, Unıversıte P Sabatıer, pp 68

3 Talou T and Kulıfaj M , 1992, Les secrets de la truffe, *La recherche*, 239(23)31-39

4 Berta G and Fusconı A , 1983, Ascosporogenesıs ın *Tuber magnatum*, *Trans Br Mycol Soc* , 80, 201-207

5 Janex-Favre M C and Parguey-Leduc A , 1976, La formatıon des ascospores chez deux truffles *Tuber rufum* Pıco et *Tuber aestıvum* Vıtt (Tuberacees), *C r Seances Acad Scı Ser D*, 283,1173-1175

6 Janex-Favre M C and Parguey-Leduc A , 1983, Etude ultrastructurale des asques et des ascospores des truffles du genre *Tuber* II Les ascospores, Cryptogamıe, *Mycologıe*, 4,353-373

7 Gross G and Schmıtt J A , 1974, Besıeungen Zwıschen Sporenvolumen v Kernal beı Eınıger Hoeren Pılze, *Z Pılzk*, 40,163-214

8 Gonzales C , Ureta T , Sanchez R and Nyemyr H , 1964, *Bıochem Bıophys Res Commun* ,16,347-359

9 Grosbard L and Schımke R T , 1966, *J Bıol Chem* , 241,3546-3561

10 Stocchı V , Magnanı M , Canestrarı F , Dacha M and Fornaını G , 1981, Rabbıt red blood cell hexokınase Evıdence for two dıstınct forms, and theır purıfıcatıon and characterızatıon from retıculocytes, *J Bıol Chem* , 256,7856-7862

11 Stocchı V , Magnanı M , Pıccolı G and Fornaını G , 1988, Hexokınase mıcroheterogeneıty ın rabbıt red blood cells and ıts behavıour durıng retıculocytes maturatıon, *Mol Cell Bıochem* , 79,133-136

12 Katzen H M , 1967, ın *Advances ın Enzyme Regulatıon*, (Weber G Ed) Vol 5, 355, Pergamon Press, New York

13 Crabtree B and Newsholme E A , 1985, *Curr Top Cell Regul* , 25,21-76

14 Kaplan N O , 1985, *Curr Top Cell Regul* , 26,371-382

15 Atkınson D E , 1977, ın *Cellular Energy Metabolısm and Its Regulatıon*, Academıc Press, New York

16 Beuchat L R , Brenneman T B and Dove C R , 1993, Composıtıon of the pecan truffle (*Tuber texense*), *Food Chemıstry*, 46,189-192

17 Beutler E , ın *Red Cell Metabolısm*, (Grune and Stratton Eds), Inc New York 3rd Ed 1984

18 Lowry O H , Rosenbrough N J , Farr A L and Randall R J , 1951, *J Bıol Chem* , 193,265-275

19 Stocchı V , Cardonı P , Ceccarolı P , Pıccolı G , Cucchıarını L , De Bellıs R and Dacha M , 1994, Hıgh resolutıon of multıple forms of rabbıt retıculocyte hexokınase type I by hydrophobıc ınteractıon chromatography, *J Chromatogr*, 676,51-63

20 Stocchı V , Masat L , Bıagıarellı B , Accorsı A , Pıccolı G , Palma F , Cucchıarını L and Dacha M , 1992, Hıgh resolutıon of multıple forms of red blood cell enzymes usıng a Toyopearl DEAE 650S, *Prep Bıochem* , 22,11-40

21 Stocchı V , Cucchıarını L , Canestrarı F , Pıacentını M P and Fornaını G , 1987, A very fast low ıon-paır reversed-phase HPLC method for the separatıon of the most sıgnıfıcant nucleotıdes and theır degradatıon products ın human red blood cells, *Anal Bıochem* , 167,181-190

22 Stocchı V , Cucchıarını L , Pıccolı G and Magnanı M , 1985, Complete hıgh-performance lıquıd chromatographıc separatıon of 4-N,N-dımethylamınobenzene-4'-thıohydantoın and 4-dımethylamınobenzene-4'-sulphonyl chlorıde amıno acıds utılızıng the same reversed-phase column at room temperature, *J Chromatogr* , 349,77-82

23 Stocchı V , Pıccolı G , Magnanı M , Palma F , Bıagıarellı B and Cucchıarını L , 1989, Reversed-phase hıgh-performance lıquıd chromatography separatıon of dımethylamınoazobenzene sulphonyl- and dımethylamınoazobenzene thıohydantoın-amıno acıd derıvatıves for amıno acıd analysıs and mıcrosequencıng studıes at the pıcomole level, *Anal Bıochem* , 178,107-117

24 Wılson, J H (1984) ın Regulatıon of mammalıan hexokınase actıvıty (Beıtner E Eds) *Regulatıon of Carbohydrate Metabolısm*, Vol I, pp 45-85, CRC Press, Boca Raton, FL

25 Palma F , Agostını D , Mason P , Dacha M , Pıccolı G , Bıagıarellı B , Fıoranı M and Stocchı V Purıfıcatıon and characterızatıon of the carboxyl-domaın of human hexokınase type III expressed as fusıon proteın, (1995) (Submıtted for publıcatıon)

26 Ceccarolı P , Fıoranı M , Buffalını M , Pıccolı G , Bıagıarellı B and Stocchı V , (1995) Rabbıt braın hexokınase evıdence for the presence of two dıstınct molecular forms, (Submıtted for publıcatıon)

17

ISOLATION AND ANALYSIS OF GENOMIC SEQUENCES FROM MYCORRHIZAL FUNGI

M. G. De Santo, S. Filosa, A. Franzè, and G. Martini

Istituto Internazionale di Genetica e Biofisica
CNR
Via G. Marconi 12
80125, Napoli
Italy

ABSTRACT

The isolation of effective probes is of paramount importance in studies aimed at elucidating the molecular mechanisms underlying the development of ectomycorrhizal fungi, as well as in the creation of biotechnological tools for improving truffle harvest and species identification. In one strategy which we are using to isolate DNA probes, recombinant plasmids are randomly isolated from genomic libraries and their nucleotide sequences compared with available data banks of nucleotide and aminoacid sequences. As a first step in this strategy, we have analyzed 56 clones from a genomic library of plasmids bearing small inserts of DNA from *Tuber albidum*. Several clones were singled out on the basis of sequence similarity with known genes. Among these was a recombinant plasmid carrying nucleotide sequences potentially coding for a protein extremely similar to the a subunit of Fatty Acid Synthase of other fungal species. The potentially coding region appeared to be interrupted by an intron and its sequence has allowed the construction of an evolutionary tree including *Tuber magnatum* and three different fungal species.

INTRODUCTION

The ascomycetous fungi *Tuberales* are able to form ectomycorrhizal symbioses with the roots of gymnosperm and angiosperm trees and have attracted attention in the scientific community due to their interesting life cycle as well as the remarkable commercial value of the fruit-bodies of some species (Martin and Hilbert, 1991; Lanfranco et al., 1994). Current molecular investigations are mostly aimed at elucidating the mechanism of symbiosis, at determining methods for species-specific identification at different developmental phases and, in the long range, at specifically altering the expression of fungal genes whose products are suspected to be involved in normal development and/or to have economic importance.

Biotechnology of Ectomycorrhizae, Edited by Vilberto Stocchi et al.
Plenum Press, New York, 1995

The accumulation and analysis of sequence tagged sequences (STS) or of partial cDNA sequences (Expressed Sequence Tags, EST) has become an important component of genome research (Green and Olson, 1990; Weber, 1990; Boguski et al., 1993). At the same time, with the rapid growth in the number of available sequences (Rice et al., 1993), it is likely that current known protein sequences may already include representatives of most protein motifs (Ancient Evolutionary Conserved Regions, ACRs) existing in nature (Green et al., 1993), so that there is a great and increasing chance of identifying new gene sequences on the basis of database comparison. Therefore, in an approach complementary to those mentioned above, it is possible to construct small-fragment genomic libraries and determine the primary sequence of randomly chosen clones to identify those bearing coding sequences by computer analysis. This approach is expected to facilitate the isolation of genes expressed at low levels, such as most master regulatory genes which may be useful in developmental studies, and at the same time to allow the identification of non transcribed repetitive sequences which may be of special value in developing tools for species discrimination. However, one important caveat on the usefulness of a strategy based on small-fragment genomic libraries comes from the absence of a thorough analysis of genome structure and sequence complexity, as is the case of most *Tuberales* species. Therefore, we decided to carry out pilot experiments, the first results of which are described in this paper.

MATERIALS AND METHODS

By comparing published primary aminoacid sequences of G6PD from different species from eubacteria to man (Jeffery et al., 1993), we determined 13 blocks identical in yeasts and humans. After back translating the aminoacidic sequences into nucleotide sequences, we were able to identify four non-degenerated oligonucleotide primers in sub-regions also conserved in *E. coli* and with their 3' ends at a codon first base or at a non degenerated second base (TU7A, 5' G G C A C C G A A G G C C G T G G C G G C T A T T T C G A; TL11A, 5' G G G C T G C A C T C T G A T G A C C A G T T C G T T; TU5A, 5' A G A A T T G A C C A T T A C T T G G G T A A A G A; TL7A, T C G A A A T A G C C G C C A C G G C C T T C G G T G C C). Other primers used in PCR amplifications are Lex4 (C T G T T C C G G G A T G G C C T T C T) and Rex5 (T C G T G A A T G T T C T T G G T G A C) already in our collection of G6PD amplimers (Calabrò et al., 1993; Filosa et al., 1993) and primers Pf8 (T T Y T C N A C C A T Y T T Y T T) and Pf2 (G A R A A R C C H T Y G G) which are short, degenerated primers successfully utilized to isolate the G6PD gene from P. falciparum (O'Brien et al., 1994). The primers described above were utilized in various combinations to optimize PCR amplifications around the reference conditions consisting of 100ng DNA, 150 of each primer, 1.25 U. Taq Polymerase (Perkin Elmer) in PCR buffer (10 mM Tris-HCl pH 8.3, 50 mM KCl, 2 mM $MgCl_2$, 800 mM dNTP) and 31 cycles (1' 94°C;.2' 53°C; 2' 72°C). For the amplification of clone 6, the primers 6L (5' T G C C G A G C T G T C T A T T G A A G G T) and 6R (5' C T C C T C T T T T T T T C A T C T C A G T) were employed.

DNA of high molecular weight was extracted (Lanfranco et al., 1994) from different samples of *Tuber magnatum* Pico and *Tuber melanosporum* obtained from Prof. Vilberto Stocchi (University of Urbino, Italy). DNA from an axenic culture of *Tuber albidum* was obtained from Prof. P. Bonfante (University of Turin, Italy).

DNA manipulations were performed according to established techniques (Sambrook et al., 1989). Nucleotide sequences were determined using a fluorescence-labelled modified cycle-sequencing method of the dideoxy chain reaction with the vector pGEM-4Z primers (Applied Biosystem sequencing manual). The reactions were electrophorised on an ABD 373A DNA sequencer. Computer assisted sequence analysis was performed locally using the GCG set of programs (Devereux et al., 1984) and remotely using NCBI's BLAST (Altschul et al., 1990) server on the World Wide Web Server (Uniform Research Locator: http://www.ncbi.nlm.nih.gov/Recipon/index.html).

RESULTS AND DISCUSSION

In our search for molecular probes to study ectomycorrhizal fungi we first attempted to use Polymerase Chain Reaction (PCR) to amplify specific genes of interest from field isolates of fruit bodies. More specifically, we attempted to isolate from *Tuber magnatum* and *Tuber melanosporum* the gene encoding the enzyme Glucose 6-Phosphate Dehydrogenase (G6PD) with the aim of better understanding the role of the Hexose Monophosphate Shunt (HMS) in these species. In fact, the HMS is a major source of NADPH, a molecule required in a variety of reactions including some reactions for nitrogen assimilation, and it has been observed in some mycorrhizal fungi that the contribution of HMS is stimulated when the fungus is associated with the root (references in Martin and Hilbert, 1991) The level of G6PD activity determines the fraction of glucose entering the HMS (Vulliamy et al., 1992) and therefore studies of the G6PD gene are expected to provide information on the regulation of the HMS as well as on the expression of a gene required at all stages of development, although at different levels.

DNA of high molecular weight was extracted from samples of *Tuber magnatum* Pico (five samples) and *Tuber melanosporum* (three samples). Interestingly, another preparation of DNA from two samples of *Tuber melanosporum*, obtained at a late stage in the season of fruiting, was degraded and when separated on agarose gels displayed a ladder-like pattern of bands having molecular weights increasing by steps of 160 bp, suggesting that degradation had occurred before cell disruption during a process resembling apoptosis (data not shown). Eight primers were used in various combinations as described in Materials and Methods to perform PCR amplifications of *Tuber* DNA.

Figure 1 shows typical results obtained with the pairs TU5A/TL7A and TU7A/TL11A. Other pairs either produced results qualitatively similar to one of those shown in Figure 1 or did not produce any PCR product at all. The nucleotide sequence of the DNA fragment produced with the pair TU5A/TL7A was identical to the corresponding segment in the *S. cerevisiae* gene; other PCR products were not sequenced.

To ascertain if the DNA segment amplified with the primers TU5A/TL7A is derived from *Tuber* DNA and not from some species contaminating the fruit bodies, we performed

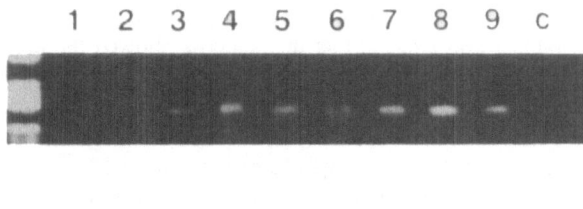

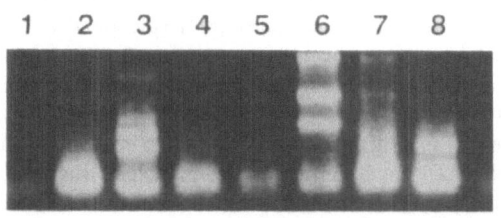

Figure 1. Agarose gel electrophoresis of PCR amplified DNA from fruit bodies of T. magnatum (lanes 1-5) and T melanosporum (lanes 5-8) using the oligonucleotide pairs TU5A/TL7A (top) and TU7A/TL11A (bottom) as primers

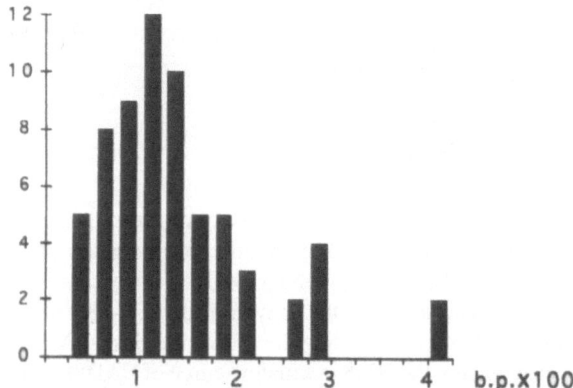

Figure 2. Size distribution of the plasmid inserts analyzed in this paper.

a Southern blot hybridization of equal amounts of HindIII-digested DNA from *S. cerevisiae* and *Tuber magnatum*, with the DNA amplified from *Tuber* as a probe. The hybridization signal was definitely stronger in the yeast then in the *Tuber* lane. Furthermore, we did not obtain any amplification product when we tried to amplify a sample of DNA obtained from mycelium of *Tuber albidum* grown as a pure culture, even after slightly varying the PCR conditions. Therefore, more work is needed to definitely prove that some, if any, of the PCR fragments obtained by amplifying DNA extracted from fruit bodies collected in the field are of *Tuber* origin.

As a second alternative, we decided to exploit the fact that conditions suitable for cultivating the mycelia of commercially important *Tuber* species *in vitro* and in pure axenic cultures have been identified (Mischiati et al., 1993). Therefore, we decided to test a strategy employing small amounts of DNA extracted from mycelium grown *in vitro* to isolate molecular probes useful for the analysis of mycorrhizal fungi.

DNA extracted from an axenic culture of *Tuber albidum* was digested to completion with the restriction enzyme MboI and inserted into the BamHI site of the plasmid vector pGEM4Z to construct a small-fragment genomic library. The primary sequence of 56 randomly chosen clones was determined.

Figure 2 shows that the size distribution of the sequenced DNA inserts, totalling 6756 base pairs, is within the 50-400 bp range. The sequences obtained were compared by means of computer programs to the collection of sequences available in different genome databases. Sequence similarities with a probability of random occurrence lower then 0.01 are reported in Table 1 and indicate that this analysis allowed the identification of clones potentially carrying the coding regions of genes of interest, including genes coding for proteins involved in regulation of transcription. The strongest similarity observed is between our clone 6 and the gene coding for the α subunit of the Fatty Acid Synthase from *Penicillium patulum* which has a probability of random occurence as low as 10^{-15}.

Fatty Acid Synthases represent one of the most functionally complex multienzymes known. In fact, nature has developed a splendid diversity in the organization of the enzymes that catalyze the synthesis of long chain saturated fatty acid from Acetyl-CoA and Malonyl-CoA and, although composed of at least six enzymatic activities, the enzymes are usually referred in the singular as Fatty Acid Synthase (FAS; Habib Mohamed et al., 1988; and references therein). In most prokaryotes and plants the complex exists as an acyl carrier protein (ACP) and a group of enzymatic activities that can be separated from one another by conventional methods of purification, while in animals all the components are organized

Table 1. Data base sequence similarities obtained with clone 6

Clone	Size		P	Accession number
clone6	190bp	Fatty acid synthase sub.alpha (Penicillium griseofulvum)	4.1xe-15	P15368
clone16	162bp	E1 protein (Papillomavirus)	0.0015	Q02261
clone3	179bp	Brain beta spectrin (Homo sapiens)	0.0018	M37885
8* seq	239bp	Spectrin beta chain (D. Melanogaster)	0.0022	Q00963
clone11	180bp	Collagen pro-alpha-1typeI (Mus musculus)	0.0024	U08020
clone32	187bp	Keratin, 64K typeII cytoskeletal (Xenopus laevis)	0.0024	M13956
clone10	200bp	AWJL 218 gene (Triticum aestivum)	0.0053	X81369
clone23*	140bp	Extensin (Lycopersicon esculentum)	0.0079	Z46675
clone29	272bp	CYTB and tRNA genes (W.makrii mitochondrial)	5.6e-05	X66594|
13* seq	94bp	Mitochondrion DNA (S.cerevisiae)	0.00088	M62622
clone13	226bp	Arginyl tRNA synthetase (argS) gene,5' end of cds and 16S ribosomal RNA promoter region (Buchnera aphidicola).	0.0011	L18930
clone36	178bp	Culmination specific prespore Dd31 gene (D.discoideum)	0.0061	X54452

in one large polypeptide chain. In the case of lower eukaryotes, such as yeasts and other lower fungi, the Fatty Acid Synthase is made of two subunits which in yeast are expressed by the genes FAS1 and FAS2 coding for the ß (MW=175.000) and α (MW= 185.000) subunit, respectively. The α subunit contains the ß-ketoacyl reductase and the ß-ketoacyl synthase activities and the ACP domain. The FAS1 and FAS2 genes are coordinately regulated to form an $\alpha_6\beta_6$ structure (Schuller et al., 1992; Schuller et al., 1994).

Figure 3 reports the nucleotide sequence of clone 6 along its conceptual 3-phase translation into amino acids. In the first reading frame, the aminoacid sequence corresponding to the region underlined is highly similar to the segment from Asp[207] to Met[251] in the *Penicillium* α FAS subunit and this segment partially overlaps the ACP domain, spanning amino acids 172-210 of the 1931-residue long α FAS polypeptide (Wiesner et al., 1988).

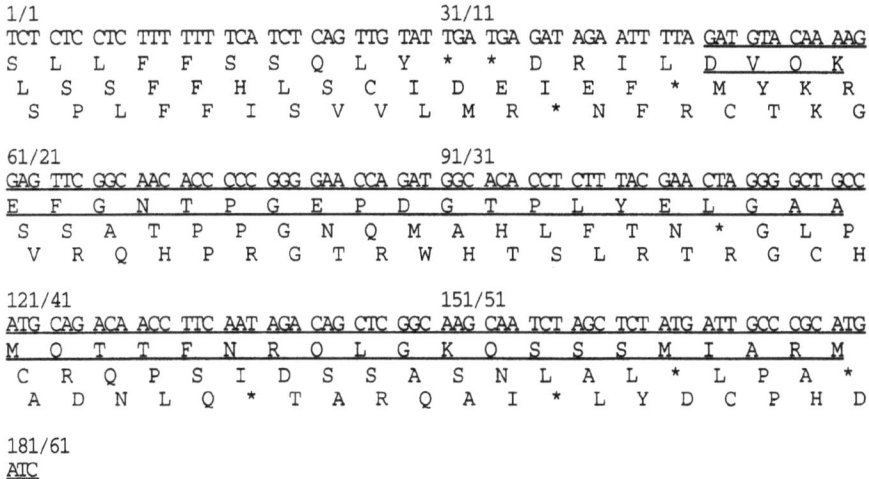

Figure 3. Nucleotide sequence of clone 6 along its conceptual aminoacid translation in the three reading frames. The region similar to FAS is underlined.

```
      C. albicans  DLGKEFGSTPEKPEDTPLEELAEQFQDSFSGQLGKTSTSLIGRLM
   S. cerevisiae   DLGKEFGTTPEKPEETPLEELAETFQDTFSGALGKQSSSLLSRLI
 P. griseifulvum   DLGKEFGSTPEKPEDVPLDELGASMQATFNGQLGKQSSSLIARMV
      T. albidum   DVQKEFGNTPGEPDGTPLYELGAAMQTTFNRQLGKQSSSMIARMI
```

Figure 4. Comparison of known fungal fatty acid synthase sequences with the sequence obtained from clone 6.

Besides *P. patulum*, the primary sequence of the α FAS subunit of two other fungal species is known. The remarkable similarity of clone 6 with the α FAS sequences from *P. patulum* (Wiesner et al., 1988), *S. cerevisiae* (Habib Mohamed et al., 1988) and *C. albicans* (Southard, S.B. and Cihlar, R.L., 1994, unpublished) is reported in Figure 4 and has allowed the construction of an evolutionary tree, as depicted in Figure 5.

In the sequence of clone 6, upstream of the region underlined in Figure 3 there is a dramatic drop in similarity with the FAS sequence, there are two stop codons in the first reading frame, and the boundary has the features of a 3' splicing junction, suggesting that an intron may be present at this position in the *Tuber albidum* gene. Interestingly, the *P. patulum* FAS2 gene has a span of 6000 bp and its coding region is interrupted by two small introns, one of which splits the Gly^{196} codon, which is located only 32 bp upstream of the position corresponding to the putative splice junction in *T. albidum*.

PCR primers based on the sequence of clone 6 were used to amplify DNA extracted from pure mycelium of *T. albidum* as well as from two different isolates of *T. magnatum* fruit bodies (Figure 6), suggesting that the sequence of clone 6 is highly conserved among *Tuber* species.

In conclusion, the experiments reported in this paper give those interested in isolating truffle genes a warning in pursuing strategies based on PCR amplification of material isolated in the field. At the same time, a strategy based on genomic libraries constructed from axenic cultures looks promising and has already allowed the isolation of interesting sequences. In comparison with the results obtained from the analysis of the DNA sequence of yeast chromosomes (Oliver, S.G. et al., 1992; Bork, P. et al., 1992; Dujon, B. et al., 1994), the fraction of *T. albidum* recombinant clones bearing sequences similar to known genes appears relatively low, suggesting a large sequence complexity for the genome of *Tuber albidum*. The clones we have isolated are expected to represent powerful tools for further studies of genome structure and sequence complexity in the *Tuberales* species.

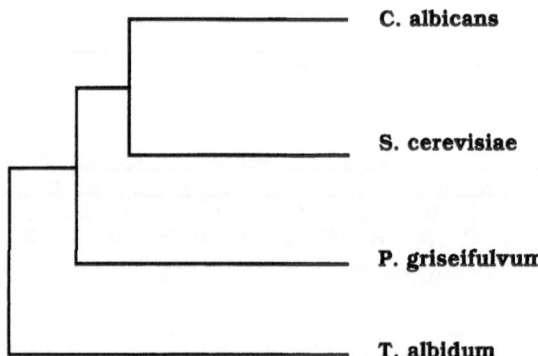

Figure 5. A branching diagram of fungal fatty acid synthases based on sequence similarity.

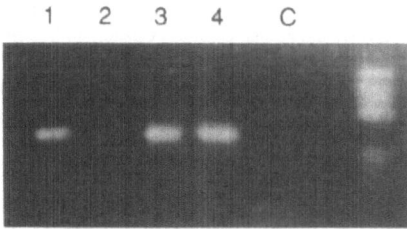

Figure 6

ACKNOWLEDGMENTS

We thank Dr. A. Ciccodicola and C. Migliaccio for their help with DNA sequencing, and Prof. V. Stocchi and Prof. P. Bonfante for providing fruit bodies and a sample of *T.albidum* DNA, respectively. The technical assistance of Ms. Maria Terracciano is also gratefully acknowledged. Work supported by Progetto Strategico Micorrizazione of the Italian National Research Council.

REFERENCES

Altschul, S. F., Gish, W., Miller, W. and Myers, E. W., 1990, Basic Local Alignment Tool, *J. of Mol. Biol.* 215: 403-410.

Boguski, M. S., Lowe, T. M. J. and Tolstoshev, C. M., 1993, dbEST-database for "expressed sequence tags", *Nature Genetics* 4: 332-333.

Bork, P., Ouzounis, C., Sander, C., Scharf, M., Schnaeider, R. and Sonnhamer, E., 1992, Comprehensive sequence analysis of the 182 predicted open reading frames of yeast chromosome III, *Protein Sci.* 1: 1677-1690.

Calabrò, V., Mason, P. J., Filosa, S., Civitelli, D., Cittadella, R., Tagarelli, A., Martini, G., Brancati, C. and Luzzatto, L., 1993, Genetic heterogeneity of glucose 6-phosphate dehydrogenase deficiency revealed by single-strand conformation and sequence analysis, *Am. J. Hum. Genet.* 52: 527-536.

Devereux, J., Haeberli, P. and Smithies, O., 1984, A comprehensive set of sequence analysis program for the VAX, *Nucleic Acids Research* 12(1): 387-395.

Dujon, B. and the European Consortium for the Yeast Genome, 1994, Complete DNA sequence of Yeast chromosome XI, *Nature* 369: 371-378.

Filosa, S., Calabrò, V., Lania, G., Vulliamy, T. J., Brancati, C., Tagarelli, A., Luzzato, L. and Martini, G., 1993, G6PD haplotypes spanning Xq28 fromF8C to Red/Green color vision, *Genomics* 17: 6-14.

Green, P., Lipman, D., Hillier, L., Waterston, R., States, D. and Claverie, J.-M., 1993, Ancient Conserved Regions in New Gene Sequences and the Protein Databases, *Science* 259: 1711-1716.

Green, E. D. and Olson, M. V., 1990, Systematic screening of yeast artificial-chromosome libraries by use of the polymerase chain reaction, *Proc. Natl. Acad. Sci. USA* 87: 1213-1217.

Habib Mohamed, A., Chirala, S., Mody, N. H., Hang, W.-Y. and Wakil, S. J., 1988, Primary Structure of the Multifunctional alpha Subunit Protein of Yeast Fatty Acid Synthase Derived from FAS 2 Gene Sequence, *J. Biol. Chem.* 263(25): 12315-12325.

Jeffery, J., Persson, B., Wood, I., Bergman, T., Jeffery, R. and Jornvall, H., 1993, Glucose-6-phosphate dehydrogenase. Structure-function relationship and the Pichia jadinii enzyme structure, *Eur. J. Biochem.* 212: 41-49.

Lanfranco, L., Wyss, P., Marzachì, L. and Bonfante, P., 1994, DNA probes for the identification of the ectomycorrhizal fungus Tuber magnatum Pico, *FEMS Microbiol. Lett.* 114: 245-252.

Martin, F. M. and Hilbert, J. L., 1991, Morphological, biochemical and molecular changes during ectomycor-rhiza development, *Experientia* 47: 321-331.

McCombie, W. R., Adams, M. D., Kelley, J. M., FitzGerald, M. G., Utterback, T. R., Khan, M., Dubnick, M., Kerlavage, A. R., Venter, J. C. and Fields, C., 1992, Caenorhabditis elegans expressed sequence tags identify gene families and potential disease gene homologues., *Nat. Genet.* 1: 124-31.

Mischiati, P., and Fontana, A. 1993, In vitro culture of *Tuber magnatum* mycelium isolated from mycorrhizas, *Mycol. Res.* 97(1): 40-44.

O'Brien, E., Kurdi-Haidar, B., Wanachiwanawin, W., Carvajal, J. L., Vulliamy, T., Cappadoro, M., Mason, P. J. and Luzzatto, L., 1994, Cloning of the glucose 6-phosphate dehydrogenase gene from Plasmodium falciparum, *Mol. Biochem. Parasitol.* 64: 313-326.

Oliver, S. G. *et al.* , 1992, The complete DNA sequence of yeast chromosome III, *Nature* 357: 38-46.

Rice, C. M., Fuchs, R., Higgins, D., Stoehr, P. J., and Cameron, G. N. 1993, The EMBL data library, *Nucleic Acids Research* 21: 2967-2971.

Sambrook, J., Fritsch, E. F. and Maniatis, T. (1989). *Molecular Cloning, a laboratory manual.* Cold Spring Harbor, Cold Spring Harbor Laboratory Press.

Schuller, H. J., Hahn, A., Troester, F., Schutz, A., Schweizer, M. and Schweizer, E., 1992, Coordinate genetic control of yeast fatty acid synthase genes FAS 1 and FAS 2 by an upstream activation site common to genes involved in membrane lipid byosynthesis, *The EMBO Journal* 11: 107-114.

Schuller, H. J., Schutz, A., Knab, S., Hoffmann, S. and Schweizer, E., 1994, Importance of general regulatory factors Rap1p, Abf1p, and Reb1p for the activation of yeast fatthy acid synthase genes FAS 1 and FAS 2, *Eur. J. Biochem.* 225: 213-222.

Vulliamy, T., Mason, P. and Luzzatto, L., 1992, The molecular basis of glucose-6-phosphate dehydrogenase deficiency, *Trends in Genetics* 8(4): 138-143.

Weber, J. L., 1990, Informativeness of human (dC-dA)n-(dG-dT)n polymorphisms, *Genomics* 7: 524-530.

Wiesner, P., Beck, J., Beck, K.-F., Ripka, S., Muller, G., Lucke, S. and Schweizer, E., 1988, Isolation and sequence analysis of the fatty acid synthetase FAS2 gene from Penicillium patulum., *Eur. J. Biochem.* 177: 69-79.

18

TESTING A SELECTED REGION OF *TUBER* MITOCHONDRIAL SMALL SUBUNIT rDNA AS A MOLECULAR MARKER FOR EVOLUTIONARY AND BIO-DIVERSITY STUDIES

J. Tagliavini,[1] A. Bolchi,[2] R. Percudani,[2] S. Petrucco,[2] G. L. Rossi,[2] and
S. Ottonello[2*]

[1] Department of Evolutionary Biology
[2] Institute of Biochemical Sciences
University of Parma
43100 Parma, Italy

SUMMARY

To set the basis for a molecular-evolutionary characterization of ascomycete ectomy-corrhizal fungi belonging to the genus *Tuber*, we have worked out PCR conditions for the amplification of a selected region of the mitochondrial small subunit rDNA (mt-SrDNA). A couple of oligonucleotides (MS1, MS2) that had previously been reported to amplify a 700 bp mt-SrDNA fragment from a variety of fungi were initially utilized as PCR primers. Based on the partial sequence analysis of a *Tuber magnatum* derived PCR fragment and on comparison with all available ascomycete sequences, two nested PCR primers (MAS1-2) have been designed. DNA fragments (380 bp) corresponding in size to the homologous mt-SrDNA region of other ascomycetes were produced by amplification reactions programmed with template DNA derived from either *T. magnatum* or *T. albidum*, in the presence of primers MAS1-2. A preliminary sequence analysis has shown that both truffle-derived fragments have high similarity (75% to 79%) with Neurospora mt-SrDNA and exhibit a characteristic distribution of conserved (U) and highly divergent (V) regions. Sequence similarity values ranged from 96% in the case of the universal region U5 to 48% for the variable region V8. This indicates that, despite its rather narrow size, the mt-DNA region we have selected can be informative for both intermediate (up to the class level) as well as for very close (intraspecific) taxonomic comparisons.

---

[*] To whom correspondence should be addressed.

Biotechnology of Ectomycorrhizae, Edited by Vilberto Stocchi et al.
Plenum Press, New York, 1995

205

INTRODUCTION

The use of molecular techniques to establish an evolutionary framework for truffles can produce a number of useful insights about these organisms. It can yield independent criteria to validate or modify existing classification schemes, as well as informative clues as to the time-frame of divergence within and outside the genus *Tuber*. Molecular data can also provide diagnostic tools to estimate biodiversity within this group of ascomycete ectomy-corrhizal fungi. In addition, defining the evolutionary relatedness of a rather poorly known group of organisms like truffles with respect to fungal species that have been extensively studied (e.g. *Saccharomyces, Neurospora, Aspergillus*) may allow meaningful comparison of gene homologues between selected groups of organisms. In some favorable cases, a specific evolutionary relatedness can even allow prediction of the existence of a particular biological process in an organism in which that process has not yet been demonstrated.

The acquisition of the above mentioned evolutionary inferences involves a wide range of comparisons, from the intraspecific to the interclass level. An important starting point for a newly addressed evolutionary analysis is thus the identification of a molecular marker that can support the broadest range of comparisons. An additional point that needs to be considered when selecting a molecular marker is the achievement of the most favorable ratio between the amount of information that can be extracted and the amount of sequence that has to be determined.

The ribosomal RNA genes, from either the nuclear or the mitochondrial (mt) genome, share a number of favorable features that make them particularly well suited for evolutionary analyses (Bruns et al., 1991). rRNA genes from distantly related organisms contain highly conserved regions that can be exploited for the design of "universal" PCR amplification primers. In addition, rRNA genes are quite well represented in the DNA database. So, there are many related sequences that can be used as references when first addressing a relatively unknown group of organisms like truffles, as well as for subsequent comparative analyses. Moreover, specific regions of the nuclear rRNA genes have been shown to evolve at distinctly different rates and they can thus provide either distant (e.g. rRNA coding sequences) or close (e.g. rDNA-Internally Transcribed Spacer sequences) evolutionary information (Gray et al., 1984; Gray, 1994; Bruns et al., 1991 and references therein).

In the case of mitochondrial rDNA, the rate of sequence divergence, and thus its most appropriate range of utilization for evolutionary studies, is extremely variable. In mammals, mitochondrial DNA variation is at least 10-fold higher than for the corresponding nuclear genes, whereas it is 5 times lower than the nuclear variation rate in some plants (Gray, 1989). Nothing is known about the mitochondrial genome of *Tuber*, nevertheless available data for other fungi indicate that the rate of sequence variation of fungal mt-DNA is generally rather high. Intraspecific variations of mt-DNA have been found in *Saccharomyces cerevisiae* and in *Neurospora crassa* (Gray, 1989; Bruns et al., 1991). More akin to truffles, a strikingly high rate of both sequence and structure variation has recently been reported for the mt-DNA genome of various ectomycorrhizal fungi belonging to the order *Boletales* (Bruns et al., 1991; Bruns and Szaro, 1992). In the case of the small subunit rRNA genes, most of this variability is concentrated in highly variable (V) regions, with only a moderate degree of variation in other portions of the sequence, the so- called universal (U) regions. The U regions of distantly related members of the *Boletales* are indeed all alignable, and the resulting evolutionary tree compares very well with that obtained from the corresponding regions of the nuclear SrRNA genes (Bruns and Szaro, 1992). Fungal mt-SrDNA thus appears to have the potential for being informative for both intermediate as well as for very close comparisons. In addition, mt-SrDNA is intronless in all the fungal species examined so far, and it is better represented in the DNA database than the large subunit mt-rDNA (72 vs 35 entries, GenBank rel. 84).

In view of all these potentially favorable features, we have worked out PCR conditions for the amplification of a selected region of *Tuber* mt-SrDNA. A preliminary sequence analysis of rDNA fragments derived from either *T. magnatum* or *T. albidum* has been carried out to evaluate the information content of this particular region of the mitochondrial genome.

EXPERIMENTAL PROCEDURES

Truffle DNA samples were obtained from the fruitbodies of *T. magnatum* and *T. albidum* (Lee et al., 1990), and were kindly provided to us by Lucia Potenza (Institute of Biological Chemistry, Urbino, Italy).

A set of previously described oligonucleotides (MS1, MS2, White et al., 1990) and a newly designed couple of oligonucleotides (MAS1, MAS2) were used as primers for PCR amplifications (see Figure 2). Amplification reaction mixtures contained 0.33 µM oligonucleotide primers, 50 µM of each dNTP, 1 U of Taq Polymerase (Perkin Elmer-Cetus) and 100 to 300 ng of template DNA in a final volume of 30 µl. DNA amplifications were carried out in a thermal cycler (Perkin Elmer mod. 480 operated in the "Step-Cycle" mode which yields a ramp time of about 1°C/2 sec) for a total of 35 cycles under the following conditions: pre-incubation at 94°C for 3 min; 1 min denaturation at 94°C; 1 min annealing at either 50°C (MS1-2) or 56°C (MAS1-2), followed by a 1 min extension at 72°C.

After amplification, one fourth of each reaction mixture was run, along with molecular size standards (ΦX174 DNA/HaeIII, Promega), on a 2% agarose, Tris-Acetate-EDTA gel containing Ethidium Bromide (0.5 µg/ml). When required, amplified DNA fragments were eluted from agarose gels with the Qiaex Gel Extraction kit (Qiagen) and cloned into the pCRII vector (InVitrogen) following manufacturer instructions.

The partial sequences of the MS1-2 amplification products were determined by direct thermal sequencing (fmol DNA sequencing system, Promega) using [33P] end-labelled MS1-2 primers. The cloned 380 bp fragments produced by MAS1-2 amplifications were sequenced on both strands with the ΔTaq Cycle Sequencing kit (United States Biochemicals) using M13/pUC universal and reverse primers.

Primers MAS1-2 were verified with the program PCRPLAN (software package PC/Gene Rel. 6.7). Homology searches were carried out on the GenBank database rel.84 using the FASTA program and sequences were aligned with the PILEUP program (GCG, Wisconsin Sequence Analysis Package).

RESULTS AND DISCUSSION

To set the basis for a molecular characterization of truffle mitochondrial SrDNA, we utilized a PCR-based approach. In the absence of any sequence information on *Tuber* mt-DNA, we initially used as PCR primers a couple of oligonucleotides that had previously been reported to amplify a 700 bp mt-SrDNA fragment from a variety of fungi (MS1, MS2, White et al., 1990). An amplification product of the expected size (marked by an arrow in Figure 1A) was obtained using these primers and total genomic DNA from the fruitbody of *T. magnatum* as template. Under these conditions, however, two slightly shorter fragments were also amplified (Figure 1A, lanes 2 and 3).

Given this heterogeneity, DNA fragments corresponding to these three major bands were eluted from a preparative gel, further purified and partially sequenced. A comparison

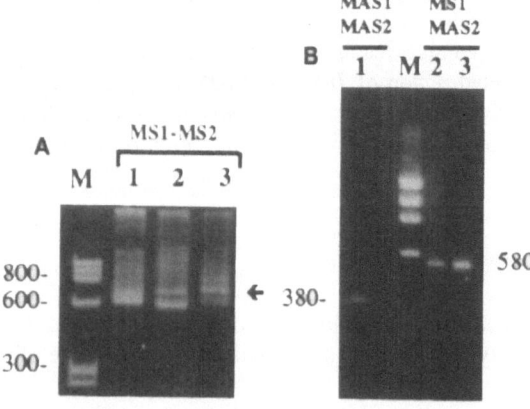

Figure 1. Electrophoretic analysis of amplification products obtained from *T. magnatum* DNA. (A) amplified fragments produced by primers MS1, MS2 using three different truffle DNA samples are shown in lanes 1-3. The arrowhead indicates a DNA band that is common to the three PCR amplification profiles. (B) amplification of *T. magnatum* DNA with primers MAS1, MAS2 (lane 1) or MS1, MAS2 (lane 2, 3). The size of the two most highly represented bands is indicated. Molecular size markers are shown in lanes M.

of their partial sequences (about 200 bp at both ends) with the DNA database showed that the lower fragments in lanes 1 and 2 are highly homologous to bacterial (Flavobacter) SrDNA and basidiomycete (Suillus) mt-SrDNA. Only the DNA fragment that was shared by all three samples exhibited a high sequence similarity to ascomycete mt-SrDNA. Both contaminants can plausibly be present in field-collected truffle specimens and, in fact, the presence of bacteria inside the sporocarps of different species of white truffles was recently reported

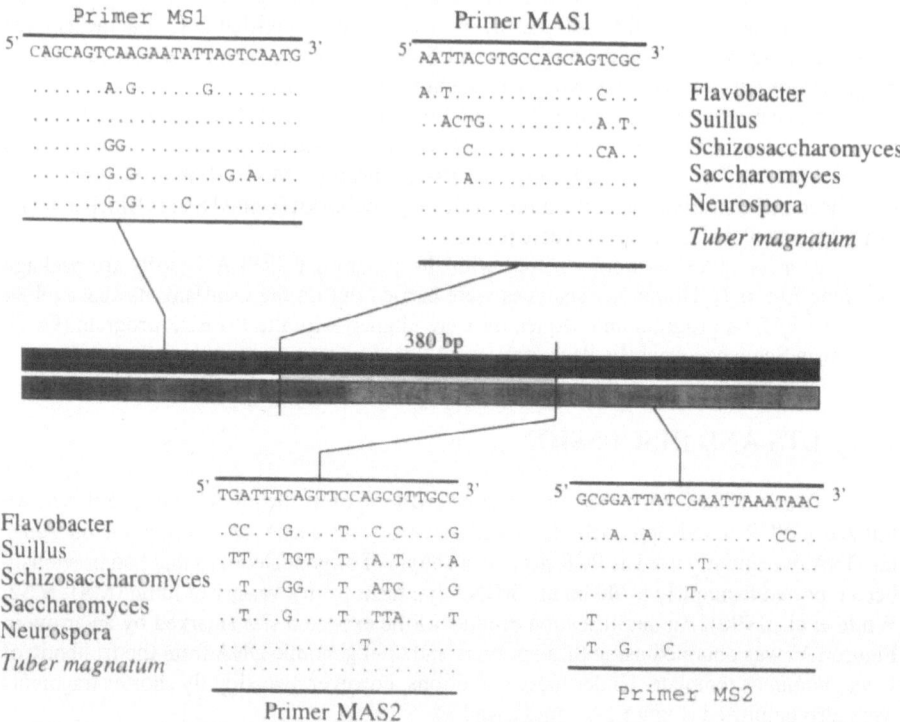

Figure 2. Outline of the oligonucleotide primers utilized for PCR amplification.

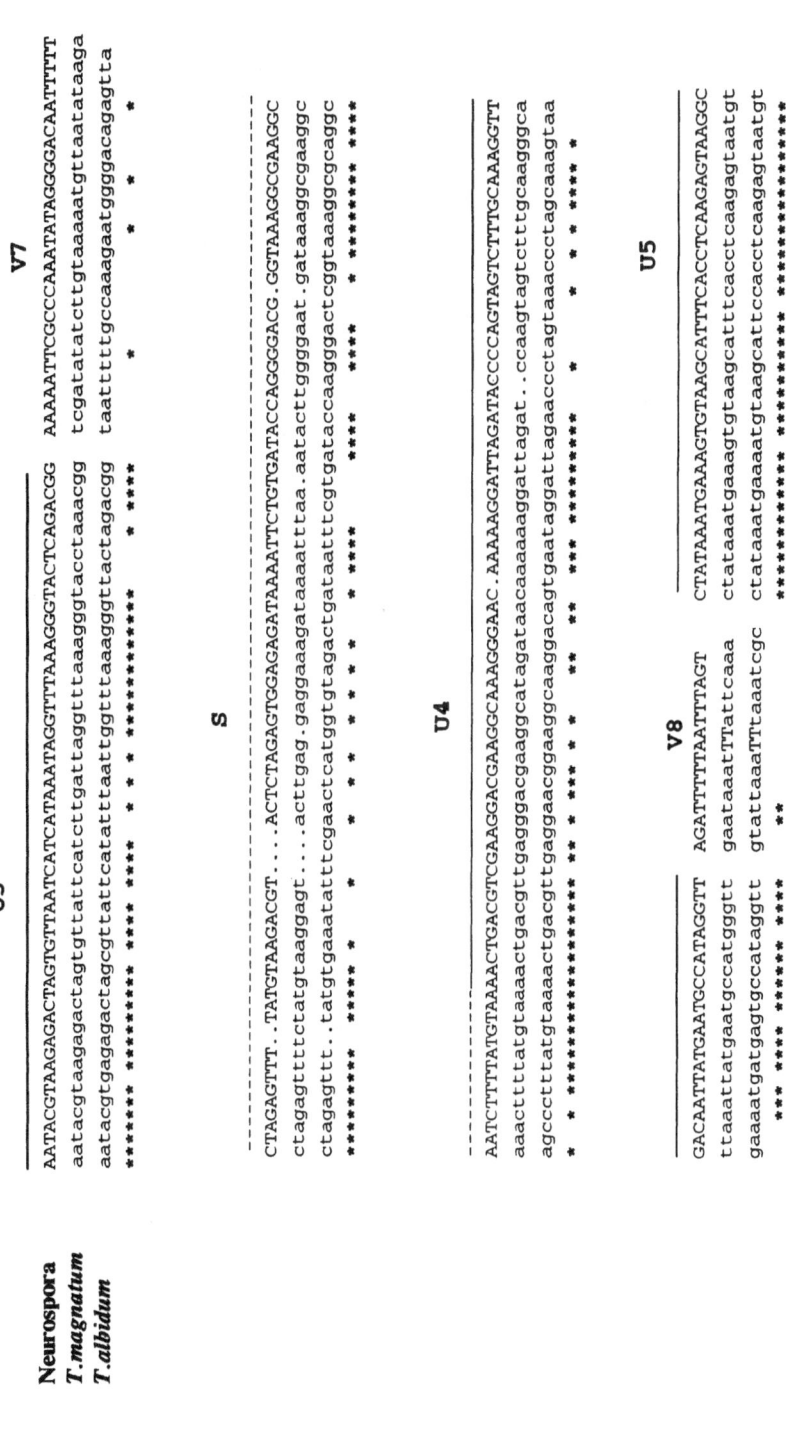

Figure 3. Alignment of *N. crassa* mt-SrDNA (GenBank accession number L33367) with the two partial sequences derived from *T. magnatum* and *T. albidum*. Conserved bases (*), universal (U), semiconserved (S) and variable (V) regions are indicated.

(Citterio et al., 1994). The above result is even less surprising if one considers the possibly small amount of intact mitochondria in the mature fruitbody and the actual discrimination capacity of primers MS1, MS2.

As shown in Figure 2, these two primers can indeed recognize various SrDNA sequences, including basidiomycete and bacterial SrDNAs. Based on the partial sequence data obtained from the previous amplification experiment, and on the alignment with available ascomycete mt-SrDNA sequences, we then designed a new couple of primers. These two primers (MAS1 and MAS2, Figure 2) anneal with two highly conserved regions of Neurospora mt-SrDNA and are expected to produce an amplification product of about 400 bp. The actual specificity of primers MAS1-MAS2 has not yet been thoroughly tested. However, as shown in Figure 1B, the amplification patterns they yield are considerably less heterogeneous than those obtained with primers MS1-MS2. Highly represented amplification products of 380 bp and 580 bp, corresponding in size to the homologous mt-SrDNA regions of Neurospora, were obtained with primers MAS1-MAS2 and MS1-MAS2, respectively. Similarly sized DNA fragments and a comparable improvement in amplification specificity were obtained when using T. albidum DNA as template (data not shown).

PCR-amplified DNA fragments (380 bp) derived from a single isolate of both truffle species were then cloned and sequenced. Both sequences exhibit the highest similarity with N. crassa mt-SrDNA: 79% and 75% for T. magnatum and T. albidum, respectively (Figure 3). Blocks of conserved bases overlap previously described universal (U3, U4, U5) and semiconserved (S) regions of Neurospora (Gray et al., 1984). A similar correspondence with Neurospora mt-SrDNA also pertains to the positioning of the two highly degenerate regions V7 and V8. Sequence homology for individual regions ranges from 96% in the case of U5 to 48% in the case of V7.

It thus appears that in spite of its rather narrow size, the mt-SrDNA region we selected is potentially informative for both intermediate as well as for very short range comparisons. The analytical approach we have described here is currently being extended to multiple isolates of T. magnatum and T. albidum and to other species of Tuber.

ACKNOWLEDGMENTS

We thank Lucia Potenza for the kind supply of truffle DNA samples and Vilberto Stocchi for helpful discussions and support. This work was supported by the National Research Council of Italy, P.S.: "Biotecnologia della Micorrizazione".

REFERENCES

Bruns, T D., White, T J. and Taylor, J W , 1991, Fungal molecular systematics, *Annu Rev Ecol Syst* 22, 525-564

Bruns, T D and Szaro, T.M., 1992, Rate and mode differences between nuclear and mitocondrial small-subunit rRNA genes in mushrooms, *Mol Biol Evol* 9, 836-855

Citterio, B., Pierotti, C , Cardoni, P, Gola, G., Potenza, L , Amicucci, A., Trillini, B and Stocchi V., Isolation of bacteria from sporocarps of *Tuber magnatum* Pico, *Tuber borchii* Vitt and *Tuber maculatum* Vitt . identification and biochemical characterization, *International Symposium on Biotechnology of Ectomycorrhizae Molecular Approaches*, Urbino 1994, pp 70-71

Gray, M W , Sankoff, D and Cedergren, R J , 1984, On the evolutionary descent of organisms and organelles: a global phylogeny based on a highly conserved structural core in small subunit ribosomal RNA, *Nucleic Acids Res* 12, 5837-5852

Gray, M V , 1989, Origin and evolution of mitochondrial DNA, *Annu Rev Cell Biol* 5, 25-50

Lee, S B and Taylor, J W , 1990, Isolation of DNA from fungal mycelia and single spores, in *PCR protocols a guide to methods and applications*, Innis, M A , Gelfand, D H , Sninsky, J J and White, T J (Eds), Academic Press, New York pp 282-287

White, T J , Bruns, T D , Lee, S and Taylor, J , 1990, Amplification and direct sequencing of fungal ribosomal RNA genes for phylogenetics, in *PCR protocols a guide to methods and applications, Innis, M A , Gelfand, D H , Sninsky, J J and White, T J (Eds), Academic Press, New York pp 315-322*

19

TRUFFLE DEVELOPMENT AND INTERACTIONS WITH THE BIOTIC ENVIRONMENT

Molecular Aspects

Giovanni Pacioni,[1] Anna Maria Ragnelli,[2] and Michele Miranda[2]

[1] Department of Environmental Sciences
[2] Department of Basic and Applied Biology
University of L'Aquila
Via Vetoio
67100 L'Aquila
Italy

SUMMARY

The nature and structure of the substances involved in black truffle morphogenetic development as well as in the interactions with the biotic environment are reviewed. The development of mycelial pellets, which form sporocarp primordia, is marked by a high L-DOPA oxidase and tyrosinase hydroxylase activity originating from a layer of cells just under the outermost one. This layer gives birth to cells which have both an outward and an inward orientation, the external ones gradually grow into isodiametric cells while the internal start branching into the ascus-bearing veins. In peridial cells and spores, tyrosinase activity produces allomelanins (dihydroxynaphtalene or cumaric derivates) which participate in cell wall structure. Some volatile compounds (alcohols and aldehydes with 2-5 C) are produced before spore formation, phenolic and sulphur compounds along with ketones and esters appear only later, when also a steady decrease in tyrosinase activity occurs. Unripe truffle substances are normal fungal metabolites produced *via* the pyruvate pathway. Some of these seem to be responsible for a strong grass-growing inhibition and a patent modification of the micro-population of the hydnosphere. Soil fungi and the plants tested are affected by three aldheydes (2-methyl propanal, 2-methyl butanal and 3-methyl butanal), the plants also by two alcohols (2-methyl butanol and 3-methyl butanol), even at very low concentrations. In contrast, a strain of truffle *Pseudomonas* showed a tolerance to much higher concentrations of these substances. While dimethyl sulphide, produced only by ripe truffles, seemed to act as an attractant of spore-spreading animals. By all accounts, truffle sporocarp morphogenesis is initiated by the tyrosinase activity and the subsequent truffle development affects all components of its biotic environment.

Biotechnology of Ectomycorrhizae, Edited by Vilberto Stocchi et al.
Plenum Press, New York, 1995

INTRODUCTION

Black truffles were the first mycorrhizal fungi to be cultivated along with their naturally mycorrhized trees and truffle farming was the first applied forestry biotechnology.

The investigations into truffle farming, begun in France more than a century ago, led to the rise of a new science focusing on mycorrhiza (Frank, 1885). Understandably, some truffle species belonging to the genus *Tuber* Mich. ex Fr. are among the ectomycorrhizal fungi most thoroughly studied. However, their distribution is restricted to just a few European countries and the difficulties in growing truffles in culture has greatly reduced the interdisciplinary interest in them. Despite the lack of insight into the very basics of truffle biology, truffle farming has brought about afforestation and the creation of nurseries as well as profitable business.

This paper is meant to discuss the results obtained thus far in research on truffle metabolism during the sporocarp morphogenetic development in connection with its effect on the biotic environment.

TRUFFLE LIFE CYCLE AND CHANGES IN THE HYDNOSPHERE BIO-ENVIRONMENT

The *Tuber* life cycle seems to go through different phases, some of which can coexist separately, without any physical connection, such as the mycorrhiza and sporocarp.

A primary mycelium produced by a germinating spore grows towards the roots of a suitable host tree, source of a still undefined chemical stimulus. It is possible that at least two mycelia are needed to produce the mycorrhiza, because the symbiotic mycelium shows two nuclei. Dikaryotic mycelia live partly as mycoclena and Hartig net and partly extramatrically, carrying out the typical functions of this symbiosis.

Two morphological types of *Tuber* mycorrhizae can be observed in the root apparatus of a truffle-producing tree: a) single or ramified mycorrhizae with characteristic mycoclena; b) coralloid clusters of mycorrhizae often covered with a mucilaginous substance, called 'glomeruli'.

The glomeruli survive winter and, in the spring, the new extramatrical mycelium, arisen from these, can produce pellets, the primordia of future truffles ('hyphenchima'). As they grow in size, a process of differentiation starts from a superficial layer of filamentous cells. The surface cells seem to produce outward the 'peridium', an external protective tissue made of generally isodiametric cells with thick, pigmented walls, while their branches grow inward, producing fertile veins, later asci and finally spores.

Several animals, wild or trained, are capable of finding and unearthing ripe truffles. In nature, these animals are active agents of spore dispersal of a peculiar category, 'hypogeous fungi', that is truffles and false-truffles, belonging to Zygomycotina, Ascomycotina and Basidiomycotina. The sporocarps of these fungi produce odours which attract truffle-eating or 'hydnophagous animals' (Pacioni,1989), which disperse spores in their faeces.

In the presence of truffle mycelium we find a strong inhibition of grasses, phenomenon commonly called 'brulé' in France or 'pianello' in Italy, and a clear modification of the micro-population. For this peculiar volume of soil, where the truffle mycelium produces a modular modification of biocoenoses, the term 'hydnosphere' has been proposed (Pacioni, 1991).

A number of bacteria live inside truffle sporocarps, mainly in their aeriferous veins and some are also present in the hyphal cells and asci (Pacioni, 1990).

Morphogenetic Development and Sporocarp Metabolism

Six different phases can be recognized in truffle sporocarp development:

1. - the 'hyphenchymatous stage'- when undifferentiated mycelial pellets are formed;
2. - the 'peridial stage'- when the outer layer transforms into round cells which then develop into the external structures (hypothecium and peridium);
3. - the 'vein stage'- when the hyphae beneath the hypothecium branch inward and evolve into fertile veins;
4. - the 'ascal stage'- when asci originate from the fertile veins;
5. - the 'sporal stage'- when spores are produced within the asci;
6. - the 'spore pigmentation stage'- when the spores are fully ripe and show both their final colour and completely developed ornamentation.

To date, the hyphenchymatous stage has not yet been studied, either from the hystochemical or the chemical point of view, and why the pellets are formed and what mechanism activates the second stage is still unknown. An external action by 'primer' bacteria may be supposed, as happens in *Agaricus bisporus* (Rainey *et al.*,1990), but physical factors, or both could be involved.

Starting from stage 2, the 'peridial stage', pigments accumulate in peridial cell walls. The black truffle peridium at the beginning is red, then goes from brown to dark brown and finally becomes black. Particularly during phase 4, a remarkable quantity of glycogen is accumulated in the asci which in phase 5 is completely utilized during spore formation, and is thus absent in phase 6. During the last two phases a new accumulation of pigments in the spore walls occurs and some new metabolic pathways are activated.

MORPHOLOGICAL-MOLECULAR CORRELATIONS OF TRUFFLE DEVELOPMENT AND MELANOGENESIS

The pigments of black truffles have been studied with a view to using them for chemosystematic purposes. Dark pigments are generally referred to as melanins.

Melanins are polymeric pigments found in bacteria, plants and animals; eumelanins and phaeomelanins are mainly present in animals while allomelanins are present in bacteria and plants, arising from polyphenol oxidation. Unique pathways for melanin biosynthesis are expressed in both Ascomycotina and Basidiomycotina (Bell *et al.*, 1975; Stussi & Rast, 1981).

Melanins appear important for the survival and longevity of fungal propagules and the inhibition of melanin synthesis prevents the penetration of plant tissues by appressorial cells (Bell & Wheeler, 1986). Moreover melanin synthesis is an important marker of the sexual differentiation of *Neurospora crassa* (Hirsh, 1954) and the tyrosinase locus is repressed during vegetative growth (Horowitz *et al.*, 1970); however no information is available as regards its function in sexual differentiation.

Fungi produce melanins that are different from those of other melanin-synthesizing organisms; in fact the allomelanins found in fungi do not contain nitrogen, being derivatives of dihydroxybenzene and/or dihydroxynaphtalene (Nicolaus, 1968). However, the tyrosinases and polyphenoloxydases of fungi are able to oxidize L-tyrosine and L-3,4-dihydroxyphenylalanine even if the former substrate is less specific than the vertebrate enzymes (Robb, 1984).

Table 1. Elemental percent composition of *Tuber aestivum* melanin (minimal formulas : Gleba $C_6H_{9.5}O_{1.8}$; Peridium $C_6H_8O_3$)

	C	H	N
Gleba	54.88	6.08	Traces
Peridium	64.46	8.39	Traces

Tyrosinase (L-tyrosine, L-3,4-dihydroxyphenylalanine: oxygen oxidoreductase, EC 1.14.18.1), a widespread enzyme in both bacteria and man, is able to oxidize monophenols and diphenols producing the black, brown, red and yellow plant and animal pigments referred to as allomelanins, eumelanins and phaeomelanins (Mason, 1965; Nicolaus, 1968; Swan, 1974; Robb, 1984). This enzyme is a binuclear copper protein (Jolley *et al.*, 1974; Malmstrom, 1982; Woolery *et al.*, 1984) showing polymorphism and different polypeptide composition in plants and animals (Robb, 1984). It is also involved in the sexual differentiation of the ascomycete *Neurospora crassa* (Hirsh, 1954; Schaeffer, 1953; Horowitz, 1970; Prade *et al.*, 1984).

Incorporation of melanins into cell walls increases both mechanical strength and antimicrobial properties enhancing survival by protecting cells from lytic enzymes and desiccation. Melanins in fact protect them from lysis by inhibiting the action of chitinase and glucanase, enzymes that have a key role in the microbial destruction of fungal propagules (Cook & Whipps, 1993).

Since truffle sporocarps need to stay several months in the soil before ripening, melanized peridium is fundamental to resist microbial attacks. Spores also can resist enzymatic aggressions in the intestines of hydnophagous animals.

Structure of Truffle Melanins

The black pigments were studied in *Tuber aestivum*. The extracts from peridium and gleba, that is spores, were analysed separately (Miranda *et al.*, 1991). Results showed that summer truffle melanins are allomelanins, because they seem to lack nitrogen (Table 1). The structures of the only identified pyrolysis products of both peridium and gleba are of the dihydroxynaphtalene and coumarine series, thus giving indication about their tyrosine origin (Fig.1, A and D). The identified structures agree with our knowledge on fungal melanin synthesis. Research, still in progress with other truffle species, has shown some differences between peridial and sporal melanins, perhaps as a result of their different chemical composition and organization.

Characteristics and Properties of Truffle Tyrosinase

Tyrosinase from several species of truffles was characterized by Miranda *et al.* (1992). Figure 2, A & B, shows the tyrosinase activities of ripe and unripe truffles of different species and Table 2 reports the effect of tyrosinase inhibitors.

The L-DOPA oxidase activity of *Tuber brumale* tyrosinase is inhibited by high phosphate concentrations, whereas tyrosinase activity is essentially independent at buffer concentrations above 20 mM. The activity of truffle tyrosinase is essentially independent of pH in the range 5-7 at the substrate concentration used, 1 mM L-tyrosine or 5 mM L-DOPA.

The effect of temperature on the L-DOPA oxidase activity of truffles is shown in Fig.3 where the Arrhenius plot of L-DOPA oxidase activity shows enzyme stability in the temperature range 20-45°C.

Table 2. Effects of PTU, DDC and mimosine on the L-DOPA oxidase and L-tyrosine 3-monooxygenase activities of truffle homogenate supernatants. The inhibitor concentration was always 10^{-4} M. Values are the percentages of the control specific activities; the mean values of at least ten measurements are reported whose S.E. never exceeded 5% of the mean

	1 mM L-tyrosine			5 mM L-DOPA		
	PTU	DDC	Mimosine	PTU	DDC	Mimosine
Tuber mesentericum	0	0	100	0	53	76
Tuber melanosporum unripe	0	0	100	0	0	80
Tuber melanosporum ripe	0	0	100	0	0	78
Tuber macrosporum	0	0	100	0	0	67
Tuber brumale	0	0	100	0	0	80

The apparent K_m for L-tyrosine was calculated to be 2.70 mM while that for L-DOPA was 0.37 mM and the V_{max} (pH 5-7, at 1 mM L-tyrosine or 5 mM L-DOPA concentration) were 0.0019 and 0.0165 respectively. The electrophoretic patterns of the L-DOPA oxidase activities present in the homogenate supernatants from *Tuber brumale*, unripe *T. melanosporum*, ripe *T. melanosporum*, *T. macrosporum* and *T. mesentericum* are reported in Fig.4. The L-DOPA oxidase activity is mainly confined in two major bands at the top of the gel in the cases of *Tuber brumale*, ripe *T. melanosporum* and *T. macrosporum*, while in the cases of unripe *T. melanosporum* and *T. mesentericum* a third band is also clearly evident, but this appears very feeble in the case of *Tuber brumale* (Miranda *et al.*, 1992).

Localization of Tyrosinase Activity

The L-DOPA oxidase and L-tyrosine 3-monooxygenase activities of truffle tyrosinase are colocalized as is apparent in the unripe truffle (*Tuber melanosporum*) where both have been tested (Fig.5). In the peridium a new melanin synthesis is not evident since it is already highly pigmented (Fig.6), while in the gleba both L-DOPA oxidase and L-tyrosine 3-monooxygenase activities (even if the L-tyrosine 3-monooxygenase reaction is less intense than the L-DOPA reaction) are clearly evident. Both sterile and fertile veins show melanin synthesis which is more intense in the sporogenic hyphae (Miranda *et al.*, 1992).

An interesting point is the significance of the black pigment within truffle tissues (peridium and gleba) (Fig. 7, A), even if some relationship seems to exist between pigment expression and reproductive differentiation (Hirsh, 1954; Schaeffer, 1953; Horowitz *et al.*, 1970; Prade *et al.*, 1984; Ragnelli *et al.*, 1990). In fact we did not find any L-DOPA oxidase (EC 1.14.18.1) activity in the ripe spores (Fig.7, D), while considerable activity was found in the sporogenic hyphae, connected or not to the asci (Fig.7, B & C), and in the unripe spores (Ragnelli *et al.*,1990) (Fig.7, D). The L-DOPA oxidase reaction was also positive in the layer of hyphae beneath the hypothecium, from which the sporogenic hyphae originate (Fig.7, A). Thus the investigation of melanogenesis in truffles might be of some interest to gain information about the reproductive differentiation of these ascomycetes and also in order to culture them.

The findings reported above indicate that a true tyrosinase occurs in truffles, which in the case of the genus *Tuber* are highly pigmented, in accordance with the occurrence of tyrosinase in Ascomycetes other than Tuberales, such as *Neurospora crassa* (Hirsh, 1954; Schaeffer, 1953; Horowitz *et al.*, 1970; Prade *et al.*, 1984; Huber & Lerch, 1987). In fact, tyrosinase activity is absent in fully ripe ascocarps of *Tuber melanosporum* while activity is found in the unripe ascocarps. The L-DOPA oxidase activities of all the species of truffles

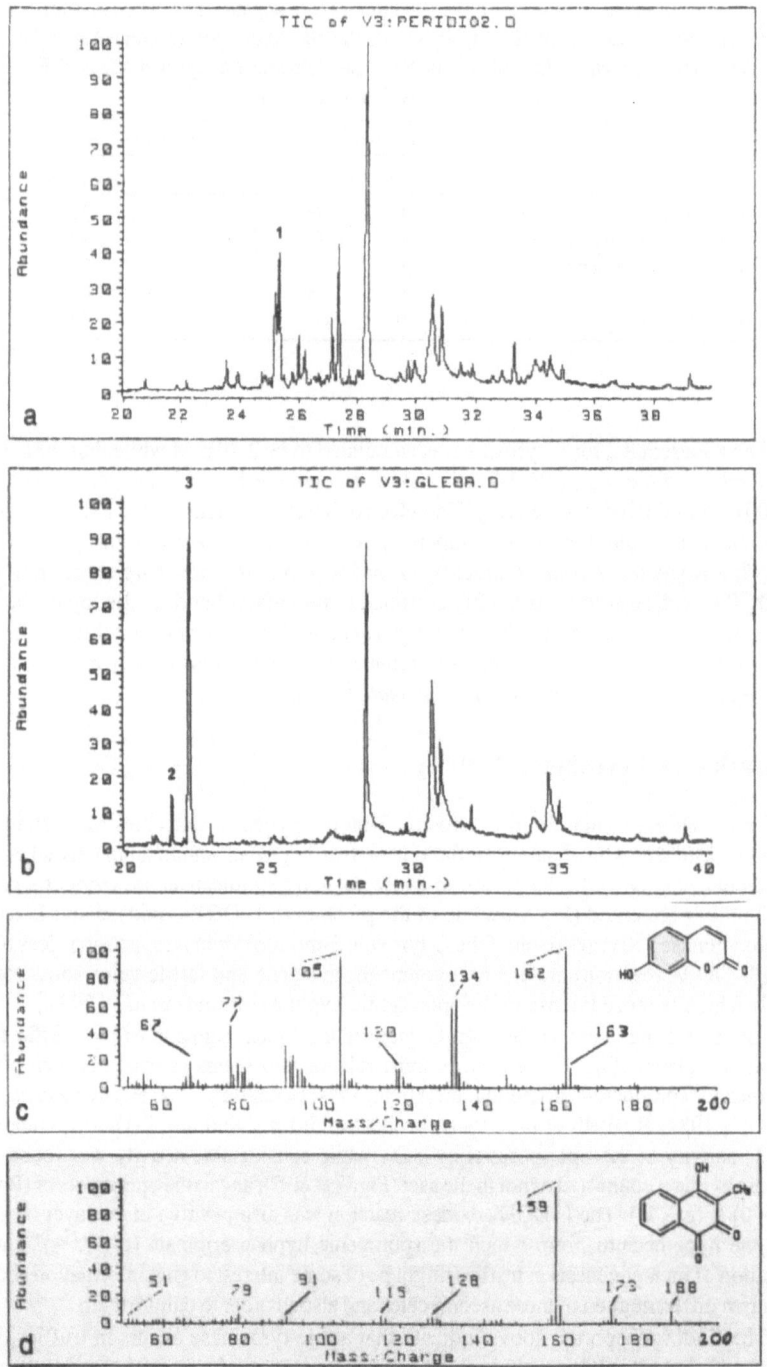

Figure 1. Gaschromatography and mass spectra of alkaline fusion products from peridium (a, c) and gleba (b, d, e) melanins of *Tuber aestivum*.

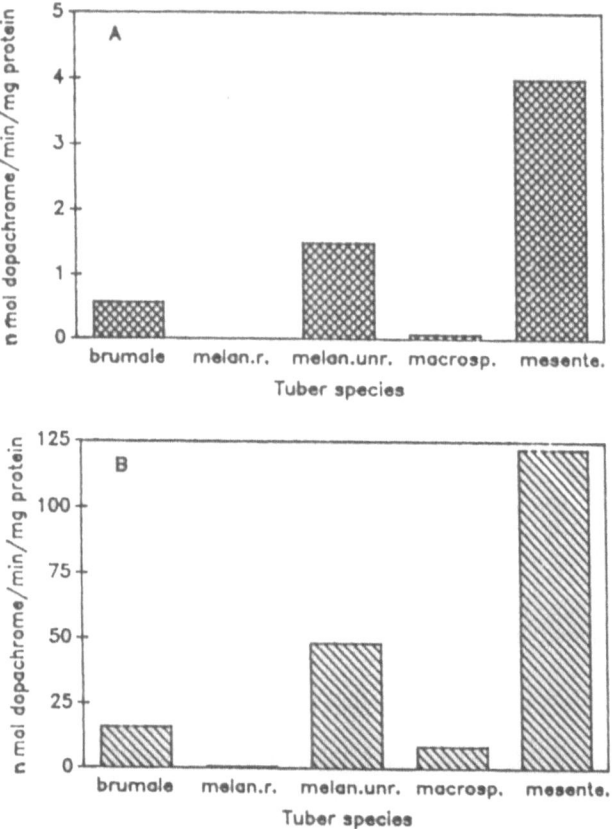

Figure 2. Specific L-tyrosine 3-monooxygenase (A) and L-DOPA oxidase (B) activities of various *Tuber* species: *T.brumale, T.melanosporum* ripe (melan. r.) and unripe (melan. unr.), *T. macrosporum* (macrosp.) and *T. mesentericum* (mesente.). When the maturation stage is not indicated it is intermediate. The values reported are the means of at least three measurements; S.E. never exceeded 5% of the mean.

investigated are about one order of magnitude lower than those of L-tyrosine 3-monooxy-genase and this is fully in accordance with what has already been described for tyrosinases of different phylogenetic sources (Robb, 1984).

The changes in specific activity versus temperature do not show important features as compared to other tyrosinases or enzymes, with the range of stability from 20-40°C.

The K_m values versus L-tyrosine and L-DOPA are in the order of magnitude of those found in mushrooms and, as regards the L-DOPA oxidase activity, the K_m is also similar to that found in vertebrates. In contrast the K_m versus L-tyrosine is one order of magnitude higher than that for L-DOPA, in accordance with the lower specificity for L-tyrosine of nonvertebrate tyrosinases when compared to the vertebrate enzymes (Swan, 1974; Robb, 1984). The higher affinity for L-DOPA may reflect the fact that Ascomycete and Basidiomy-cete melanins are mostly allomelanins produced by catechol or 1,8-dihydroxynaphtalene oxidation (Mason, 1965; Nicolaus, 1968; Swan, 1974; Robb, 1984; Wheeler, 1983).

L-DOPA oxidase activity is distributed in many electrophoretic bands, due to protein polymorphism, or to aggregates of the enzyme protein (Robb, 1984).

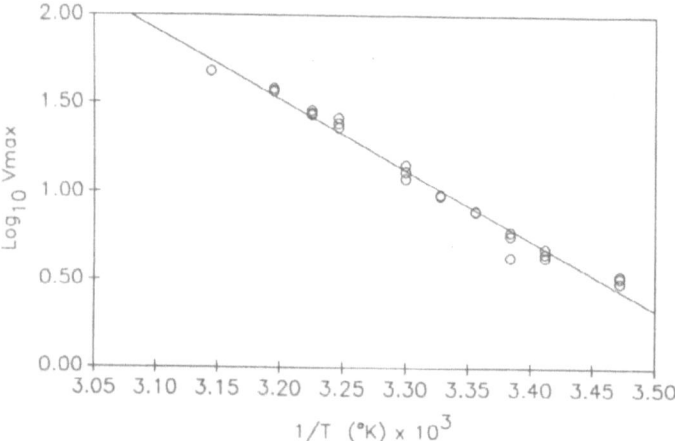

Figure 3. Arrhenius plot of *Tuber brumale* L-DOPA oxidase activity, the curve is the first-order linear regression of the experimental points

In conclusion, the tyrosinase activity present in the Ascomycetes of the order Tuberales is in some way correlated with the age, life cycle and reproductive differentiation of their sporocarps

PRODUCTS OF TRUFFLE METABOLISM DURING DEVELOPMENT

Related to tyrosinase activity and melanin production, the different phases of truffle development are characterized by a different composition of volatile substances, a group of

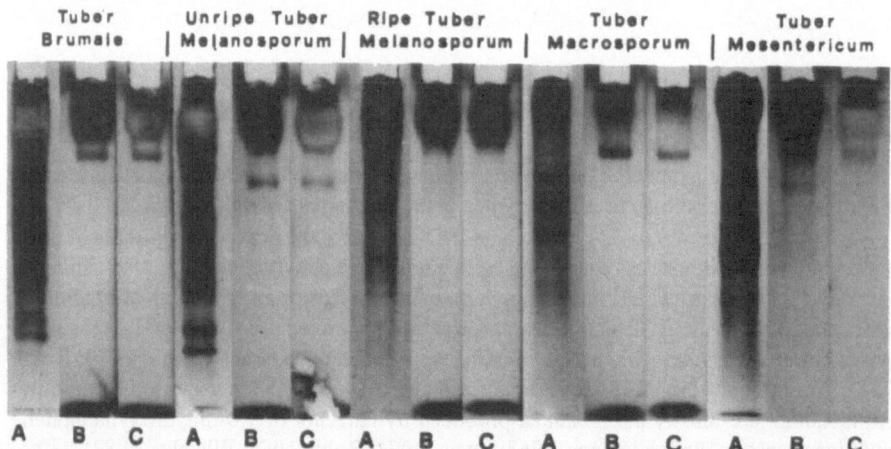

Figure 4. Polyacrylamide slab-gel electrophoresis of the homogenate supernatants from different species of truffles The different species are indicated in the Figure (A) Coomassie staining, (B) L-DOPA + PTU staining

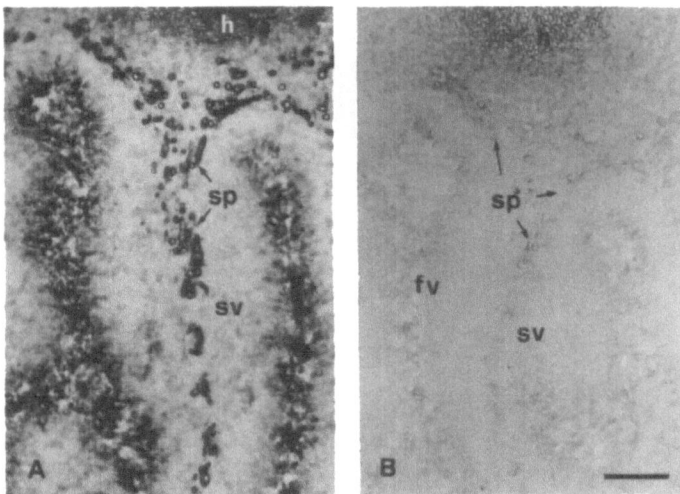

Figure 5. L-DOPA oxidase (A) and tyrosine hydroxylase (B) activities in serial sections of unripe *Tuber melanosporum* ascocarp h, hypothecium, sp, sporogenic hyphae, sv, sterile vein, fv, fertile vein Bar = 100 μm

compounds regarded as an evolutionary character that also makes truffles so highly appreciated and prized

Because of the economic importance of these compounds several attempts to identify them were made with different techniques and tools before the reliable results obtained with gas-chromatography coupled with mass spectrometry using direct (Bellina-Agostinone, D'Antonio & Pacioni, 1987, Pacioni, Bellina-Agostinone & D'Antonio, 1990) or dynamic head-space (Talou, Delmas & Gaset, 1987) analyses

Volatile Compound Analysis Methods

In our research, direct analysis was preferred so as not to interfere with truffle metabolism a very complex system which formerly made it impossible to distinguish the metabolism of truffle hyphae from that of guest bacteria

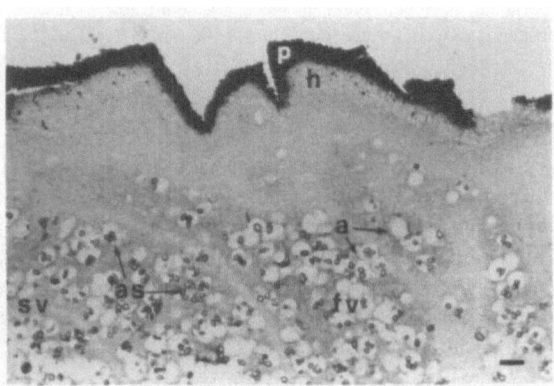

Figure 6. Giemsa staining of *Tuber aestivum* sub-adult ascocarp p, peridium, h, hypothecium, sv, sterile vein, fv, fertile vein, as, asci, a, ascospores Bar = 100 μm

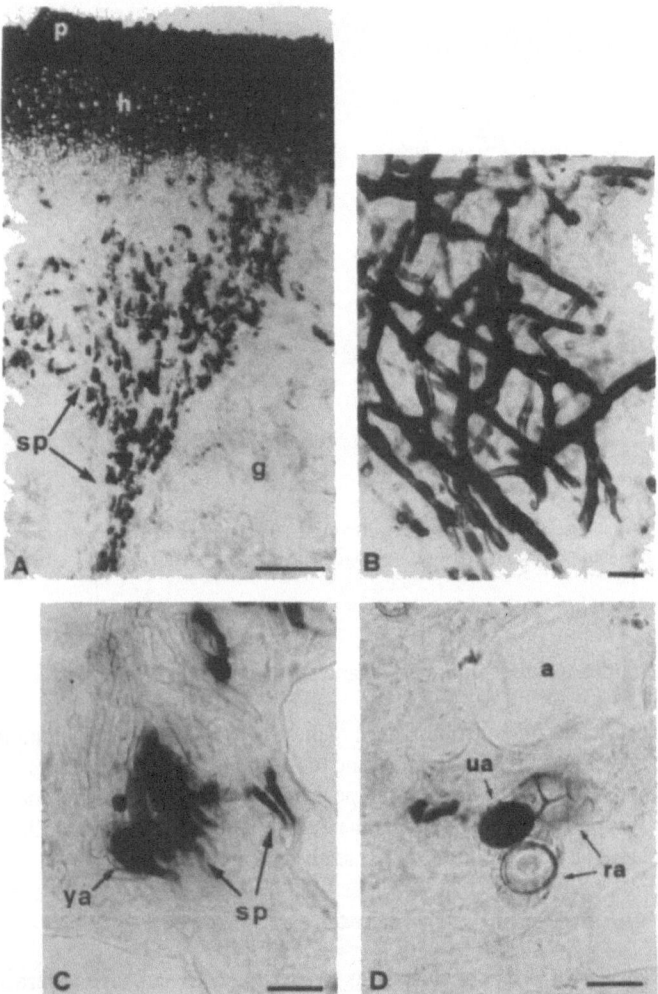

Figure 7. DOPA oxidase reaction in *Tuber aestivum* A and B young ascocarp, C and D sub-adult ascocarp p, peridium, h, hypothecium, g, gleba, sp, sporogenic hyphae, a, ascus, ua, unripe ascospores, ra, ripe ascospores, ya, young ascus. Bars in A = 100 μm, in B, C and D = 20 μm

Clean and whole specimens were stored in small air-tight bottles as soon as possible after collection. Air in the bottles was sampled with a gas syringe at room temperature or after heating the bottles quickly to a temperature of 50°C with the aim of increasing the peaks of higher-boiling substances. The sampled air was injected into the helium stream of our analyser system: GC-MS Hewlett-Packard Model 5986, computerized with HP 9825B, using a 6 ft 0.1 SP 1000 on 80/100 Carbopack-C column programmed from 40 to 220°C, rate 10°/min. Chromatogram time was 30 min.

The use of a helium stream was necessary to prevent, in part, a natural oxidation of both alcohols and thio-methyl compounds.

Volatile Substances from Truffles

The characteristic odours of truffles are the result of mixtures of several substances. The majority are oxygen-containing organic compounds with molecular masses between 46-125 AMU accompanied by one of two different sulphur methylated compounds: thiobismethane (dimethylsulphide) or bis-methylthiomethane, in black and white truffles, respectively. Generally they are aliphatic molecules coming from the primary metabolism. Aromatic compounds with one ring, derived from the first steps in the shikimate-chorismate pathway (Towers, 1976), were found in only a few species. In the taxonomic complex of *Tuber mesentericum* , singly or in addition to aromatic compounds, 2-methyl-1,3 butadiene (isoprene) can also be produced. This is not usually free, being as a rule a basic compound of certain secondary fungal metabolites (McCorkindale, 1976; Pacioni, Bellina-Agostinone & D'Antonio, 1991)

The main compounds are shown in Table 3. Qualitatively few differences were observed among the different taxa examined. Generally the relative amounts of some compounds changed. A remarkable variation occurs according to the development stage.

Before spore formation and pigmentation only some oxygen containing organic compounds (alcohols and corresponding aldehydes) are produced, while after pigmentation ketones and esters, phenolic and sulphur compounds appear (Pacioni, Bellina-Agostinone & D'Antonio, 1990), even as the tyrosinase activity starts decreasing steadily (Miranda *et al.*, 1992).

CHEMISTRY OF THE INTERACTIONS WITH THE BIOTIC ENVIRONMENT

'Quod terra ubi Tubera crescunt, herbis et graminibus plane sit destituta' wrote Marsigli (1714) describing the phenomenon of 'brulé', that is, the circular area around the

Table 3. Main volatile substances identified with GC-MS from several species of black and white truffles. Names from *Chemical Abstracts*

Alcohols	Ether
Ethanol	Anisole
1-Propanol	*m*-Methyl anisole
2-Butanol	
2-Methyl-1-propanol	Esters
2-Methyl-1-butanol	Formic acid isopropyl
3-Methyl-1-butanol	ester
3-Octanol*	Formic acid 1-methyl
1-Octen-3-ol*	propyl ester
2-Nonanol*	
	Diene
Aldehydes	2-Methyl-1,3-butadiene
Acetaldehyde	
2-Methylpropanal	Sulphur compounds
2-Methyl-1-butanal	Thiobismethane
3-Methyl-1-butanal	Bis-methylthiomethane
Ketones	
Acetone	
2-Butanone	

*from heated samples.

Table 4. The LC$_{50}$ in ppm of the some *Tuber* metabolites tested on higher plants and soil micro-organisms.

Species	Substances				
	2-methyl propanal	2-methyl butanal	3-methyl butanal	2-methyl butanol	3-methyl butanol
Triticum vulgare	—	—	26.029	22.684	23.259
Lens caulinare	21.227	20.724	15.074	11.250	15.528
Penicillium vinaceum	1.009	16.044	0.252	—	—
Aspergillus alliaceus	12.984	12.050	2.379	—	—
Pseudomonas sp.	74.952	152.493	44.543	—	—

truffle mycorrhizal tree where very few grasses can survive. The grass coenoses of these 'burned' areas were extensively studied in Italy (Montacchini & Caramiello-Lomagno, 1977; Montacchini, Lo Bue & Caramiello-Lomagno, 1977). The microbial changes in the soil flora were also accurately examined (Luppi-Mosca, 1972; Luppi-Mosca, Gribaldi & Jaredi-Sodano, 1970; Marletto, 1969; Ozino-Marletto & Sartoris, 1978-79). The phytotoxic effect was reproduced with water extracts in the laboratory by Fasolo-Bonfante, Fontana & Montacchini (1971), and Papa & Porraro (1978-79) tried to characterize the substances responsible.

The main spread among the previously identified substances in Table 3 were tested in sealed desiccators against wheat, *Triticum vulgare*, lentil, *Lens caulinare*, *Penicillium vinaceum*, *Aspergillus alliaceus* and *Pseudomonas sp.* According to Luppi-Mosca (1972), the two fungi should be sensitive, in a different manner, to truffle metabolites. Instead, the bacteria strain was isolated from the soil stuck around a truffle. Various species of *Pseudomonas* live in large numbers in the truffle habitat (Mamoun, Poitou & Olivier, 1985) as well as inside the truffle sporocarp (Pacioni, 1990) demonstrating a great tolerance of truffle metabolites.

The results obtained are reported in Table 4. All the organisms tested appeared to suffer a great inhibition when exposed to some of these substances evaporated into desiccators. Three aldehydes (2-methyl propanal, 2-methyl butanal and 3-methyl butanal) produced effects on the two soil fungi tested. In addition to some of these, two alcohols (2-methyl butanol and 3-methyl butanol) are also active on plants. 2-Methyl butanol generally affects plants to a high degree, but the strongest effects on microfungi are caused by 3-methyl butanal. However, the strain *Pseudomonas* tolerated much higher concentrations of these three aldehydes. These substances affecting the micro-aerobic environment of the soil, where seeds germinate and roots live, are normal compounds of fungal metabolism produced *via* the pyruvate pathway. These substances are produced by unripe truffles and probably by the mycelium.

Their toxic action could be a consequence of alterations caused both to the cellular membrane and the nucleic acids.

The same substances tested on plants and microorganisms and the pheromone 5-α-androst-16-en-3-a-ol, isolated by Claus, Hoppen & Karg (1981) from some truffles, were also tested as to their ability to attract truffle-eating insects and mammals.

Pit-fall traps, containing a satured solution of potassium bichromate, with microreaction vessels were used. The vessels are provided with a valve (Mininert® by Supelco) that, when open, permits a slow and continuous release of volatile substances. The traps were placed in a truffle habitat for one year starting in June 1989 and sampled every 15 days.

Arthropods were trapped by the thousand in all traps, but the truly mycetophagous insects, as well as those species strictly bound to truffles in their biology, appear clearly attracted only by thiobismethane (dimethylsulphide). (Pacioni, Bologna & Laurenzi, 1991). A successful application of this attractant capacity of thiobismethane was carried out by Bratek *et al.* (1992) to record hydnophagous insects in Hungary.

Pigs and trained dogs were also significantly attracted by thiobismethane.

CONCLUSIONS

Technologies and research related to truffle-farming are ultimately aimed at truffle production, a process which involves a complex of environmental conditions as well as genetic controls. Even though seasonal microclimate and bacteria primers can promote sporocarp initiation, its differentiation and ripening seem to be regulated by tyrosinase locus activity, which partly resembles what happens in the case of *Neurospora crassa*.

The sporocarp metabolism is independent from that of the mycorrhiza or the extramatrical mycelium and some volatile substances have been proved to be important as regulators of its relation life. The production of volatiles follows the morphogenetic process and some simple primary metabolites, also produced by the mycelium, perturb the biotic environment inducing a decrease in competition and a modulation of neighbouring coenoses. The action of these simple alcohols and aldehydes on the soil micro-environment is highly effective and selective. Only when ripening is completed does the dimethyl sulphide, an attractant of spore-spreading animals, appear.

Truffle sporocarp biology represents a noteworthy case of coevolution. Many of its fundamental steps have been chemically clarified but the mechanism of tyrosinase locus activation and the pathway and regulation of sulphate reduction are still obscure. A genetic study with *Neurospora crassa* could be instrumental in providing insight into the former problem, while the latter could be solved with direct truffle investigations. In fact, in truffle sporocarps a high bioaccumulation of sulphur takes place, and sulphur metabolism seems to be a limiting production factor in truffle-farming (Pacioni, 1992).

ACKNOWLEDGMENTS

We are very grateful to P. Aimola, O. Zarivi and C. Visca for their help. We are also indebted to Prof. Edmondo Grilli for his assistance with the English manuscript. This study was supported by the M.U.R.S.T. 40% Grant 'Biology and ecology of fungi and lichens'.

REFERENCES

Bell, A.A., Stipanovich, R.D., Puhalla, J.E., Wheeler, M.H. and Tolmsoff, W.J. ,1975, Pathway of melanin biosynthesis from (+)- scytalone in *Verticillium dahliae. Proc. Am. Phytopathol. Soc.*, 2:55.

Bell, A.A. and Wheeler, M.H. ,1986, Biosynthesis and functions of fungal melanins. *Ann. Rev. Phytopathol.*, 24:411-451.

Bratek, Z., Papp, L., Merkl, O., Adam, L., and Takas, V., 1992, Insects living in truffles. *Micol. Veget. Medit.*, 7:103-107.

Claus, R., Hoppen, H.O., and Karg, H., 1981, The secret of Truffles: a steroidal pheromone?. *Experientia*, 37: 1178-1179.

Cook, R.C., and Whipps, J .M., 1993. Ecophysiology of Fungi. Blackwell Sc. Publ.: Oxford, U.K.

Fasolo-Bonfante, P., Fontana, A:, and Montacchini, F., 1971, Studi sull'ecologia del *Tuber melanosporum*.I. Dimostrazione di un effetto fitotossico. *Allionia*, 17: 47-54.

Frank, A.B., 1885, Ueber die auf Wurzelsymbiose beruhende Ernahrung gewisser Baume durch unterirdische Pilze. *Ber. Deut. Bot. Gesell.*, 3:128-145.

Jolley Jr., R.L., Evans, L.H., Makino, N., and Mason, H.S.,1974, Oxytyrosinase. *J. Biol. Chem.*, 249:335-345.

Hirsh, H.M. (1954) Environmental factors influencing the differentiation of protoperithecia and their relation to tyrosinase and melanin formation in *Neurospora crassa*. *Physiol.Plant.*, 7:72-97.

Horowitz, N.H., Feldman, H.M. and Pall, M.L., 1970, Derepression of tyrosinase in *Neurospora crassa* by cycloheximide, actinomycin D and puromycin. *J. Biol. Chem.*, 245:2784-2788.

Huber, M. and Lerch, K. ,1987, The influence of copper on the induction of tyrosinase and laccase in *Neurospora crassa*. *FEBS*, 219:335-338.

Luppi-Mosca, A.M., 1972, La microflora della rizosfera delle tartufaie.III. Analisi micologiche dei terreni tartufiferi francesi. *Allionia*, 18 :33-40.

Luppi-Mosca, A.M., Gribaldi, L., and Jaredi-Sodano, G., 1970, La micoflora della rizosfera nelle tartufaie.II. Analisi micologiche di terreni tartufiferi piemontesi. *Allionia*, 16:115-132.

Malmstrom, B.G. (1982) Enzymology of oxygen. *Annu. Rev. Biochem.*, 51:21-59.

Mamoun, M., Poitou, N., and Olivier, J.M., 1985, Etude des interactions entre *Tuber melanosporum* et son environment biotique. In *Mycorrhizae:physiology and genetics* (ed. V. Gianinazzi-Pearson & S. Gianinazzi), pp.761-765. I.N.R.A.: Paris.

Marletto, F., 1969, La micoflora della rizosfera delle tartufaie.I. Blastomiceti dei tartufi e della rizosfera delle tartufaie. *Allionia*, 15: 155-171.

Marsigli, L.F., 1714, De generatione fungorum. Roma.

Mason, H.S. (1965) Oxidases. *Annu. Rev. Biochem.*, 84:595-634.

McCorkindale, N.J., 1976, The biosynthesis of terpenes and steroids. In *The Filamentous Fungi.II.Biosynthesis and Metabolism* (ed.J.E.Smith and D.R. Berry), pp.369-422. Arnold:London.

Miranda, M., Bonfigli, A., Zarivi, O., Ragnelli, A.M., Pacioni, G., and Botti, D., 1992, Truffle tyrosinase: properties and activity, *Plant Science*, 81: 175-182.

Miranda, M., De Angelis, F., Barbarulo, M.V., Arcadi, A., Marinelli, F., Pacioni, G., and Botti, D., 1991, Struttura delle melanine dei tartufi: caratterizzazione attraverso la spettrometria di massa. *La Chimica & L'industria, Quaderni del Laboratorio di Spettrometria di Massa, suppl.2* : 17-19.

Montacchini, F., and Caramiello-Lomagno, R., 1977, Studi sull'ecologia del *Tuber melanosporum*.II. Azione inibitrice su specie erbacee della flora spontanea. *Allionia*, 22:81-85.

Montacchini, F., Lo Bue, G., and Caramiello-Lomagno, R., 1977, Studi sull'ecologia del *Tuber melanosporum*.III. Fenomeni di inibizione nell'ambiente naturale nell'Italia centrale. *Allionia*, 22: 87-104.

Nicolaus, R.A. ,1968, The Melanins. Hermann: Paris.

Ozino-Marletto, O.I., and Sartoris, A., 1978-79, Studi sull'ecologia del *Tuber melanosporum*.V. La blastoflora delle 'aree bruciate' nell'Italia centrale. *Allionia*, 23:91-94.

Pacioni, G., 1989, Biology and ecology of the truffles. *Acta Med. Rom.*, 27:104-117.

Pacioni, G., 1990, Scanning electron microscopy of *Tuber* sporocarps and associated bacteria. *Mycol. Res.*, 94:1086-1089.

Pacioni, G., 1991, Effects of *Tuber* metabolites on the rhizospheric environment, *Mycol. Res.*, 95: 1355-1358.

Pacioni, G., 1992, Ruolo dello zolfo nel metabolismo dei tartufi. *Micologia Italiana*,21: 71-76.

Pacioni, G., Bellina-Agostinone, C., and D'Antonio, M., 1990, Odour composition of the *Tuber melanosporum* complex, *Mycol. Res.*,94: 201-204.

Pacioni, G., Bellina-Agostinone, C., and D'Antonio, M., 1991, On the odour of *Tuber mesentericum*. *Mycol. Res.*, 95: 1016-1017.

Pacioni, G., Bologna, M., and Laurenzi, M., 1991, Insect attraction by *Tuber*: a chemical explanation, *Mycol. Res.* , 95: 1359-1363.

Papa, G., and Porraro, G., 1978-79, Studi sull'ecologia del *Tuber melanosporum*.VI. Analisi spettrofotometriche di estratti di terreni tartufigeni ed azione inibente la germinazione. *Allionia*, 23: 95-102.

Prade, R.A., Cruz, A.K. and Terenzi, H.F. ,1984, Regulation of tyrosinase during the vegetative and sexual life cycles of *Neurospora crassa*. *Arch. Microbiol.*, 140:236-242.

Ragnelli, A.M., Pacioni, G., Aimola P., Lanza, B., and Miranda, M., 1992, Truffle melanogenesis : correlation with reproductive differentiation and ascocarp ripening. *Pigment Cell Research*, 5: 205-212.

Ragnelli, A.M., Pacioni, G. and Miranda M. ,1990, Histochemical investigation of truffle (*Tuber aestivum*) melanogenesis. *XIVth Int. Pigment Cell Conf., Kobe, Japan, Oct. 31 - Nov. 4 1990*, Abs. DB-9.

Rainer, P.B., Cole, A.L.., Fermor, T.R., and Wood, D.A., 1990, A model system for examining involvement of bacteria in basidiome initiation of *Agaricus bisporus*. *Mycological Research*, 94: 191-195.

Robb, D.A. ,1984, Tyrosinase. In *Copper proteins and Copper Enzymes*, Vol. 2(ed R.Lontie),, pp. 207-234. CRC Press: Boca Raton, FL, U.S.A.

Schaeffer, P. ,1953, A black mutant of *Neurospora crassa*. Mode of action of the mutant allele and action of light on melanogenesis. *Arch. Biochem. Biophys.*, 47:359-379.

Stussi, H. and Rast, D.M. ,1981, The biosynthesis and possible function of γ-glutaminyl-4-hydroxybenzene in *Agaricus bisporus*. *Phytochemistry*, 20:2347-2352.

Swan, G.A. ,1974, Structure, chemistry and biosynthesis of the melanins. *Fortschr. Chem. Org. Naturst.*, 31:521-582.

Towers, G.H.N., 1976, Secondary metabolites derived through the shikimate-chorismate pathway. In *The Filamentous Fungi.II.Biosynthesis and Metabolism* (ed.J.E.Smith & D.R. Berry), pp.460-474. Arnold:London.

Wheeler, M. ,1983, Comparisons of fungal melanin biosynthesis in ascomycetous, imperfect and basidiomycetous fungi. *Trans. Br. Mycol. Soc.*, 81:29-36.

Woolery, G.L., Powers, L., Winkler, M., Solomon E.I., Lerch, K. and Spiro, T.G.,1984, Extended X-ray absorption fine structure study of the coupled binuclear copper active site of tyrosinase from *Neurospora crassa. Biochim. Biophys. Acta*, 788:155-161.

20

METHODS FOR STUDYING SPECIES COMPOSITION OF MYCORRHIZAL FUNGAL COMMUNITIES IN ECOLOGICAL STUDIES AND ENVIRONMENTAL MONITORING

J.-E. Nylund, A. Dahlberg, N. Högberg, O. Kårén, K. Grip, and L. Jonsson

Department of Forest Mycology and Pathology
Swedish University of Agricultural Sciences
Box 7026
S-750 07 Uppsala
Sweden

INTRODUCTION

Descriptions of the species composition of mycorrhizal communities has up to recent times almost exclusively been made using fruitbody inventories (Vogt et al. 1991). However, recent studies (Dahlberg & Stenlid, 1994) have shown that sporocarp biomass constitutes only a few percent of the total (annual accumulative) biomass of ectomycorrhiza (Taylor and Alexander, 1990; Danielson and Visser, 1989). More importantly, sporocarps only poorly reflect the composition of the mycorrhizal community. Efforts have been made to use morphotyping of mycorrhizal roots in order to better describe ectomycorrhizal communities (ECM), but this has had only limited application and success (Egli et al., 1993). Compared to the advances of plant ecology, our knowledge of the fungal communities, both mycorrhizal and saprophytic, is rudimental.

Meanwhile, not only the demands of basic science make intensified efforts in this field desirable. Alarming reports on declining occurrence of macromycete fruitbodies, including popular edible species, from the European continent (Arnolds, 1991) have created awareness of the effects of environmental degradation on fungal communities. Changes have to be quantitatively and qualitatively described both in planned experiments and during environmental monitoring. Landowners and companies may be required to document the environmental consequences of their forest management in order to sell their products.

The advent of molecular biology tools has opened new pathways to the study of fungal communities and populations (White et al., 1990; Bruns et al., 1991), and these are rapidly being explored in mycorrhiza research (Taylor & Alexander, 1990; Gardes et al,

Biotechnology of Ectomycorrhizae, Edited by Vilberto Stocchi et al.
Plenum Press, New York, 1995

1991; Egger et al., 1991; Gardes & Bruns, 1992; Erland et al., 1994, etc). By the polymerase chain reaction (PCR; Mullis & Faloona 1987) large amounts of specific target DNA can be produced using crude preparations of total DNA template. The key factor is the design of specific primer pairs identifying and limiting the target gene sites. Using suitable primers, questions at various taxonomic levels can be studied (Taylor & Bruns 1988), or selected taxa can be identified in mixed materials, such as mycorrhizal fungi mixing with soil and plant tissue (Gardes & Bruns, 1993). By digesting the PCR products with restriction endonu-cleases and examining the products by gel electrophoresis, distinctive band patterns will show polymorphisms in the amplified material (Restriction fragment length polymorphism, RFLP). Alternatively, the base pair sequence of the entire DNA fragment can be determined (sequenced) and analysed.

This article will broadly examine some issues relevant to the species determination of mycorrhiza, as well as matters related to sampling and interpretation of the material obtained. Most research reported is ongoing and incomplete; the paper should thus not be seen as a full-fledged review, but rather as a voice in an ongoing research debate, based on our present experience.

In our ongoing work, our primary target is to establish a reference library of RFLP patterns of taxonomically well described common mycorrhizal fungi. The basis of this is a selection of geographically widely separate specimens, collected from the Nordic countries and available at the public fungal herbaria. To establish a reference data base, we first selected primers and restriction endonucleases, and tested their discriminating power regarding the ITS polymorphisms (Gardes & Bruns, 1992). These data are then used to identify mycorrhi-zal samples from various field surveys or experiments, some of which are done in plots where macrofungal fruitbody inventories have been carried out for a longer period. Results from the library work will soon be published, while an early report of the application has already been presented (Kårén et al, 1995).

In this paper, the following subjects will be brought up: Sampling methods; the usefulness of morphotyping as a tool to identify species; PCR procedures for tree roots; inter- and intra- species polymorphism using the ITS region of ribosomal DNA as seen in a RFLP library of some 40 species; recent experiences from some ongoing field experiments.

MORPHOTYPING

A number of previous studies of mycorrhiza in the field have comprised various schemes of morphotyping the mycorrhizal rootlets (i.e. to establish types based on macro- and microscopic characters, assuming that one morphotype to a reasonable degree coincides with a species or species group. Very precise such schemes have been published by Agerer (1987) and Ingleby et al. (1990). A combined morphotyping and PCR approach would be attractive, not least for cost reasons: first establishing the morphotypes occurring in a set of samples, and identifying them by PCR-RFLP, then making estimates of the morphotype frequencies in the individual samples, assuming that type recognition would be a relatively easy process. Occasionally, the accuracy of the recognition would be verified by PCR. By this procedure, relatively large samples could be scanned, and accurate proportions of the fungal species be calculated.

In a field study of a Norway spruce stand recently reported (Kårén et al., 1995), we examined this approach in considerable detail. We encountered several difficulties.

- Morphotyping following the scheme of Agerer, whatever its inherent qualities when used by highly trained experts, was too complicated for us to use, considering the large number of roots examined and the skill of the staff available. But in line with other colleaugues (cf. Egli et al, 1993; Gardes and Bruns, 1994), we also question some underlying assumptions (cf. below).

- Using a simplified scheme, mainly based on macroscopical characters such as colour, character of the extramatrical mycelium, surface of the mantle and general appearance, some types could be defined which were very stable, while others were hard to describe precisely, resulting in lack of replicability of the classification.

- between 40 and 70% of all mycorrhizas, and increasing with the degree of environmental stress, consisted of a uniform, smooth brown, thin-mantled type with no evident extramatrical mycelium. This could upon quick inspection be considered to be non-mycorrhizal, and we suppose that many assumedly non-mycorrhizal roots in various studies may represent this type. However, inspection of a large sample of thin (4µm) sections, selected to represent most probably non-mycorrhizal roots, showed that every single specimen was mycorrhizal, with a thin but perfectly regular mantle, and a Hartig net which extended several cell layers deep.

Cross-matching morphotypes and RFLP patterns (Tab. 1), it was found that types and patterns by no means formed unique combinations. This was the case particularly with the smooth brown type (type 1), which contained a number of species also occurring as other morphotypes. Based on our previous results (Wallander & Nylund, 1992) that increased nitrogen availability leads to reduced soil mycelium and lower fungal biomass in mycorrhizas, we hypothesize that a number of species under nitrogen load change their morphology, developing similar macroscopic characters. If this can be verified, it means that any morphotyping exercise must take into account not only species characters but also environment-induced variations. Egli et al. (1993) also questioned morphotyping, the major problems being the poor comparability of results from different authors. Most of the macro- and microscopic features are in their opinion not well enough differentiated or stable.

The same lack of unique combinations of morphotype characters and RFLP patterns was found by Mehmann et al. (in this volume). We thus conclude, in line with Egli and coworkers, that morphotyping is often unusable as a way of reducing the number of samples for PCR analysis (yet, in one case this was very successful, cf. below). Also, field studies based on coarse morphotypes must not be given too much importance when it comes to estimating number of species present, or to tentatively matching fruitbody occurrences with associated morphotypes.

SAMPLING PROCEDURES

Both basic statistics and the innate variation in soils complicate the sampling procedure. This matter has so far largely been ignored in the discussion of the PCR tool in mycorrhiza ecology research.

Looking completely away from the spatial variation, the number of sample mycorrhizas required to determine frequencies varies greatly with the frequency itself. Assuming (unrealistically) that the mycorrhizal species are randomly distributed over the roots tips in uniformity over larger areas, simple calculations can roughly show us what number of samples need to be taken to ensure a desired error at various expected frequencies. It seems to us that this type of error calculation has seldom been done in past morphotyping studies; we currently don't know how to handle this problem in PCR-RFLP field studies.

What complicates the picture is of course the combined effects of patchiness and spatial variation. Patchiness: most species colonize distinct areas of roots and soil. A fungal

Table 1. Correlation between RFLP pattern and morphotype in a pilot material from Skogaby, matched against the full present database (40 spp). Patterns designated *Nx* have been tentatively identified. Patterns designated by running letters have not been identified so far. The figures give the number of observations of a specific combination. Data from Kårén et al., 1995

RFLP pattern	Brownish							Black		White		Other					
	1	2	3	4	5	6	7	8	9	10	11	12	13	14	15	16	17
N1	2	11	3	1	1												
N2			1														
N3			1														
N4										3							
N5	3	1					1	1		1							
N6										1							
N7										1							
A													2				
B	2							1				1					
C			2		1									1			
D	1						1	1	3								
E								1									
F		1															
G	1																
H		1															
I	1				1												
J		2	1														
K	1	1															
L	2																
M			1														
N	1																
O					1												
P			1														
Q			1														
R	1																
S	1																
T													3				
U															2		

species frequently colonizes a cluster of rootlets, not just single root tips; several clusters may belong to the same genet (at least in *Suillus bovinus;* Dahlberg & Stenlid 1994); this pattern seems to be highly species-dependent. In some situations even mycorrhizal mats (cf. Unestam, 1991) may develop, covering areas from five to fifty centimetres, and excluding most other species. But even otherwise there is a heterogeneity in the forest soil, partly depending on the irregular distribution of litter items, logs, stones, and the varying thickness of the organic soil layer - and as a consequence of the microenvironments created by vegetation. We have found distinct variation between ectomycorrhizal communities under patches of different vegetation (lichens, *Cladina spp,* and ericaceous vegetation, *Empetrum hermaphroditum* with *Pleurozium schreberii)* in Scots pine forest, only 25 cm apart (Dahlberg, unpublished).

This variation is a general problem in all soil biology. We suppose that it also occurs on a microscale. We are currently addressing this issue by making model samplings from a restricted area. Our current strategy is to increasingly take many very small soil cores over

Table 2. Number of samples required, at various species frequencies and at various error levels

error of estimation (% units)	frequency of sp. (p)						
	0.5	0.250	0.125	0.06	0.03	0.015	0.008
1%	9900	7440	4360	2340	1210	620	310
2%	2490	1870	1090	590	300	150	80
3%	1110	830	490	260	140	70	30
4%	630	470	270	150	80	40	20
5%	400	300	180	100	50	30	10

Equation for the calculation: $n = (Npq)/((N-1)D + pq)$, where p = species frequency; $q = 1 - p$; $D = B^2/4$; $B = 2\sqrt{V(\hat{y})}$; $\hat{y}$ = average of species y_i; and V = variance (Schaeffer et al., 1990, p 74.)

a larger area, and extracting just a few mycorrizas from each core, rather than taking a few larger cores and analyzing them thoroughly.

The limitation will in all cases be the economy. Recent calculations of total sample processing cost, from extracted root tip to ready gel, and including all materials, minor equipment amortizations, and labour (lab technician hours) was recently estimated at approximately 50 USD per sample. Using present technology, sample costs may be cut to some extent, but not enough to make really large-scale screening possible. There is a risk that the economic constraints of a project will make required levels of statistical precision impossible to achieve in inventories of field experiment effects on the mycoflora; this aspect has to be taken into consideration when planning a project.

PCR: PRIMERS AND PROTOCOLS

We struggled for a long time to make the PCR procedure work in a replicable manner with root material, and succeeded only after generous advice and assistance from the teams of Monique Gardes and Tom Bruns in Berkeley, and of Francis Martin in Nancy. Besides the general need to optimize all parts of the process (being particularly sensitive to Mg concentrations), it seems that forest-grown conifer roots contain contaminants which inhibit the polymerase reaction. In our early steps, broadleaf tree roots (even silica gel dried dipterocarps from Java) worked well, as did peat-grown conifer seedling mycorrhiza. Only by diluting the samples was it possible to make the procedure run without problems. Another essential matter was the fast dehydration of samples. Drying of root tips in silica gel (good for field material to be mailed), storage in alcohol or lyophilization all prevented nuclease activity and preserved the DNA.

As discussed in the next section, we have almost directly focused on the ITS region of ribosomal RNA, using the standard primers ITS1-ITS4. In repeated runs, using all kinds of mycorrhiza samples, as well as sterile roots of *Pinus sylvestris* and *Picea abies*, we detected no amplification of the ITS region of host DNA. This sometimes seems to cause problems, particularly with angiosperm hosts, but can be avoided by using a fungus-specific primer, ITS1F (Gardes & Bruns 1992) instead of the universal primer, ITS1.

Regarding extraction protocols, most procedures tried in our lab have worked well. There is no need to extensively purify the DNA template material. Our experience is, as stated above, that dilution is more efficient than purification, if the amplification process fails. As the rDNA is a multicopy region of the genome, dilution in the range of 10^{-4} to 10^{-5} still yields satisfactory PCR products.

A RFLP LIBRARY OF FUNGAL SPECIES

In order to create the required reference material, the short way would be to establish a collection of fruitbodies from a research area, and make as many matches as possible. This approach was taken in the pilot study referred to below and in Tab. 1. Out of 28 RFLP patterns discerned, only 7 could be matched with fruitbody patterns (less frequent fruiting species have yet to be analyzed). While several less frequent fruitbodies are still unprocessed, we expect to improve the record, but ultimately sequencing of relevant genes will be the best way to identify unmatched RFLP patterns.

Thus, it stood quickly clear that we needed to establish a large data base comprising a selection of important mycorrhizal species, distributed over most genera occuring in our coniferous forests. It is presently estimated that Sweden harbours more than 800 species, probably 1000, of mycorrhizal fungi, to which may be added a number of hypogeous and corticiaceous species which may be recognized as mycorrhizal only in the course of root tip analysis (such as *Tylospora fibrillosa* or *Amphinema byssoides*). It was also evident that such a data base had to contain information about intraspecific variation as well. Finally, a strategic decision had to be made about what kind of data to record: sequences or RFLP digestion patterns. For cost and convenience reasons, and considering the good discrimination between species in the pilot work, we decided to set up the system using RFLP patterns.

The library presently contains some 40 species from 27 genera, and at least 3 collections of each, coming from sites at least 500 km apart, preferably 1000 km or more. Most of the material consists of herbarium specimens, and fresh collects have been deposited in public herbaria. In a few cases, well defined fungal cultures have been used. Thus, should our work reveal discrepancies with the species determination of the reference material, this can easily be reexamined. We have chosen only well-defined species; even within the *Cortinarius* subgenera *Telamonia* discussed below, all species analysed are considered to be taxonomically well distinct. However, as the current taxonomy is based on morphological characters, it is likely that DNA data will add to the current classification. This library will of course be continuously expanded, and should also include identified soil saprophytes which may contaminate the mycorrhiza samples.

After screening a large number of restriction endonucleases for the RFLP, we have selected three, *Hinfl, Cfol and Mbol* for testing on larger fruitbody material.

In most studies presented up to now, critical samples have been compared on the same gel. Our approach, however, requires accurate base pair lengths to be read on different gels. Up to now, all DNA material has been separated on agarose, which after staining with ethidium bromide were recorded on Polaroid prints, and measured with 0.1 mm precision. To correct the readings from skewness and transform them into base pair lengths, we developed a procedure for SAS for Windows software, based on a non-linear algorithm taking two neighbouring molecular markers into account. This procedure is currently being published (Olsson and Kårén), and will later be available from the authors over e-mail.

In spite of various agarose formulas and electrophoresis set-ups, there was a disturbing variation in the data when running the same DNA product on different gels. Examination of polyacrylamide gels showed the same degree of variation. This variation must be handled in the computerized matching of unknown samples with the RFLP library data base. In order to enhance the accuracy, we are currently exploring the possibilities of capillary electrophoresis as presented by Henrion et al (1992), and of sequencing gels.

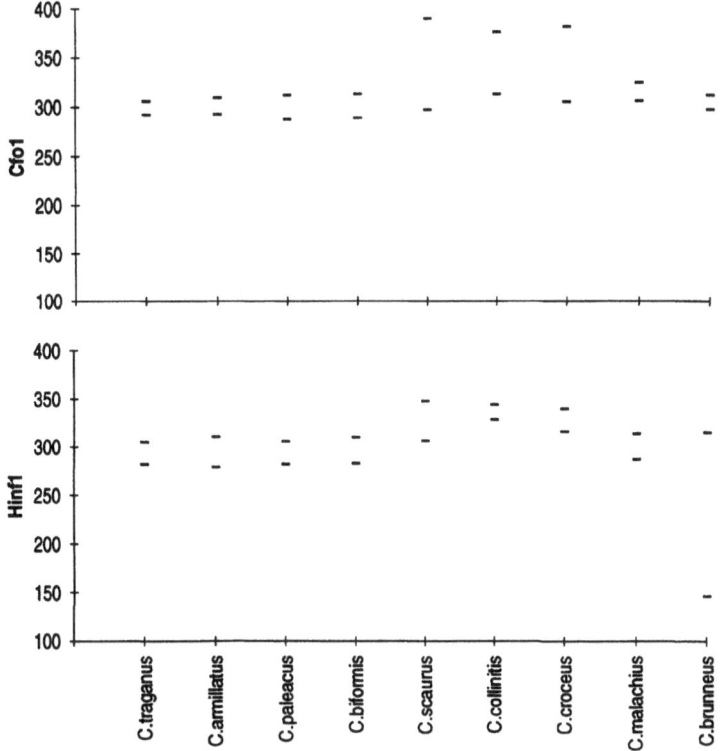

Figure 1. Nine *Cortinarius* species, digested with *Hinfl* and *Cfo1. C. traganus, armillatus, paleacus* and *biformis* all belong to the subgenus *Telamonia*. Printout from base pair data (means of at least 3 isolates).

Performing a cluster analysis of this data set (SAS for Windows), we made the following observations (cf. Figs 1 and 2):

- Overall, there is a good separation between species. Only taxonomically closely related subspecies groups, such as the *Telamonia* group of *Cortinarius*, seemed to coincide with all restriction enzymes (Fig 1).

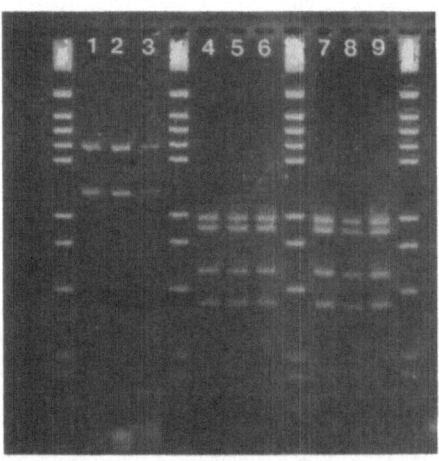

Figure 2. Agarose gel of *Hinfl* digestions of fruit-body DNA, amplified using the primers ITS and ITS4. (1-3) Xerocomus subtomentosus: (1) Uppland, East Svealand, (2) Halland, Soutwest Götaland, (3) Lapland, Northern (alpine) Sweden. (4-9) Suillus bovinus: (4) Skövde, West Götaland, (5) Laxå, Central Svealand, (6) Björklinge, East Svealand, (7) Skövde,West Götaland, (8) Laxå, Central Svealand, (9) Björklinge, East Svealand.

- The RFLP patterns lead mostly directly to species. In the cluster analysis, we usually don't find genus-level groupings of species.
- With most species, there was no discernible polymorphism even between geographically distant specimens (eg. *Suillus bovinus and Xerocomus subtomentosus*, Fig 2). But there are cases where we are detecting possible polymorphisms. Some species, such as *Chroogomphus rutilans* clearly seem to be characterized by geographically distinct RFLP patterns, which may indicate restricted gene-flow and present speciation. This variation points at the need for collecting reference material for each study site and controlling the congruence with the library data. In other cases, the intraspecies variation follows no pattern. Our data are still too limited to point out genera where this is the case. It is, however, obvious that species show various levels of polymorphism in different taxa, and the variable taxa have to be identified. In these cases, collection of reference material from the study site is of particular importance.

Species identification is currently done by introducing the unknown pattern into the complete library file and rerunning the cluster analysis.

RECENT APPLICATIONS

Drought, Pollution and Fertilization Experiments in Skogaby, S. Sweden

The Skogaby experimental area comprises plots of 40 year-old Norway spruce treated with ammonium sulphate (NS) to accelerate the acidification and nitrogen deposition, irrigation and irrigation + balanced full fertilization, drought (D) (temporary roofs), wood ash application and "vitalization fertilization" (V), i.e. addition of all macro - and micronutrients except nitrogen, plus lime to compensate for acidification. This replicated field trial, situated near Halmstad in SW Sweden, and subject to considerable acidification and nitrogen deposition, has been subject to multidisciplinary investigation by a number of Swedish research groups. Here, we have sampled the NS, D and V plots as well as the control (C). A report on the comparison between D and C is being published (Kårén et al 1995), the core of which being the lacking correspondence between morphotyping and PCR-RFLP quoted above, and between fruitbodies and underground species. The V and NS treatments have so far been only preliminarily reported (Nylund 1993). Of the total number of RFLP patterns found (28), only 7 have been identified, but these cover almost half of all root tips. The Skogaby material presented the greatest difficulties at the PCR stage.

Mapping of Nursery Mycorrhiza on Peat-Grown Seedlings

Seventy-five percent (400 million) of all outplanted seedlings in Sweden (Scots pine and Norway spruce) are cultured as containerized peat grown seedlings. They are usually grown in greenhouses for 3 months, then kept in the open for some months, and either planted in the autumn or cold-stored until the following spring.

We have examined seedlings from more than half of all Swedish forest nurseries. The seedlings were normally fully mycorrhizal when delivered, in spite of conditions adverse to mycorrhiza development during the greenhouse phase (mycorrhiza-free substrate, high nitrogen supply, ample irrigation and high temperatures). We also attempted to identify the more frequent mycorrhiza species and estimate their frequency. *Telephora terrestris* was clearly the most frequent one, followed by *Laccaria laccata*. But we also found many other species at low frequencies, even species which are normally found only in mature forests, such as *Suillus variegatus*. The material is still not fully processed (Dahlberg et al., unpublished).

Mycorrhiza in Dipterocarp Plantations on Java

In this pilot study, mycorrhizal root tips collected from dipterocarps in plantations near Bogor, Java (Indonesia) were compared with fruiting body material collected in the same stands. The hypothesis was that the plantations would show a narrower range of species, being established on former farmland, slightly outside the natural distribution of the host trees. The material was immediately placed in silica gel and mailed to Sweden. The PCR procedure worked very well on these collections; thus, we have here a simple and inexpensive way of handling field material under difficult conditions. Just as in other studies, the match between roots and fruit bodies was poor. We are currently trying to identify the mycorrhiza from herbarium material (Nylund, unpublished). It seems evident, however, that some abundantly fruiting species used for nursery inoculations are infrequent on the roots, an observation which calls for a changed inoculation strategy.

Mycorrhiza Synecology

The success of natural establishment of seedlings in different boreal forest biotopes and plant communities varies considerably, a well known fact.Traditionally, abiotic factors have been claimed as an explanation. However, recently biotic factors have received more attention.

We are now surveying four sites in northern Sweden where more than 1500 seedlings have been sown in either *Empetrum hermaphroditum* or *Cladina* spp. vegetation. The seedling roots have been examined regarding morphogenesis, and representative samples of each mycorrhizal morphotype have been subjected to PCR-RFLP. In this study, we could distinguish two stable morphotypes, a *Knollenmykorrhiza* and a light yellow one, both of which were monospecific: the first one was *Suillus variegatus* the other either *Cortinarius croceus* or *C.cinnamomea*. The proportion between the two types was clearly distinct for each type of field cover (Dahlberg, in preparation).

In other ongoing work, we are mapping the mycorrhizal species on plots in undisturbed forest in south central Sweden, where detailed fruitbody collections have been done over many years.

CONCLUSIONS

- 1. Fruitbody inventories correlate poorly with mycorrhiza community structure, seen as species distribution on mycorrhizal root tips
- 2. Morphotyping is no alternative to molecular identification
- 3. Polymorphisms in the ITS region of the ribosome RNA are of a magnitude suitable for species identification
- 4. RFLP pattern data, using two restriction enzymes, are enough for species identification except for closely related species groups
- 5. Intraspecific variation in this region seems to be relatively small, but has to be taken into account
- 6. For field inventories, sampling methods and statistical considerations are crucial for the success of any study, and have to be given due consideration.

ACKNOWLEDGMENTS

The development of the application of PCR to mycorrhiza has been funded by the Swedish Environmental Protection Agency, which has also been sponsoring the Skogaby study as well as ongoing work on nitrogen and silviculture effects The synecology project is funded by the National Council for Forestry and Agriculture Research For the field work in Java, Dr Yadı Setıadı made an important contribution, assisting some Swedish forestry students in material collection Contributions for the nursery study have been received from Swedish forest companies

The authors want to express their great indebtedness to Monique Gardes, Tom Bruns and Francis Martin for generous assistance during the development phase of the programme They are also grateful to a number of unmentioned fungal taxonomists for providing indispensable herbarium material

REFERENCES

Agerer, R , 1987, *Colour atlas of ectomycorrhizae* Munich Einhorn-Verlag

Arnolds, E , 1991, Decline of ectomycorrhizal fungi in Europe *Agriculture Ecosystems and Environment* 35 209-244

Bruns, T D , White, T J , and Taylor, J W , 1991, Fungal molecular systematics *Annual Revue of Ecological Systems* 22 525-564

Dahlberg, A , and Stenlid, J , 1994, Size, distribution and biomass of genets in populations of *Suillus bovinus* (L Fr) Roussel revealed by somatic incompatibility *New Phytologist* 128 225-234

Danielson, R M , and Pruden, M , 1989, The ectomycorrhizal status of urban spruce *Mycologia* 81 335-341

Egger, K N , Danielson, R M , and Fortin, J A , 1991, Taxonomy and population structure of E-strain mycorrhizal fungi inferred from ribosomal and mitochondrial DNA polymorphisms *Mycological Research* 95, 866-872

Egli, S , Amiet, R , Zollinger, M , and Schneider, B , 1993, Characterization of Picea abies (L) Karst ecto-mycorrhizas discrepancy between classification according to macroscopic versus microscopic features *Trees* 7 123-129

Erland, S , Henrion, B , Martin, F , Glover, L A , and Alexander, I J 1994, Identification of the ectomycorrhizal basidiomycete *Tylospora fibrillosa* Donk by RFLP analysis of the PCR amplified ITS and IGS regions of ribosomal DNA *New Phytologist*, 126 525-532

Gardes, M , White, T J , Fortin, J A , Bruns, T D , and Taylor, J W , 1991, Identification of indogenous and introduced symbiotic fungi in ectomycorrhizae by amplification of nuclear and mitochondrial ribosomal DNA *Canadian Journal of Botany* 69 180-190

Gardes, M , and Bruns, T D , 1992, ITS primers with enhanced specificity for basidiomycetes - application to the identification of mycorrhizae and rusts *Molecular Ecology* 2 113-118

Gardes, M , and Bruns, T , 1995, ITS-RFLP matching for identification of fungi In *Methods in Molecular Biology* Humana Press (in press)

Henrion, B , LeTacon, F , and Martin, F , 1992, Rapid identification of genetic variation of ectomycorrhizal fungi by amplification of ribosomal genes *New Phytologist* 122, 289-298

Ingleby, K , Mason, P A , Last, F T , and Fleming, L V , 1990, *Identification of ectomycorrhizas* Institute of Terrestrial Ecology, Res Publ 5

Kåren, O , Hogberg, N , Dahlberg, A , Grip, K , and Nylund, J -E , 1995, Influence of drought on ectomycorrhizal species composition - morphotype versus pcr identification *Proceedings from the 4th ESM Granada 1994* (in press)

Mullis, K B, and Faloona, F A , 1987, Specific synthesis of DNA *in vitro* via polymerase-catalysed chain reaction *Methods Enzymology* 155 335-350

Schaeffer, R L , Mendenhall, W , Lyman, O , 1990, *Elementary survey sampling* 4th edition Wadsworth Publishing Company p 74 ISBN 0-534-92185-X

Taylor A F S , Alexander I J (1990)a Demography of ectomycorrhizas of sitka spruce fertilised with nitrogen *Agriculture Ecosystems and Environment* 28 493-496

Taylor, A F S and Alexander I J (1990b) Ectomycorrhizal synthesis with Tylospora fibrillosa, a member of the Corticiaceae *Mycological Research* 95, 381-384

Unestam, T., 1991, Water repellency, mat formation, and leaf-stimulated growth of some ectomycorrhizal fungi. *Mycorrhiza* **1**, 13-20.

Wallander, H., and Nylund, J.-E., 1992 Effects of excess nitrogen and phosphorus starvation on the extramatrical mycelium of mycorrhizas of *Pinus sylvestris* L. *New Phytologist* **120**, 495-503

White, T.J., Bruns, T., Lee, S., and Taylor, J., 1990, Amplification and direct sequencing of fungal ribosomal RNA genes for phylogenetics. In Innis M.A., Gelfand D.H., Sninsky J.J., White T.J., eds. *PCR Protocols. A Guide to Methods and Applications.* San Diego: Academic Press, 315-322.

Vogt, K., Bloomfield, J., Ammirati, J. F., and Ammirati, S. R., 1991, Sporocarp production of basidiomycetes, with emphasis on forest ecosystems. In: *The Fungal Community - its organization and role in the ecosystem, 2nd ed., G C Carroll and D T Wicklow (eds), Marcel Dekker Inc. New York. pp 563-582.*

21

ISOLATION OF BACTERIA FROM SPOROCARPS OF *TUBER MAGNATUM* PICO, *TUBER BORCHII* VITT. AND *TUBER MACULATUM* VITT.

Identification and Biochemical Characterization

B. Citterio,[1] P. Cardoni,[2] L. Potenza,[2] A. Amicucci,[2] V. Stocchi,[2]*
G. Gola,[3] and M. Nuti[4]

[1] Dipartimento di Tecnologie Alimentari, Università di Udine
Udine, Italy
[2] Istituto di Chimica Biologica "Giorgio Fornaini", Università di Urbino,
Via Saffi, 2
61029 Urbino, Italy
[3] Stazione Sperimentale per l'Industria delle Conserve Alimentari
Parma, Parma, Italy
[4] Dipartimento di Biotecnologie Agrarie e CRIBI Biotechnology Centre,
Università di Padova
Padova, Italy

SUMMARY

The presence of bacteria inside the fruitbodies of the *Tuber* spp. of mycorrhizal fungi has been found, however the characteristics of these bacteria and the mechanisms by which they can influence the interaction between fungi and the root of the host plant have not yet been elucidated. In this study the presence of bacteria in sporocarps of different species of the white truffles *Tuber magnatum* Pico, *Tuber borchii* Vitt. and *Tuber maculatum* Vitt. is reported. Bacterial strains were isolated and biochemically identified as different species of *Micrococcus*, *Moraxella*, *Pseudomonas* and *Staphylococcus*. The strains most frequently recurring were subsequently characterized at the protein level by sodium dodecyl sulphate polyacrylamide-acrylamide gel electrophoresis (SDS-PAGE) and at the DNA level by random amplified polymorphism (RAPD) analysis. The presence of similar banding patterns suggests inter-strain similarities at both the protein and DNA levels.

Biotechnology of Ectomycorrhizae, Edited by Vilberto Stocchi et al.
Plenum Press, New York, 1995

INTRODUCTION

A wide array of organisms is present on roots or mycorrhizae, taking advantage of the various organic compounds released by the plant. These organisms may in turn affect the plant growth by various means: by providing protection against pathogens; by inducing drought tolerance; by supplying vitamins or growth factors or by contributing to plant nutrient uptake. These organisms include bacteria which are able to enhance ectomycorrhizal development under both laboratory and greenhouse conditions (1-7). These bacteria have been collectively named Mycorrhizal Helper Bacteria (MHB) (3). There are many possible mechanisms for this stimulation: the helper microorganisms could produce a substrate used by the mycorrhizal fungus, or mycorrhizal fungi can accumulate secondary metabolites in their growth medium which are toxic to the fungus itself and limit its growth (8). Furthermore, some rhizosphere bacteria are able to break down these molecules, and thus enhance the fungal growth (9). Another mechanism could be an increase in the susceptibility of the root to penetration by the mycorrhizal fungus. In fact, some strains of *Pseudomonas* spp. release cellulolytic and pectinolytic enzymes, and possibly phytohormones (10-11). All the stimulation mechanisms suggest that the inoculation of seeds, root systems or seed-beds with beneficial ectomycorrhiza-associated microorganisms could be of practical interest in the improvement of mycorrhizal inoculation techniques, particularly for species of economic interest. In this study we examined different species of white truffle: *Tuber magnatum* Pico, *Tuber borchii* Vitt. and *Tuber maculatum* Vitt.. They have fruitbodies with very peculiar organoleptic properties that make them of considerable commercial interest. The study was aimed at gaining a better understanding of the tripartite interaction of MHB/fungus/roots by isolating and characterizing the bacterial components.

METHODS

Isolation of Bacteria

Bacterial strains were isolated from sporocarps of *T. magnatum* Pico, *T. borchii* Vitt. and *T. maculatum* Vitt. collected in different soils of the Umbria and Marche regions of central Italy. Sporocarps were washed in running tap water and their surfaces sterilized in 1.5% (v/v) NaClO for two minutes, then rinsed in sterile water. Only the inner fragments, potter blended in physiological solution (0.9% w/v NaCl), were used for the isolation procedures. Serial solutions of the suspensions from sporocarps were plated on Tryptic Soy Agar medium (TSA) (Difco, Detroit, MI, USA). Representative samples were isolated and subcultured on the same medium. The ten most frequently recurring isolates were biochemically identified using the API-System (bioMérieux, Marcy-l'Etoile, France).

SDS-PAGE Analysis

The selected isolates were cultured in Tryptic Soy Broth (TSB) (Difco) at 30°C until the stationary phase of growth. It should be remembered that protein expression can change in relation to the culture, physiological and metabolic conditions of organisms. Therefore, all these variables were rigourously controlled in the samples used. The samples were centrifuged at 10,000 rpm for 10 minutes then washed 4 times with 50 mM Tris-HCl buffer, pH 7.6, containing 10% (w/v) sucrose, 1 mM DTT and 1 mM EDTA. The bacterial pellets were lysed overnight at 30°C in the same buffer containing 0.5% (v/v) Triton X-100 and 25 U/ml lysostaphin (Sigma Chemicals, St. Louis, MO, USA). The lysed samples were centrifuged at 14,000 rpm for 10 minutes and the supernatants were charged onto the gel. SDS-PAGE was carried out in 10% polyacrylamide slab gel containing 0.1% SDS according to the method of Laemmli (12). The following standard proteins (Bio-Rad, Richmond, CA, USA) were used to calibrate the gels: phosphorylase B (97,400 kDa), serum albumin (66,200 kDa), ovalbumin (45,000 kDa), carbonic anhydrase (31,000 kDa), trypsin inhibitor (21,500 kDa) and lysozyme (14,400 kDa). The gels were stained with Coomassie brilliant blue R-250.

DNA Preparation of Bacterial Strains

Genomic DNA was extracted from 5 ml suspensions of cells in the stationary phase of growth. The cells were pelleted and resuspended in 0.2 ml of TE 1-0.1 (1 mM Tris-HCl, pH 8.0, and 0.1 mM EDTA) with lysozyme (1 mg/ml) and incubated at 37°C for one hour; 0.2 ml proteinase K solution (0.5 mg/ml proteinase K, 1% sarkosyl, 200 mM EDTA, 1 mM calcium chloride) was then added and incubated for 1 h at 50°C. The clear lysates were extracted with phenol/chloroform, precipitated with ethanol, dried briefly and resuspended in TE (13). The final concentration was estimated by spectrophotometric analysis and DNA integrity was verified using agarose gel stained with ethidium bromide.

RAPD Analyses

The oligonucleotide primers used in this study were obtained from Operon (Alameda CA, USA). Amplification was carried out in a total volume of 50 µl with 100 ng of template DNA, 0.2 µM primer, 100 µM of each dNTP, 3 mM MgCl$_2$, 10 mM Tris-HCl, pH 8.8, 50 mM KCl, 0.1% (v/v) Triton X-100 and 0.8 U SuperTaq DNA polymerase (Sthelin, Basel, Switzerland). The reaction was overlaid with mineral oil and cycled through the following temperature profile: 94°C for 5 min, followed by 45 cycles of 30 sec at 94°C, 1 min. at 36°C, 2 min at 72°C. The final cycle was 72°C for 7 min (14). Amplifications were carried out in a Cetus DNA Thermal Cycler (Model 480) (Perkin-Elmer, Foster City, CA, USA). Reaction products were analyzed by electrophoresis in a 1.4% agarose gel stained with ethidium bromide.

RESULTS AND DISCUSSION

The isolation of bacteria from sporocarps of *Tuber* spp. gave rise to approximately one hundred isolates. Based on morphotypes, groups were formed and subjected to bio-chemical tests (API-System). The following species were identified among the most frequently recurring strains: *Micrococcus kristinae, M. roseus, M. varians, Pseudomonas fluorescens, P. vescicularis, Staphylococcus lugdunensis, S. xylosus, S. warneri*. In Table I the bacterial species are reported in relation to the *Tuber* species they were isolated from.

In addition, using the approach specified above, one isolated from *T. borchii* was tentatively assigned to *Moraxella* spp., and another one from *T. maculatum* to the *Pseudomonas* group. For the latter isolate, comparative analysis of protein pattern in 10% polyacry-lamide gel electrophoresis gave essentially the same profile as *Pseudomonas fluorescens* isolated from *T. borchii* (data not shown), and the strain was tentatively assigned to the species *fluorescens*. This assignment was further supported by DNA fingerprinting using random amplification of polymorphic DNA (RAPD) reported in Figure 1A lane 5. This technique can allow the identification of different species within a genus (14,15), and was used in this study as a confirmatory test of the biochemical identification. All the strains

Table I. Bacterial species isolated from different fruitbodies of *Tuber magnatum*,
Tuber borchii and *Tuber maculatum*

Tuber magnatum Pico	*Tuber borchii* Vitt.	*Tuber maculatum* Vitt.
Pseudomonas fluorescens	*Pseudomonas fluorescens*	*Pseudomonas fluorescens*
	Pseudomonas vescicularis	*Pseudomonas spp.*
Micrococcus varians		
Micrococcus roseus		*Micrococcus roseus*
Staphylococcus xylosus		*Staphylococcus xylosus*
	Staphylococcus lugdunensis	
	Staphylococcus warneri	
	Moraxella spp	

The bacterial strains were isolated from the internal part of the fruitbodies, which were
washed and sterilized. From the numerous strains isolated, those most frequently
recurring were biochemically identified and used for the experiments.

listed in Table I were analyzed by RAPD using several 10-mer primers for amplification
(Table II).

Amplified sequences ranged from 0.2 to 3 Kbp. Figure 1A shows the DNA fingerprints obtained with primer OPE-08 and *Pseudomonas* strains. The banding patterns reveal
the presence of two constant fragments, the upper one specific for the genus *Pseudomonas*,
the lower one for the species *fluorescens*. RAPD analyses obtained with primer OPE-20
showed similar patterns in the strains of *Staphylococcus xylosus* isolated from *T. magnatum*

Table II. Sequences of DNA oligonucleotide primers screened for production of RAPD profiles

Primers	*Pseudomonas* spp.	*Staphilococcus* spp.	*Micrococcus* spp.
OPE-01 5'-CCCAAGGTCC-3'	+	+	+
OPE-02 5'-GGTGCGGGAA-3'	+	+	+
OPE-03 5'-CCAGATGCAC-3'	+	+	+
OPE-05 5'-TCAGGGAGGT-3'	+	+	+
OPE-08 5'-TCACCACGGT-3'	+	+	+
OPE-10 5'-CACCAGGTGA-3'	–	+	+
OPE-11 5'-GAGTCTCAGG-3'	+	+	+
OPE-12 5'-TTATCGCCCC-3'	+	+	+
OPE-14 5'-TGCGGCTGAG-3'	+	+	+
OPE-15 5'-ACGCACAACC-3'	+	+	+
OPE-16 5'-GGTGACTGTG-3'	–	+	+
OPE-17 5'-CTACTGCCGT-3'	+	–	–
OPE-18 5'-GGACTGCAGA-3'	+	–	–
OPE-20 5'-AACGGTGACC-3'	+	+	+
OPZ-01 5'-TCTGTGCCAC-3'	–	+	–
OPZ-02 5'-CCTACGGGGA-3'	–	+	–
OPZ-03 5'-CAGCACCGCA-3'	–	+	–
OPZ-04 5'-AGGCTGTGCT-3'	–	+	–
OPZ-05 5'-TCCCATGCTG-3'	–	+	–
OPZ-06 5'-GTGCCGTTCA-3'	–	+	–

+, primers screened; –, no amplification.

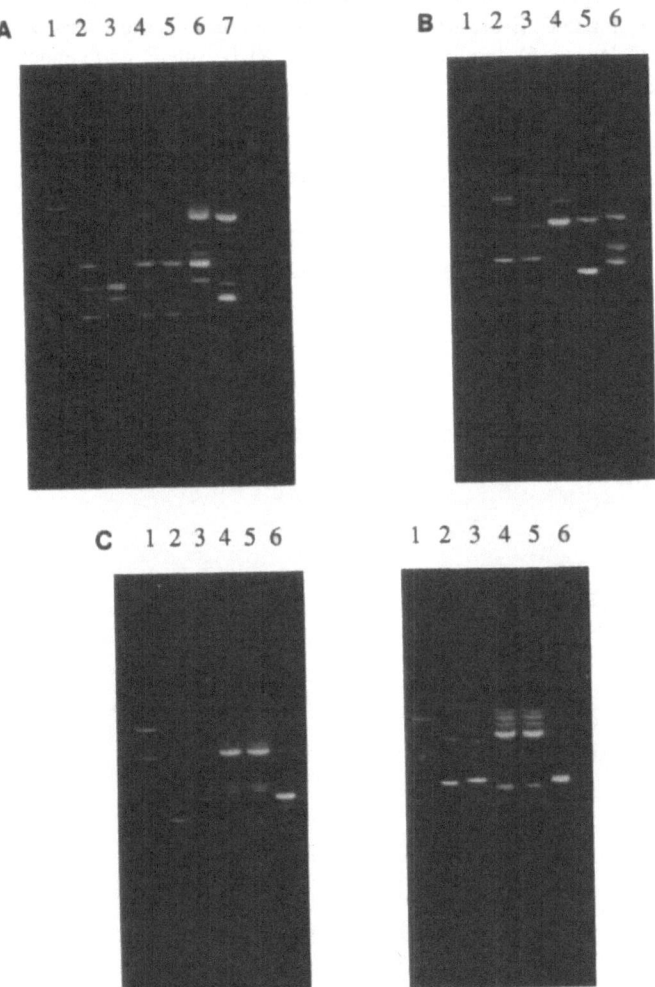

Figure 1. RAPD fingerprints of three different genera of bacteria: *Pseudomonas* spp., *Staphylococcus* spp. and *Micrococcus* spp. The amplification products were electrophoresed through 1.4% agarose gels stained with ethidium bromide. RAPD conditions and amplification were performed as described in materials and methods. **A:** Lane 1: Taq I digested pEMBL 8 DNA; Lanes 2-5: RAPD DNAs from *Pseudomonas fluorescens*; Lane 6: *P. vescicularis*; Lane 7: XL1-Blue *E.coli* strain. The *Pseudomonas* samples are characterized by the presence of a specific marker, using primer OPE-08 (5'-TCACCACGGT-3'). A marker is also evident among the *Pseudomonas fluorescens* samples. **B:** RAPD fingerprints of *Staphylococcus* spp. DNAs generated by primer OPE-20 (5'-AACGGTGACC-3'). Lane 1: Taq I digested pEMBL 8 DNA; Lanes 2-3: *Staphylococcus xylosus;* Lane 4: *S. lugdunensis;* Lane 5: *S. warneri;* Lane 6: the XL1-Blue *E.coli* strain. **C:** Lane 1: Taq I digested pEMBL 8 DNA; Lanes 2: *Micrococcus kristinae;* Lane 3: *M. varians;* Lanes 4-5: *M. roseus;* Lane 6: the XL1-Blue *E.coli* strain. The DNA fingerprints of two *Micrococcus roseus* strains, isolated from two different truffle species and amplified with primers OPE-05 (5'-TCAGGGAGGT-3') and OPE-11 (5'-GAGTCTCAGG-3'), show a common electrophoretic profile.

and *T. maculatum*, as reported in Figure 1B (lanes 2-3), while no common fragments were found for isolates of *Staphylococcus*.

Analyzing the species of *Micrococcus* isolated from *T. magnatum* and *T. maculatum*, with primers OPE-05 and OPE-11, genetic markers were found for the *Micrococcus roseus* species (Figure 1C, lanes 4-5). In all experiments, the XL1- Blue *E. coli* genome was also amplified with each single primer for comparison with a different genus.

Truffle and bacterial RAPD profiles obtained with the same primers were compared to evaluate whether significant contamination by bacterial DNA had taken place during the identification of truffle markers (16). Figure 2 reports the RAPD profiles of both the bacterial strains and the *Tuber* species from which they were isolated. The comparison shows that bacterial and truffle RAPD profiles are clearly different and rules out the possibility of bacterial contamination in truffle amplification products.

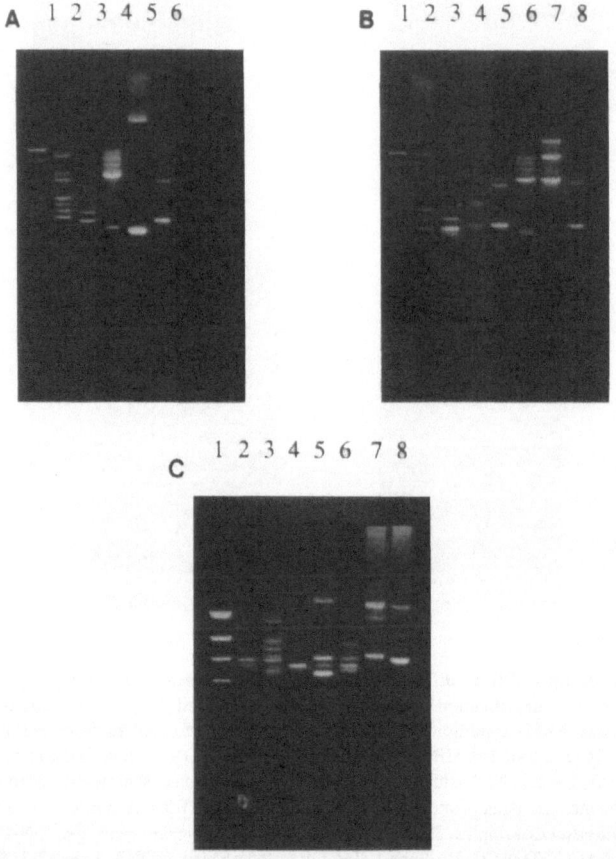

Figure 2. Comparison of RAPD profiles of bacterial strains with those of the *Tuber* species which they were isolated. The analyses of *Tuber* and bacteria fingerprints showed no significant overlapping. **A:** Lane 1: pBR322 digested with Ava I and II/Eco RI; Lane 2: *Tuber magnatum*; lane 3: *Pseudomonas fluorescens*; Lane 4: *Micrococcus roseus*; Lane 5: *Staphylococcus xylosus*; Lane 6: *Micrococcus varians*. **B:** Lane 1: pBR322 digested with Ava I and II/Eco RI; Lanes 2-3: *T. maculatum*; Lanes 4-5: *P. fluorescens*; Lane 6: *M. roseus*; Lane 7: *S. xylosus*; Lane 8: *M. kristinae*. **C:** Lane 1: Taq I digested pEMBL 8 DNA; Lanes 2-3: *T. borchii*; Lane 4: *P. fluorescens*; Lane 5: *Moraxella*; Lane 6: *P. vescicularis*; Lane 7: *S. lugdunensis*; Lane 8: *S. warneri*.

CONCLUSIONS

The evaluation of fruitbodies of *T. magnatum*, *T. borchii* and *T. maculatum* allowed the isolation of different bacterial genera, all of which are described as common colonizers of the rhizosphere and mycorrhizosphere (17). This finding extends and strengthens recent data by other authors (18.19) on the interactions between soil microflora and mycorrhizal infection. Overall, the mycorrhization helper bacteria are to be considered common inhabitants of this particular environment. In fact, they were found every time they were looked for, under very different site conditions and in various plant-fungus association. Moreover, they seem to be closely associated with mycorrhizal fungi in symbiotic organs (20). Among the bacterial genera isolated from the *Tuber* examined, we only tested the species most frequently recurring. The *Pseudomonas* species were found in most of the samples examined and in all *Tuber* species. This finding was reported also by Dupannois and Garbaye (21) who showed that the associated bacteria stimulate mycorrhiza formation by *Laccaria laccata* increasing the mycorrhizal index of the plants.

The molecular approaches utilized to characterize the strains isolated led us to observe that plasmid and restriction analyses provide a rather limited contribution to the identification of the species. The information provided by RAPD banding patterns is more informative. RAPD fingerprints of the three different genera of bacteria considered showed overlapping profiles for samples belonging to the same species, thus confirming the biochemical identifications made. The same approach allowed us to rule out the possibility of significant bacterial DNA contamination in the truffle samples studied with the selected arbitrary primers (16). Further studies are in progress in order to understand the interaction mechanisms between *Tuber* and bacteria.

ACKNOWLEDGMENTS

This work was supported by CNR, P.S. "Biotecnologia della micorrizazione".

REFERENCES

1 Bowen, G.D. and Theodoran, C., Interactions between bacteria and ectomycorrhizal fungi, *Soil Biol. Biochem.*, 11 (1979) 119-126.
2 De Olivera, B.L. and Garbaye, J., Les microorganismes auxiliaires des l'établissement des symbioses ectomycorrhiziennes (revue bibliographique), *Eur. J. For. Path.*, 19 (1989) 54-64.
3 Duponnois, R. and Garbaye, J., Mycozzhization helper bacteria associated with Douglas fir-*Laccaria laccata* symbiosis: Effects in vitro and in glasshouse conditions, *Ann. Sci. For.*, 48 (1991) 239-251.
4 Garbaye, J. and Bowen, G.D., Effects of different microflora on ectomycorrhizal inoculation of *Pinus radiata*, *Can. J. For. Res.*, 17 (1987) 941-943.
5 Garbaye, J. and Bowen, G.D., Stimulation of ectomycorrhizal infection of *Pinus radiata* by some microorganisms associated with the mantle of ectomycorrhizas, *New Phytol.*, 112 (1989) 383-388.
6 Duponnois, R. and Garbaye, J., Some mechanisms involved in growth stimulation of ectomycorrhizal fungi by bacteria, *Can. J. Bot.*, 68 (1991) 2148-2152.
7 Garbaye, J., Duponnois, R and Wahl, J.L., The bacteria associated with *Laccaria laccata* ectomycorrhizas or sporocarps: Effect of symbiosis establishment on Douglas fir, *Symbiosis*, 9 (1991) 267-273.
8 Duponnois, R. and Garbaye, J., Some mechanisms involved in growth stimulation of ectomycorrhizal fungi by bacteria, *Can. J. Bot.*, (1990) 2148-2152.
9 Azcon-Aguillar, C., Diaz Rodriguez, R.M. and Barca, J.M., Effect of soil microorganisms on formation of vescicular-arbuscular mycorrhizas, *Trans. Br. Mycol. Soc.*, 84 (1985) 536-537.
10 Strzelczyk, E. and Rozycki, M., Production of B-group vitamins by bacteria isolated from soil, rhizosphere and mycorrhizosphere of pine (*Pinus silvestris*)., *Zbl. Mikrobiol.*, 140 (1985) 293-301.

11 Strzelczyk, E., Kamper, M. and Michalsky, M, Production of cytokinin-lyke substances by mycorrhizal fungi of pine (*Pinus silvestris*) in cultures with and withouth metabolites of actinomicetes., *Acta Microbiol. Pol.*, 34 (1985) 177-186.

12 Laemmli, U.K., Cleavage of structural proteins during the assembly of the head of bacteriophage T4, *Nature*, 227 (1970) 680-685.

13 Welsh, J., Mc Clelland M., Fingerprinting genomes using PCR with arbitrary primers, *N.A.R.*, 18 (1990) 7213-7218.

14 Williams, J.G.K., Kubelik, A.R., Livak, K.J., Rafalskj, J.A. and Tingey, S.V., DNA polymorphism amplified by arbitrary primers are useful as genetic markers, *N.A.R.*, 18 (1990) 6531-6535.

15 Barral V., This P., Imbert-Establet D., Combes C. and Delseny M., Genetic variability and evolution of *Schistosoma* genome analysed by using random amplified polymorphic DNA markers, *Molecular and Biochemical Parassitology*, 59 (1993) 211-222.

16 Potenza, L., Amicucci, A., De Bellis, R., Cardoni, P. and Stocchi, V., Identification of *Tuber magnatum* Pico DNA markers by RAPD analysis, *Biotechnology Techniques*, 8 (1994) 93-98.

17 Florenzano G. in *Fondamenti di Microbiologia del Terreno*, REDA Ed., (1993), 2° Ed.

18 Garbaye J. and Dupannois R., specificity and function of mycorrhization helper bacteria (MHB) associated with the *Pseudotsuga menziesii- Laccaria laccata* symbiosis, *Symbiosis*, 14 (1992) 335-344.

19 Garbaye J., Dupannois R. and Wahl J.L., The bacteria associated with *Laccaria laccata* ectomycorrhizas or sporocarps: effect on symbiosis establishment on Douglas-fir, *Symbiosis*, 9 (1990) 267-273.

20 Garbaye J., Tansey Revue No. 76 Helper bacteria: a new dimension to the mycorrhizal symbiosis, *New Pytol.*, 128 (1994) 197-210.

21 Dupannois R. and Garbaye J., application des BAM (bactères auxiliaires de la mycorhization) à l'inoculation du Douglas par *Laccaria laccata* en pépinière forestière, *Rev. For. Franç.*, 44 (1992) 641-650.

INDEX